D0161618

Fisheries Conservation and Management

MICHAEL R. ROSS
University of Massachusetts, Amherst
Department of Forestry and Wildlife Management

PRENTICE HALL
Upper Saddle River, New Jersey 07458

Library of Congress Cataloging-in-Publication Data

Ross, Michael R.
    Fisheries conservation and management/Michael R. Ross.
      p. cm.
    Includes bibliographical references and index.
    ISBN 0–02–403901–2
    1. Fisheries. 2. Fishery conservation. 3. Fishery management. I. Title.
SH331.R635 1997   95–41533
333.95'6—dc20   CIP

Editor: Teresa K. Ryu
Senior Editor: Sheri L. Snavely
Editorial/Production Service: Electronic Publishing Services Inc.
Manufacturing Manager: Trudy Pisciotti
Page Compositor: Eric Hulsizer
Cover Designer: Bruce Kenselaar
Cover Photograph: National Marine Fisheries Service, Alaska Fisheries Science Center

© 1997 by Prentice Hall, Inc.
Simon & Schuster / A Viacom Company
Upper Saddle River, New Jersey 07458

Printed in the United States of America.

All rights reserved. No part of this book may be reproduced or transmitted in any form or by any means, electronic or mechanical, including photocopy, recording, or any information storage and retrieval system, without permission in writing from the Publisher.

1 2 3 4 5 6 7 8 9 10

ISBN 0-02-403901-2

Prentice-Hall International (UK) Limited, *London*
Prentice-Hall of Australia Pty. Limited, *Sydney*
Prentice-Hall Canada Inc., *Toronto*
Prentice-Hall Hispanoamericana, S.A., *Mexico*
Prentice-Hall of India Private Limited, *New Delhi*
Prentice-Hall of Japan, Inc., *Tokyo*
Simon & Schuster Asia Pte. Ltd., *Singapore*
Editora Prentice-Hall do Brasil Ltda., *Rio de Janeiro*

Contents

Contents

In memory of my parents,
who first took me fishing

Preface

A number of textbooks have been written about the use, conservation, and management of fisheries resources. Many of these provide technical detail that is well suited for advanced undergraduates or beginning graduate students who have broad backgrounds in the natural sciences, mathematics, and the social and behavioral sciences. Successful entry into the profession of fisheries conservation and management demands such broad training. Indeed, in an opinion article J. Hard (1995) proposed that the entire curriculum of fisheries programs at the undergraduate level be focused on this background training, and that courses that deal specifically with fisheries conservation and science generally be offered at the graduate level.

Although many fisheries educators believe that some fisheries-specific training must be integrated into undergraduate curricula of fisheries programs, we are faced with a dilemma. When background training is coupled with the general education requirements of many universities, students majoring in a fisheries-related field are often unable to complete the academic training necessary to delve fully into the technical nature of the "science" of fisheries conservation before their junior or even senior year. As relevant as the argument for broad training may be, many students who have selected fisheries as an undergraduate major remain most interested and motivated to pursue this choice if they do not have to wait until the final year of their degree program before taking some coursework specifically in their major field.

I have written this textbook for use in introductory fisheries conservation and management courses at the sophomore or junior level. Because such a course occurs reasonably early in the students' undergraduate training, I have chosen to introduce them to the breadth of the fisheries conservation and management process rather than the depth and detail of specific fisheries conservation and management settings. Knowing that background coursework in the natural sciences, the social and behavioral sciences, and mathematics may not have been completed by the time a course such as this is taken, I have minimized coverage of the mathematical approaches of fisheries science and have included chapters providing basic background information on aquatic environments, the ecology of fishes, and characteristics of humans who harvest fisheries resources or indirectly interact with them through land- and water-use activities.

The history of fisheries conservation and management is a mixed bag of successes and failures. Throughout this book, I have included examples of both successes and failures and have discussed the strengths and weaknesses of particular conservation and management strategies. By doing so I hope to instill in students (1) an understanding of the complexity of the fisheries conservation and management arena and (2) the importance of making wise decisions based on sound information to conserve our wild fish resources and the benefits that people gain through interactions with fishes and other aquatic life.

ACKNOWLEDGMENTS

I am grateful to a number of people who contributed to the preparation of this book. Among those individuals who were particularly helpful were Steve Murawski, John Boreman, and Jay Burnett of the Northeast Fisheries Science Center (NMFS:NOAA); Kurt Fausch of Colorado State University; J. Ellen Marsden of the Illinois Natural History Survey; Ron Essig, Regional Office, Region 5, U.S. Fish and Wildlife Service; John Merriner of the Southeast Fisheries Science Center; David Beauchamp of Utah State University; Karsten Hartel of the Harvard Museum of Comparative Zoology; David Ross, Ken Paxson, and Roger Knight of the Ohio Department of Natural Resources; Christine Carr of the International Pacific Halibut Commission; Francis Juanes, David Loomis, and Robert Muth of the University of Massachusetts; and Steve Quinn of the "In Fisherman."

The following people reviewed drafts of various chapters of the book: Cynthia Annette, University of Kansas; Willard E. Barber, University of Alaska; David H. Bennett, University of Idaho; Paul B. Brown, Purdue University; Lynn Decker, USDA Forest Service; Kurt D. Fausch, Colorado State University; Ronald A. Fritzsche, Humboldt State University; Gary D. Grossman, University of Georgia; Daniel B. Hayes, Michigan State University; Thomas K. Hill, University of Tennessee; Donald C. Jackson, Mississippi State University; Brett Johnson, Colorado State University; William Kelso, Louisiana State University; Charles Knight, Southeastern Louisiana University; John J. Ney, Virginia Polytechnic Institute and State University; and William D. Youngs, Cornell University. I am grateful for the constructive comments of these reviewers.

Finally, I am extremely grateful to my wife, Peg, who spent countless hours proofing what must have seemed like endless drafts of book chapters and literature cited sections.

Some materials in this book were either modified or reprinted with permission from the handbook *Recreational Fisheries of Coastal New England* by Michael R. Ross with contributions by Robert C. Biagi (Amherst: University of Massachusetts Press, 1991), copyright (c) 1991 by the University of Massachusetts Press. I thank the University Press for allowing the use of these materials.

1

Introduction

Fishes and other living aquatic resources have served as a source of food, commerce, and recreation for people since ancient times. For example, the Chinese were raising carp and other freshwater fishes for food as long as 4,000 years ago (Johnson 1989). Harvesters of Sumaria, near the Persian Gulf, were organized into government fishing fleets as early as 2300 B.C., and rulers of Egypt fished recreationally—that is, for pleasure rather than to provide their families with food or income—more than 3,400 years ago (Royce 1987). Fisheries resources played an important role in the European settlement of North America, being harvested first as a family food source and then for commercial purposes, in many areas that later became part of the United States. Fishes and shellfishes were so important a food staple to Native Americans that their harvest became an integral part of the cultural activities and ceremonies of some tribes (Johnson 1989).

Today, living aquatic resources are intensively harvested around the world as a source of protein in the family diet, for commercial gain, or for recreation. The worldwide average yearly harvest of fisheries resources ranged from 70 to 90 million metric tons (154 to 198 billion pounds) in the 1970s and 1980s (Miller and Johnson 1989). In the United States, aquatic resources are harvested commercially for marketing as human foods, or for use in such diverse industrial products as animal feeds, clothing accessories (for example, buttons made from shells of clams or other bivalves), and pharmaceutical medications (many baby boomers remember the less-than-tasty cod liver oil tablets they were given daily as children in the 1950s and 1960s, long before the health benefits of eating fish were well understood). Coastal commercial fish harvesting, processing, and marketing operations produce billions of dollars of economic activity annually and employ over 300,000 people (Royce 1987). Although U.S. citizens eat markedly less seafood than do people of some other regions of the world, the per capita yearly consumption of seafood products in the United States has steadily risen in recent decades. An increase in consumption was particularly rapid during and after the mid-1980s, when the value of eating fish to reduce cardiovascular disease became widely publicized. Each U.S. citizen consumes an average of more than 16 pounds of fish and other seafood products per year (Miller and Johnson 1989).

In addition to their value in commercial enterprises, fishes are the focus of one of the most avidly pursued of outdoor activities in the United States: recreational fishing, or angling. About 36 million people 16 years and older fished recreationally in 1990 (USDI and USDC 1991). The money that they spent while fishing produced the equivalent of 600,000 full-time jobs in tourism and recreation-based industries (Miller and Johnson 1989).

Some fishes and other aquatic organisms are enjoyed by humans in a non-consumptive manner (consumptive uses include all harvesting activities of humans, whereas nonconsumptive uses encompass all interactions between humans and resources that do not include harvesting, or keeping, the resources). For example, many fishes are caught by anglers simply for the enjoyment of landing them; fishing regulations may require that the fishes be released after they are netted or brought to the boat. People value the opportunity simply to watch some resources, such as whales and coral reef fishes, because these animals fascinate them. Other aquatic organisms, although seen by very few people—such as cavefishes or desert pupfishes—are considered beneficial by some because they represent unique elements of the earth's diverse living communities.

Substantial, long-term increases in the harvest of fishes and other living aquatic resources, and the intensive manner in which humans use our waterways, have led to a significant decline in the condition of an ever-growing number of aquatic resources. The fisheries conservation and management process, carried out by public agencies or commissions appointed by public officials, is the means that our society has created to protect these resources and the benefits that humans receive from them.

WHAT IS A FISHERY?

Simply defined, a fishery is the union of aquatic organisms and humans (Miller and Johnson 1989). There are three basic elements to a fishery: the resource itself, its aquatic environment, and the people who harvest the resource or change the condition of its aquatic surroundings. Fishes are not the focus of all fisheries. Laws that have facilitated the formation of public fisheries conservation and management agencies or commissions (see Chapter 6) have assigned these groups broad responsibilities in overseeing human interactions with a variety of aquatic life-forms. Thus fisheries agencies or commissions may find themselves developing conservation and management programs that focus on fishes, other vertebrate groups such as marine mammals and sea turtles, molluscan bivalves and snails, crustaceans such as lobsters and crabs, sea urchins, sponges, or even marine algae.

Historically, the label *fishery* was restricted to specific consumptive interactions between humans and aquatic resources. In this book, the term *harvest* will be used in describing aquatic resources that are either eaten by humans or used to produce industrial products. Humans affect and interact with aquatic organisms much more broadly than simply harvesting them. Our use of waterways and the lands that surround them has caused numerous aquatic resources to suffer severe declines in abundance and distribution and, in some instances, to become extinct. In the latter half of the 20th century U.S. society, through its elected officials, passed federal and state laws (such as the Endangered Species Act and the Marine Mammal Act of the 1970s; see Chapter 6) that defined our responsibility to protect forms of life threatened with extinction. Such actions broadened the fisheries conservation and management process from that focused only on species of some consumptive importance to humans to a much wider concern for the wise use and protection of all aquatic life. Today, the fisheries conservation and management process may focus upon harvested resources; specific, rare forms of aquatic life; or the diversity of life in aquatic habitats and the quality of aquatic habitats themselves.

WHY CONSERVATION *AND* MANAGEMENT?

Fisheries conservation and management can be defined as the manipulation of human interactions with living aquatic resources in a manner that allows humans to gain some sustainable benefit from these resources (Nielsen 1993). This manipulation may be directed toward the behavior of people—for example, controlling harvest with regulations, or controlling the manner in which nonconsumptive uses such as whale watching are conducted; aquatic habitats that the resources live within—for example, preventing deterioration of aquatic habitats caused by pollution; or the resources themselves—for example, introducing new species into lakes or reservoirs to create fishing opportunities or to provide new food resources for fish species deemed to be important.

Too often, laypeople equate conservation with preservation. In fact, the first conservation efforts in the United States were established to gain some sustainable use from natural resources, not to provide absolute protection. Fisheries conservation can be defined simply as *the wise, sustainable use of wild (naturally produced) fisheries resources. Sustainable use* is an important element of conservation, as the benefit humans receive from living resources can be sustained only if the viability of those resources and the quality of the environments they live within are protected and maintained.

Prior to the mid-20th century, in most instances the wise use philosophy included only resources that were harvested, because people generally viewed fishes and other aquatic resources as beneficial only if they could be used consumptively. Indeed, fisheries conservation and management "was born [from] a need to balance the supply-demand equation" (Nielsen 1993)—in other words, to control harvest so that the resources provided by nature could supply the harvest demand of people. As mentioned above, in the last several decades U.S. society has broadened its view of benefit to include organisms that have little marketable or recreational value. Thus, in some instances wise use of a resource might be synonymous with no use. In this text, fisheries conservation efforts include all programs that support wise, sustainable use of wild fisheries resources, including controlling harvest to prevent overfishing and protecting or enhancing habitats to improve the condition of populations of fishes or other aquatic organisms.

In one sense, because fisheries agencies and other entities manipulate or manage interactions between resources and people, all conservation efforts can be considered management. However, not all agency efforts to manipulate the interaction between humans and fisheries resources include natural, or wild, resources. For example, public agencies have developed expansive hatchery systems that provide millions of fish each year for capture in recreational fisheries. Many of these fisheries depend wholly upon hatchery-reared, rather than naturally produced, fishes. Thus agencies may establish effective fisheries resource management programs that have little or nothing to do with conservation of natural resources but provide humans with an interaction with, and benefit from, fishes. In this context, *fisheries management* applies to programs that manipulate people-fish interactions, when the fish are a product of human rearing and stocking activities rather than a product of nature.

WHAT'S AHEAD?

This book is organized into two parts: Chapters 2 through 4 provide general information concerning the three basic elements of a fishery: Chapter 2, aquatic environments; Chapter 3, ecology of fishes that live within aquatic systems; and Chapter 4, the people who harvest these resources or affect them through land- and water-use practices. These chapters are not a substitute for the much more detailed training that students receive in courses in limnology, oceanography, fish ecology, or the human dimensions of resource conservation; they are intended to provide some basic information to students who have not yet taken specialized courses in some or all of the above subject areas. Chapter 5 provides an overview of major fisheries resources of the United States, and the remainder of the text focuses on the fisheries conservation and management process.

2
Aquatic Habitats

Water covers about 70 percent of the earth's surface, with over 99 percent of the volume of surface waters contained in ocean basins and polar ice caps (Wetzel 1983). Although freshwater systems constitute less than 1 percent of the earth's surface waters, they represent an array of different physical, chemical, and biological conditions. This chapter reviews the basic types of aquatic systems where fish or other taxa that support fisheries live. This chapter is not intended to provide a detailed description of aquatic environments and the biotic communities that inhabit them; numerous textbooks have been written that provide such reviews for freshwater and coastal systems, and college curricula often include semester-long courses focused on each of these basic types of aquatic environments. Provided here is a brief review of aquatic environments to be used by students who might be taking an introductory course in fisheries conservation prior to completing coursework in freshwater and marine ecology. The focus is on some of the basic characteristics of aquatic systems that are critical to the occurrence and health of particular fisheries resources, and that are particularly susceptible to modification and deterioration as a result of human activities within basins or the watersheds that surround them.

PHYSICAL AND CHEMICAL CHARACTERISTICS OF SURFACE WATERS

Water is an unusual natural compound. At its freezing point, 0 degrees Celsius (°C), it is about 8.5 percent lighter as a solid (ice) than as a liquid at the same temperature. It reaches its greatest density at 4°C. Compared with most substances, water has a high specific heat—it must gain or lose relatively substantial amounts of heat energy before a change in its temperature occurs (Wetzel 1983). These unique physical properties of water strongly influence the biological communities that inhabit aquatic systems, whereas chemical properties influence the productivity of these communities. This section reviews the effects of several basic physical and chemical characteristics on aquatic life. Unless otherwise cited, the basic properties of water reviewed here are referenced from Reid and Wood (1976) and Wetzel (1983).

Light Penetration into the Water Column

A small proportion of the light that strikes the surface of water is reflected back into the atmosphere. The percentage that is reflected varies seasonally and daily. On still, summer days about 5 to 6 percent is reflected, whereas in the winter in temperate regions about 10 percent is reflected because light strikes the water at a lower angle of incidence—that is, the angle formed between the direction of light movement and the water's surface. On a windy day, waves constantly change the angle of incidence, causing a modest increase in the percentage of light that is reflected by surface waters.

Over half of the light that is not reflected at the surface is absorbed and transformed into heat energy within the first meter (3.3 feet) of water (Ruttner 1974). Shorter wavelengths of light penetrate the water column more deeply than longer wavelengths. All red and orange is absorbed within the first 4 and 17 meters of water column, respectively. Even in very clear water, 70 percent of blue light (the shortest of visible wavelengths) is absorbed within 70 meters of water. Turbidity—the level of opaqueness caused by materials suspended in the water column—can significantly reduce light penetration through a water column by absorbing or scattering light rays. In glacial-fed lakes that are highly turbid due to suspended clay particles, most light is absorbed within a few centimeters below the surface (Moss 1988).

The relatively limited penetration of light into deep bodies of water produces two effects that significantly influence these aquatic systems. In most systems, heat energy that is added to lakes is derived from the surface. This greatly influences temperature cycles, circulation patterns, and productivity of standing waters. Secondly, all photosynthesis is restricted to shallow inshore areas or to some limited depth below the surface of deeper, open waters. Thus production of phytoplankton—buoyant photosynthetic microorganisms—occurs near the surface, and the distribution of rooted aquatic plants is restricted to areas shallow enough to allow leaves of the plants to receive sufficient light. Generally, there is an inverse relationship between levels of turbidity and depth of the photic zone—the zone of photosynthesis.

Temperature Cycles

Generally, any gain or loss of heat that occurs in water takes place at the surface. Gains arise from solar radiation when water temperatures are lower than air temperatures, and losses occur when air temperatures are lower. Daily and seasonal temperature fluctuations are affected by water's high specific heat. On a daily basis, temperatures fluctuate less in surface waters than in the air over the water. Seasonally, water temperatures fluctuate less than air temperatures, and temperature changes in large bodies of water such as lakes and oceans occur more slowly than seasonal increases or decreases in atmospheric temperature.

Temperature Characteristics of Flowing Waters. Within a river basin, temperature tends to increase from the headwaters to the mainstem (Winger 1981). Headwaters are generally cooler than downstream areas because water entering the stream channel from underground sources has not had time to be warmed by sunlight, particularly if the riparian zone has a tree canopy that shades the water's surface from sunlight (Hynes 1970; Winger 1981). Tributary waters are significantly cooler than daytime air temperatures in the warmer months of the year. As water flows through a basin, surface warming increases water temperature until it is close to the mean ambient air temperature. Once it reaches this level it tends to fluctuate with weather conditions and daily warming and cooling cycles (Hynes 1970). In winter, streams fed largely by underground runoff may be warmer than ambient air temperatures in northern areas of the United States.

Many streams will not form a stable layer of surface ice in the winter, particularly in turbulently flowing areas, even when air temperature drops below the freezing point of water. This is due in part to the warming effect of water percolating into the stream channel from underground, and the resistance to freezing of rapidly moving water. In smaller streams with reduced winter flows, underwater ice may occur in riffles in the form of soft slush or anchor ice—ice that forms on the substrate at the bottom of the water column when no ice forms at the surface. Such ice typically forms when dark-colored substrates underneath water cooled to 0°C reach significantly lower temperatures due to radiant loss of heat (Hynes 1970). Drag created at the surface of rocks in a riffle slows flow to the point that allows ice crystals to form on the cold rock surfaces. When it releases during warming from solar radiation, anchor ice can scour the substrates of riffles in the same manner that high flood waters do at other times of the year.

Temperature Cycles in Standing Waters. Figure 2.1 illustrates the relationship between the temperature and density of water. Note that water is most dense at 4°C, and the differences in density that occur between any two consecutive temperatures from 0 to about 8 to 10°C are relatively smaller than differences that occur between any two consecutive higher temperatures. This unique characteristic influences the annual temperature cycles in deep bodies of standing water and seasonal patterns of water circulation within these aquatic systems.

In temperate regions, late winter–early spring water temperatures in most large basins may vary little from the surface to the deepest waters, usually ranging from about 1 to 3°C. As day length and air temperatures increase in the spring, surface waters are warmed toward 4°C, causing the density at the surface to become greater than that of waters below. At this time, surface waters mix into deeper areas of the water column through two processes:

1. Denser surface waters sink into less dense waters, creating convection currents.

2. Wave-induced surface currents created by winds fold downward across the bottom when they approach shorelines, increasing the downward movement initiated by convection.

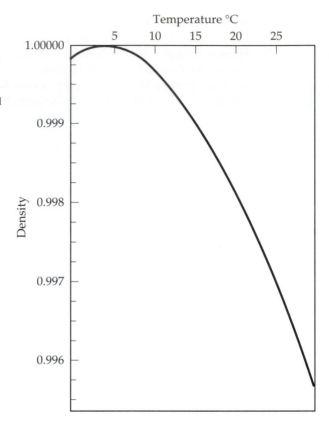

Figure 2.1 The relationship between temperature and density of water. Water reaches its maximum density at 4°C. (From: J. R. Vallentyne, "Principles of Modern Limnology;" *American Scientist*, 45: 218–244. Copyright 1957. Reprinted with permission.)

This process of mixing from the surface to the bottom is called the turnover. Complete mixing of the water column continues even after the surface waters become warmer than 4°C. Because the density differential between temperatures within the range of 4 and about 10°C is relatively slight, winds in the spring are generally strong enough to produce wave-induced currents that will continue downward mixing of warmer, less dense water.

As summer approaches, the strength of winds typically lessens, and the density differential between surface waters warming to higher temperatures than waters below becomes greater. Thus at some time surface-to-bottom mixing ceases, initiating the period of summer stratification. During this time of year, the water column is separated into three temperature regions. The surface waters continue to warm, mixing to some depth below the surface based on how deep wave-induced currents can force lighter water downward against the density differential. This near-surface region of warm, mixed waters is called the epilimnion. Immediately below this is the thermocline, a vertically narrow region where temperature declines rapidly from that characteristic of the surface to that characteristic of deeper waters that do not warm after stratification is initiated. Below the thermocline is the hypolimnion, which throughout the summer remains at the temperature that had been reached prior to stratification. During the summer, hypolimnion waters are completely separated and do not mix with epilimnion waters. At this time, no nutrients rise from deep-water sediments to the surface. Thus, planktonic primary production may not be as high during the summer as during spring turnover in many systems. Similarly, because all oxygen in lakes is either dissolved at the surface or produced by photosynthesis largely restricted to the epilimnion, dissolved oxygen used by living organisms in the hypolimnion will not be replaced during the length of summer stratification. During summer stratification, aerobic microbial decomposition of organically enriched bottom sediments in some lakes can deplete dissolved oxygen levels in the hypolimnion. In extreme conditions, dissolved oxygen concentrations of bottom waters can become so depleted that fish and many invertebrates die or move to other areas of a basin.

As air temperatures and hours of solar radiation decline in the fall, surface waters cool, increasing in density to the point where currents can once more drive water downward through the thermocline. Thus, as in the spring, complete circulation of the water column occurs with the fall turnover. By early winter, water temperature can reach 4°C throughout the water column. At this time, further surface cooling may result in a "reverse stratification"—cooler, lighter surface waters do not mix downward into deeper waters due to differences in density. Because the density differences at these temperatures are slight, late fall or winter wind-driven waves can mix surface waters downward. If winter air temperatures drop below zero, ice will form on the surface, insulating the water below from further cooling and preventing further circulation.

The probability of summer stratification and the depth at which a thermocline forms depend in part on the depth of water and the *fetch* of a system—that is, the distance over which winds can blow across a water's surface. The size of waves is dependent on the force of winds and the fetch: For any given wind speed, the size of the wave will be directly related to the fetch. Thus thermoclines may form just a few feet below the surface in some small lakes if the basin is surrounded by hills protecting at least part of the lake's surface from wind and the fetch is small. Conversely, even deep lakes with large surface areas may be resistant to stratification. For example, the western basin of Lake Erie, with a maximum depth of nearly 15 meters (50 feet), rarely stratifies even for short periods of time. Even moderate winds working over the great surface area of this elongate lake will create waves of sufficient size to mix surface waters downward to the substrate. Shallower continental shelf areas of the oceans may vertically mix in all seasons due to a combination of waves and tidal currents. For example, areas of Georges Bank (off New England) that are 60 meters (about 200 feet) or shallower mix thoroughly from surface to substrate throughout the year (Csanady and Magnell 1987), whereas adjoining deeper waters stratify in the summer.

Chemical Characteristics of Aquatic Systems

Nutrient Levels.   Numerous ions are present in surface waters. Among these, nitrogen and phosphorus ions are often the most commonly limited plant nutrients (Moss 1988). Nitrate, the nitrogen compound most readily taken up by algae and other primary producers, occurs in low concentrations in unpolluted waters. Soluble phosphates also are typically present in minute concentrations because this form of phosphorus is readily absorbed by planktonic and macroscopic plants.

Land- and water-use activities strongly influence the concentrations of these nutrients in river systems. Both nitrate nitrogen and soluble phosphate may occur in substantially higher concentrations in waters receiving sewage treatment plant effluents or agricultural runoff. Generally, the concentration of these nutrients is directly related to the percentage of a watershed engaged in agriculture and inversely related to the percentage of land that is forested rather than farmed (Winger 1981).

In standing waters such as lakes, coastal estuaries, and bays, nitrogen compounds may be washed in, or they may be formed within the system by nitrifying microorganisms. Lakes tend to act as nutrient "sinks," accumulating nitrogen, phosphorus, and other nutrients in bottom sediments. Nutrients accumulate through decomposition of organic material either transported into the lake from the surrounding watershed or from dead aquatic organisms produced by the lake's food webs. Due to these processes, lakes normally progress from an oligotrophic, "poorly nourished" (Reid and Wood 1976) condition with low biological productivity to a more eutrophic condition characterized by relatively high concentrations of available nutrients and high levels of biological productivity. In many eutrophic systems, spring and fall turnover are times of high biological productivity, because nutrients in bottom sediments are flushed upward in the water column where phytoplankton can utilize them. The process of eutrophication—the accumulation of usable nutrients and the increase in biological production—is normally a very slow process in unpolluted systems. In many waters the rate of nutrient buildup is greatly accelerated by the influx of soluble nutrients and organic materials from human sources, particularly domestic sewage and agricultural fertilizers. The increase in the rate of the eutrophication process due to human activities is called *cultural eutrophication.*

Buffering Capacity of Natural Waters.   Many atmospheric gases dissolve in water. Because carbon dioxide chemically combines with water droplets in the atmosphere to form the weakly acidic carbonic acid H_2CO_3, the pH of precipitation is about 5.6 (Moss 1988): pH is a measure of the logarithm of the reciprocal of the concentration of free hydrogen ions—the lower the pH, the greater the concentration of hydrogen ions, and thus the acidity, of water (Wetzel 1983). Surface waters exhibit wide variation in levels of acidity. Water can become acidic due to production of acids from the decomposition of organic matter. However, acid precipitation, which occurs as a result of the burning of fossil fuels, is often responsible for extreme acidity of surface waters. Sulfur dioxide, which comes largely from industrial burning of sulfur-rich coal, and nitrogen oxides, released into the atmosphere by automobiles and industrial burning of petroleum products, dissolve in atmospheric water to produce the highly acidic sulfuric and nitric acids (Moss 1988).

Aquatic systems have different capacities to buffer acidity through chemical combination of free hydrogen ions with naturally occurring compounds such as carbonates, bicarbonates, and hydroxides. For example, calcium carbonate (limestone) combines with sulfuric acid to produce calcium sulfate and the more weakly acidic carbonic acid:

$$CaCO_3 + H_2SO_4 \rightarrow CaSO_4 + H_2CO_3$$

Magnesium, sodium, and potassium compounds deriving from bedrock and soils also can buffer the acidity of waters. In many regions of the country, limestone in the soils and bedrock has protected waterways from increasing acidity caused by acid precipitation. However, in other regions underlain by granite or other calcium-poor bedrocks, the buffering capacity of the water is quite limited. In recent decades, the importance of buffering capacity of soils and bedrock has become increasingly apparent as the pH of river systems drops due to increasingly acid precipitation falling on watersheds. Acidification of surface waters has led to the loss of numerous fisheries resources in these regions (see Chapter 9).

STANDING FRESHWATER SYSTEMS

Standing (also called lacustrine) waters are small to large bodies of water with no evident flow throughout much of their expanse, and with areas of water that are deep enough to prevent the growth of emergent herbaceous or woody vegetation. Standing waters can be divided into lakes and ponds that occur naturally and reservoirs that are created either by construction of a dam within a river basin or within a valley subject to periodic, seasonal water flow. Definitions that describe the differences between lakes and ponds have varied considerably. Some authors (such as Reid and Wood 1976) define ponds as small bodies of water that are sufficiently shallow to allow growth of either emergent or submergent aquatic vascular vegetation on the entire substrate underlying the water, and lakes as natural standing bodies of water that have a proportion of the substrate deep enough to prevent the growth of rooted vegetation. Other authors classify these two systems according to size: Standing bodies of freshwater less than 4, 5, or even 10 acres have been called ponds, whereas other systems larger than these arbitrary surface area measures have been considered lakes (Bennett 1970).

The Biological Community of Standing Freshwaters

Two major plant communities occur in lakes and ponds. Rooted vascular plants and attached algae are restricted to shoreline areas where the water depth is shallow enough to allow light to penetrate to the substrate. In this part of the lake, called the littoral zone, photosynthesis is conducted by both attached and planktonic organisms. The profundal zone includes all substrates in water sufficiently deep that attached photosynthetic organisms do not occur due to insufficient light penetration. In these deeper-water areas, photosynthesis is restricted to planktonic organisms in the limnetic or pelagial zone—the portion of the water column under the surface that has sufficient light to support photosynthetic activity.

Filamentous algae and rooted vascular plants support a diverse invertebrate community in the littoral zone. Numerous fishes utilize this inshore habitat not only for the available invertebrate food resources, such as aquatic insects, amphipods, and snails, but also for the protection from predators that they can gain by hiding in the dense stands of aquatic vegetation. Primary production by plankton in the limnetic zone supports a wide variety of zooplankters that in turn serve as the food base for numerous larval and juvenile life stages, and some adult stages of fish species. Many benthic—bottom-dwelling—invertebrates including insects, oligochaetes, bivalve mussels, and snails either filter feed or graze on organic detritus.

Eutrophication can lead to robust food webs and high biomass production of fishes. However, in systems where cultural eutrophication is occurring, habitat conditions can change significantly, causing declines in some fish stocks that support very important fisheries. For example, coldwater-adapted trouts and whitefishes, which must reside in the hypolimnion or thermocline of many lakes in the summer to avoid physiologically stressful high temperatures of surface waters, may decline in organically enriched systems. Chronic oxygen depletion that occurs below the epilimnion during summer stratification reduces the quality of deepwater habitats for such fishes (see Chapter 9). Oxygen depletion can also cause substantial changes in the benthic invertebrate community, which serves as an important food base for many lacustrine fish species (Moss 1988; Chapters 9 and 12).

Reservoirs

The manner in which reservoir systems are constructed and managed strongly influences the quality and longevity of fisheries in these systems. Many reservoirs differ substantially from natural lakes in characteristics that have significant impact on fisheries resources. These differences, reviewed by Kimmel and Groeger (1986), are summarized below.

Reservoirs typically have a higher ratio between the drainage area of the basin to water surface area than do lakes of similar size. These large drainage basins produce relatively high seasonal and yearly flows of water into a reservoir. Thus the volume of water flowing into and out of a reservoir per unit time compared with the reservoir's volume is substantially higher than that ratio characteristic of natural lakes from the same geographic region.

The preparation of a watershed for construction of a dam and flooding of a basin and subsequent land-use patterns of unflooded slopes often include deforestation or other changes in the watershed's natural vegetation that increase erosion. Thus reservoirs receive high levels of sediments, generally causing faster filling of the basin than is characteristic of natural lakes. Small reservoirs may become more than 30 percent filled with deposited sediments in less than 30 years. Some large reservoirs may fill relatively quickly, whereas others may not fill for several centuries.

Although reservoirs collect sediments, they tend to lose nutrients rather than accumulate and recycle them through biotic communities as in natural lakes. Nutrient levels in reservoirs are extremely high after filling, due to organic detritus produced when the watershed was deforested, decomposition of herbaceous vegetation in flooded areas of the basin, and soluble ions leached from newly flooded soils. High flushing rates of the water in reservoirs cause this initial flux of nutrients eventually to be lost downriver, either as organic detritus and soluble ions or as phytoplankton and zooplankton that are washed from the system. Thus, instead of nutrient levels building through time as in lakes, early nutrient concentrations in reservoirs decline with time to more stable but much lower levels.

The flow into and out of many reservoirs is managed in a manner that produces reasonably stable water levels. Such reservoirs may possess well-developed littoral zones with expansive stands of rooted vegetation. However, in reservoirs managed for the generation of electricity, for municipal water supplies, for irrigation, or for flood control, water levels may rise or decline greatly as water is stored or released depending on the season and the uses of the water. Because nearshore substrates are alternately inundated and exposed to air, reservoirs with fluctuating water levels may not have well-developed stands of rooted aquatic plants, nor do they possess the littoral animal community that is associated with such habitats. In some highly fluctuating systems with very deep open-water areas, the benthic invertebrate community that serves as a food base for many freshwater species of fishes is largely missing or highly modified from that characteristic of systems with stable water levels (Ploskey 1986).

Sedimentation and early nutrient fluxes that decline with time produce fisheries resources that progress through an early stage of rapid population expansion and high productivity and then decline to a lower, more stable biomass and level of productivity. This normally occurs regardless of the fish species supporting the fisheries. The initial stage of high productivity is called the *trophic upsurge.* The inevitable decline to a lower level of productivity, the *trophic depression,* is caused by decreased nutrient loading, reduced availability of organic detritus, and changes in habitat conditions caused by sedimentation. The duration of the trophic upsurge varies greatly among reservoirs, depending on differing levels of nutrient loading and flushing, characteristics of watershed use, management of reservoirs and the nature of water releases, and the types of fishes present. However, in many systems the quality of fisheries and the productivity of fish populations supporting them may decline and stabilize at lower levels within 5 to 10 years after flooding.

RIVERINE HABITATS

The Morphology of River Basins.   A river system consists of a series of tributaries that coalesce into larger streams, forming a dendritic pattern of channels that drain a watershed. Small headwater tributary channels form in the highest elevations along the basin. As they move down slopes, headwater tributaries merge. Such mergers continue within the basin until the mainstem of the river system is formed. Because headwater systems are at the highest elevations in the basin, they receive water from limited drainage areas. Thus relatively low volumes of water flow through headwater tributaries. During dry periods of a year, many headwaters stop flowing and become a series of disconnected, small pools.

Larger tributaries carry more water because the area of the watershed that drains into these sites is greater. The mainstem, which is the "collector" of all surface water flowing through the watershed, generally has the largest volume of water passing through it of any section of the river system. Within many basins, once headwaters merge into larger tributaries, flow volumes may vary during a year, but flow will not cease. Generally, within a basin, the larger the area from which any particular reach of stream or river receives water, the greater the volume of water flowing at any given time, and the lower the relative variability of flow among days and seasons (Hynes 1970).

Source of Water.   Although precipitation provides the water that flows through river basins, only a fraction of the rain or snow that falls in an area enters river channels. The rest evaporates, is taken up and transpired by terrestrial vegetation, or sinks into underground aquifers (Hynes 1970). Water may enter the river channel as surface runoff during periods of precipitation or snow melt, or it can enter the stream from underground sources after it sinks into soils. The greatest proportion of a river's flow in semiarid regions with sparse vegetation comes from surface runoff. In less arid regions, rivers flowing through basins covered with native, uncut vegetation receive most water even during storms from underground sources (Winger 1981). Extensive removal of vegetation within a watershed will generally increase the input from surface runoff and decrease input from groundwater percolation into a river channel.

Gradient, Velocity, and Flow.   Stream gradient—the slope of the landscape—is defined as the vertical descent of the streambed over some horizontal distance (Reid and Wood 1976). Gradient is usually measured in meters descent per kilometer, or feet descent per mile, of river channel. Gradient is generally highest in a river system's headwaters and lowest near the mouth of the system. Gradient influences or is correlated with numerous characteristics of riverine habitats (Winger 1981). Gradient determines the velocity of water flowing through a river channel, which in turn influences features such as the types of substrates in the streambed, the frequency and size of pools and riffles, and the morphology of banks (sloped or undercut). Because gradient influences numerous stream habitat features, specific taxa of riverine plants and animals tend to be associated with particular gradients.

Water velocity at a specific site on a river channel varies according to the position within the water column. Friction between moving water and substrates of the streambed causes a reduction in velocity near the substrate. Lesser levels of drag are produced at the water's surface due to surface tension—the adhesion of water molecules to each other at the interface between air and water (Wetzel 1983). Thus, in a symmetrical river channel, the greatest velocity occurs just below the surface and equidistant between the stream banks (Hynes 1970). In asymmetrical streambeds, the highest velocity occurs closer to the deeper side of the stream (Figure 2.2). Varying velocities within a site always create some turbulence in flow (Hynes 1970). The amount of turbulence will vary with speed of flow and unevenness of substrate that the water is flowing over. Water flowing over a rock and boulder substrate will display markedly greater turbulence than water flowing at similar velocities over a smoother substrate.

Channel Morphology and River Channel Substrates.   Turbulently flowing water tends to erode substrates, cutting the streambed toward some base level (Reid and Wood 1976). High-gradient, high-velocity streams tend to erode vertically, creating a steep-sided, deep channel. Low-gradient systems with low-velocity flows tend to erode the sides of a stream's banks, creating a wide channel that is relatively shallow in relation to the volume of water passing through it (Hynes 1970). Through time erosion and deposition that occur as flow levels change seasonally will create a wide, flattened floodplain.

Pool habitats alternate with riffle habitats in reasonably consistent and predictable intervals in moderate- to high-gradient streams. Riffles—areas of faster-flowing, shallow water where the surface is noticeably turbulent—normally are

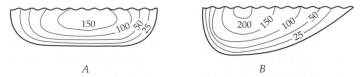

A B

Figure 2.2   Velocity patterns in a streambed that is symmetrical and one that is asymmetrical in cross section. Lines connect sites of equal velocity, here represented as centimeters per second, in the channel. (Reprinted by permission of H. B. N. Hynes, *The Ecology of Running Waters*. Copyright © H. B. N. Hynes 1970.)

spaced along a stream reach at regular distances of five to seven times the average width of the stream (Hynes 1970). They form less consistently in low-gradient, slow-flowing rivers whose substrates consist of sand, silt, or clay particles. However, slow-flowing, low-gradient rivers whose substrates are dominated by these small-particled substrates show a similar pattern of distribution of meanders or bends (Winger 1981).

Lateral erosion leads to the formation of bends in a stream when one bank is more susceptible to erosion than the other. Once bends are formed in a stream channel crossing a broad floodplain, erosional forces will increase the size of the bend, causing the stream to "meander" across the floodplain. This occurs because water flowing around a bend tends to hug the outside bank. The higher velocities on the outside of a bend create greater erosional forces, deepening that bank. A bend becomes more sharply angled with time because the outside bank is an area of high velocity and turbulence—thus net erosion—and the inside bank is an area of low velocity and net deposition of silt, gravel, and other materials (Hynes 1970; Figure 2.3). Once the angle of a bend becomes too sharp, flood waters may erode across the base of the bend, forming a straight channel and cutting the bend off from the river. When completely separated from the river proper, such bends are called oxbows (Figure 2.4). Widening and meandering of the channel across a floodplain creates other types of backwaters—side channels, sloughs, and marshes—which collectively provide a variety of nursery, shelter, and feeding habitats for fishes as well as provide sites of high biological productivity that contribute to the river system's food webs (Figure 2.5).

Numerous channelization projects have been carried out to prevent a stream from eroding floodplains whose rich soils are being used for agriculture. Similarly, dikes and levees have been constructed to prevent water from leaving a river's main channel during high water periods of the year. These activities destroy natural habitat conditions of river channels, cut off access of fishes and other aquatic organisms to all backwater habitats, and affect the productivity of food webs of such modified systems (see Chapter 9).

The average size of particles in the streambed decreases in a downstream direction due to the general relationship between gradient, velocity, and particle size (Hynes 1970). Boulder-, rock-, and gravel-dominated substrates are most typical of streams of moderate to high gradient. Only in the lower courses of river systems are

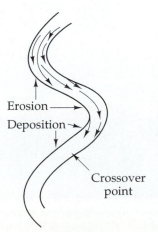

Figure 2.3 Patterns of flow and of erosion and deposition in a stream bend. The arrows depict direction of current. (Redrawn by permission of H. B. N. Hynes, *The Ecology of Running Waters*. Copyright © H. B. N. Hynes 1970.)

Erosion

Deposition

Crossover
point

Figure 2.4   An oxbow on the Connecticut River, Massachusetts. (Photograph by Alex S. MacLean; courtesy of Peter Arnold, Inc.)

substrates composed mostly of silt, clay, and organic debris (Reid and Wood 1976). Even in many low-gradient, slow-flowing streams, large volumes of silt occur mostly in backwaters and oxbows; the substrates of the main river channel will consist largely of sand particles (Hynes 1970).

Figure 2.5   An overview of the Platt River, Missouri, illustrating side channels and backwater habitats. (Photograph by Craig Aurness; courtesy of Westlight.)

Temperature.   The distribution of many fishes in river systems is influenced by seasonal temperature regimes. Trout and other taxa adapted to coldwater conditions are very susceptible to changes in seasonal temperatures brought about by the removal of tree canopies in the riparian zone. The removal of shade increases solar radiation that strikes the surface of the water, potentially warming it well above standard summertime temperatures, and well beyond the temperature ranges that are best suited for these taxa (chapter 9).

Turbidity and Sedimentation.   Excessive siltation, or the infusion of eroded sediments into river systems, is a major factor affecting fish habitat in the majority of U.S. rivers (Rabeni 1993). Some rivers are constantly turbid. For example, glacier-fed streams of Alaska and Canada and some rivers of the prairies of the mid-continental United States carry high concentrations of soil particles through high and low flows (Hynes 1970; Moss 1988). However, water in many North American river systems is relatively clear during times of moderate to low flow, becoming highly turbid only during high-water periods. As a rule, there is a close relationship between flow volumes and relative loading of silt and clay particles into a stream. Excessive, persistent turbidity is usually the result of improper land use practices that cause excessive erosion (Winger 1981). The transport of excessive amounts of sediments in rivers can affect fish communities in several ways (Rabeni 1993):

1. High turbidity limits light penetration into a water column, thus inhibiting photosynthesis and plant production.

2. Sediments can degrade spawning substrates in riffle systems, increasing egg and larval mortality.

3. Sediments can reduce production of aquatic insects and other invertebrates that serve as the food base for many stream fish species.

Dissolved Oxygen.   Due to natural turbulence at the water's surface, oxygen levels in unpolluted streams are usually high due to constant dissolution at the water's surface from atmospheric oxygen (Winger 1981). Dissolved oxygen concentrations in streams typically will not become severely depleted unless heavy loads of organic materials from sources such as sewage treatment plants, certain types of industrial wastes, or tree and leaf litter from heavily logged hillsides are released into streams. In such systems, decomposition of the organic load by aerobic microbes may cause oxygen depletion (Winger 1981). However, oxygen levels typically return to suitable concentrations downriver from the zone of organic decomposition.

Biology of River Systems

Primary Production and Heterotrophy.   Dissolved and particulate organic material from decomposing vegetation is abundant in river systems. Much of this organic material is allochthonous—it is derived from leaves and other vegetative material from the riparian plant community that are washed into the river. Allochthonous organic material supports a major proportion of the biological production that takes place in the food webs of stream and river systems (Moss 1988).

Primary production is limited in many river systems, in part due to the low concentrations of nutrients typical of many waters not culturally "enriched." In fast-flowing headwaters, primary producers are often limited largely to benthic diatoms and filamentous algae. Vascular aquatic rooted plants and phytoplankton occur more commonly in lower-gradient rivers with relatively large and slow flows. However, high turbidity in many major rivers can limit the distribution of these plant taxa to shallow backwater areas (Sheehan and Rasmussen 1993). The food webs of these systems receive only limited energy input from primary producers. Such food webs depend more on heterotrophic processes, where energy is derived from decomposition or consumption of organic materials, than on autotrophic processes, where energy is derived within the system through photosynthesis (Hynes 1970).

Consumers.   Few species of fishes in temperate-region rivers graze on plant material. Many of the invertebrate "grazers" or primary consumers eat leaf litter or the decomposer microbes that are biochemically breaking down this detritus. For example, about half of the invertebrate food eaten by young coho salmon in two Oregon streams came from food webs supported by allochthonous plant materials (Chapman and Demory 1963).

As with phytoplankton, zooplankton communities are most fully developed in slow-flowing rivers, and even in these systems they may be limited due to low phytoplankton production resulting from high levels of turbidity. The types of invertebrates dominating the benthic riverine habitats depend on the types of substrates present. Sand supports low biomasses of invertebrates because it is a constantly shifting, unstable substrate. Gravel and stone substrates support a variety of invertebrates including mollusks (both snails and mussels), crayfishes and gammarids, and adult and immature stages of numerous taxa of aquatic insects such as stoneflies, mayflies, caddisflies, and midges and other dipterans that form the food base of many fish species. Mud substrates of slow-flowing systems also house numerous invertebrates, although the taxa are different from those in rocky substrates (Hynes 1970).

The distribution of particular fish species in river systems is correlated to gradient, because this characteristic influences numerous habitat features of streams (Trautman 1942; Hocutt and Stauffer 1975). Other factors such as temperature patterns, flow rates, sedimentation, dissolved oxygen concentrations, and abundance of food also affect the general patterns of distribution and abundance of fish species within river basins (as in Trautman 1981; Berkman and Rabeni 1987; Cada et al. 1987; Chapman 1988; Harris et al. 1991; see Chapter 9). The diversity of the fish community is related to the relative position within a river basin. Headwaters house relatively few species of fishes. The number of species inhabiting particular sites increases consistently from the headwaters to the mainstem of a river system (Winger 1981).

THE IMPORTANCE OF RIPARIAN ZONES AND BACKWATER HABITATS TO FRESHWATER SYSTEMS

Both aquatic and terrestrial habitats adjacent to the margins of standing waters or stream channels have significant effects on the productivity of freshwater food webs and the suitability of habitats for particular species of fishes.

The riparian zone and associated wetland and deepwater habitats serve several functions in the maintenance of particular habitats in riverine and lacustrine systems. Wetlands include all seasonally or permanently flooded areas with emergent herbaceous vegetation, shrubs, or trees, and deepwater habitats include all back-waters, oxbows, or other flooded sites that are too deep for emergent vegetation. Tree canopies shade the surface of small streams, reducing direct solar radiation and preventing substantial warming of the waters (see the section on logging in Chapter 9). The root systems of vegetation growing along the edge of rivers stabilize stream bank soils, reducing the potential for excessive erosion and silt loading into the river channel (Wesche 1993). Root systems and fallen trees provide cover for fishes. Streamside vegetation provides leaf litter and other organic detritus that supports heterotrophic production, and serves as a source of terrestrial and aerial invertebrates, particularly insects, that drop to the water's surface and become food for fishes (Wesche et al. 1987).

During high waters, the velocity of water that overflows the stream channel is slowed by herbaceous vegetation and the stems and trunks of woody plants of the riparian zone and associated wetlands. This tends to slow the discharge of water, increasing seepage into underground aquifers and reducing the impact of flooding in downriver sites. Increased seepage into underground aquifers produces greater discharge of water down the river channel during subsequent dry periods of the year (Golet et al. 1993; Wesche 1993). Stable discharges, reduced silt loading, and stabilized banks enhance the spawning success of substrate spawning fishes and production of benthic invertebrates in rivers (Wesche 1993).

Wetlands and deepwater habitats provide critical spawning and nursery grounds for a variety of fishes. Over 40 species of fishes are associated with coastal marshes of western Lake Erie (Herdendorf 1987). About 50 percent of the fish species present in the lower Mississippi River use floodplain wetlands as nursery grounds, 7 of the 10 most common species spawn in backwater habitats (Wharton et al. 1982). Backwaters provide not only shelter but also substantial food resources to fishes. The loss of backwaters substantially reduces fish production in river systems (see the section on channelization in Chapter 9).

COASTAL HABITATS

Estuaries and Delta Salt Marshes

Estuaries and coastal delta salt marshes form at the mouths of rivers and coastal streams. Thus they are sites where freshwater mixes with and dilutes saltwater. Estuaries form in bays where barrier beaches, built up by the deposition of sand and gravel, develop where the bay opens into the coastline, producing a largely enclosed, protected body of water. Delta marshes project out from the coastline, occurring where sediments deposited by a river build a large expanse of substrate that is alternately flooded and drained by rising or falling tides. Both of these systems typically exhibit high levels of primary production. This supports active food webs within these ecosystems and in adjacent coastal areas through tidal movement of plant detritus (Odum 1980; Teal 1986). As they move along the coastline, many coastal fishes feed within estuarine or delta marsh habitats. Eighty percent of recreationally and commercially important finfish and shellfish species of the Atlantic coast, and nearly all those of the Gulf coast, depend on estuaries or delta salt marshes to complete at least one critical phase of their life cycles (Arnett 1983;

Gosselink 1984). Although commercially and recreationally important species of the Pacific coast of the United States generally are less estuarine dependent, some taxa including Pacific salmon species, striped bass, and flounder and sole species do use these habitats (Zedler 1982; Seliskar and Gallagher 1983; Simenstad 1983).

Impact of Tides and River Flows on Estuarine Habitats. Barrier beach ridges protect the estuary from the full force of tidal currents and waves that strike coastal shore areas. This protection makes estuaries quiet bodies of water in which the incoming flow of inland streams and tides is slowed. Because of this, estuaries are sites of sediment deposition, from both incoming tides and outflowing streams. As rising tides carry over barrier beaches, their flow is slowed to the point where large particles such as sand and small gravel can no longer remain suspended in the moving water, thus settling to the substrate. As tidal currents continue into the estuary, their flow is further slowed due to the drag caused by the bottom of the shallow basin and to the counterdirectional flow of freshwater entering from streams. Because of this further decrease in the rate of flow, smaller silt and clay particles that can no longer be carried by the water's movement sink to the bottom. Likewise, the velocity of streams flowing into the estuary is slowed by the opposing force of tides and the width of the area the streams flow into as they leave the stream channel and enter the broader estuary basin. This process also causes settling of suspended particles. Sediment deposited at the mouth of an estuary is dominated by coarse sand and gravel; sediment farther into the estuary is dominated by silt and clay, forming soft-mud substrates (Nixon 1982).

Tidal intrusions affect other characteristics such as salinity and temperature of an estuary, creating markedly different habitat conditions from site to site within the estuarine system. In addition, the oscillating nature of tides creates constantly changing conditions at any particular site. Tides cycle from one high tide to the next about every 12 hours and 25 minutes or about every 25 hours (Stout 1984). Tidal amplitude—the change in water level between high and low tide—is determined by the shape and bottom contour of the estuary, the size of the estuary's mouth, and other physical factors of the area and region, as well as the phase of the moon. Gravitational forces of the moon and sun create greater tidal forces, thus higher tides, during full and new moon phases than at other times of the lunar cycle. In some areas the two semidiurnal tides may display different amplitudes within any daily tidal cycle (as in Seliskar and Gallagher 1983). Tidal amplitude typically varies among regions of the country (Table 2.1). Unless dammed, streams flowing into estuaries do not display major changes in discharge within a day but will do so seasonally. These daily and seasonal changes in the amount of freshwater and saltwater input create extremely variable salinity and temperature conditions in an estuary, not only from site to site but also from time to time.

In many estuaries, freshwater flows seaward over denser, inflowing tidal saltwater, because saltwater has a higher density due to the dissolved mineral salts (Reid and Wood 1976). Thus differences in salinity occur vertically in the water column. Incoming high tides flow along the estuary's bottom farther inland than do low tides. Thus the salinity above the sediment at particular sites shifts markedly during the tidal cycle. In other estuaries, vertical differences in salinity may be less marked due to mixing of the fresh and salt waters. In such instances, the salinity throughout the water column at many sites within the estuary is constantly shifting.

TABLE 2.1 Tidal amplitude of areas of the Atlantic and Gulf coasts of the U.S.

Area	Average Tidal Amplitude in Feet (Meters)	Source
Northern Maine	maximum of 20 (6.1)	Whitlach 1982
New England, north of Cape Cod	9–12 (2.7–3.7)	Whitlach 1982
New England, south of Cape Cod	3–5 (0.9–1.5)	Whitlach 1982
Southern North Carolina	nearly 0–6 (0–1.8)	Bahr and Lanier 1981
South Carolina and Georgia	6–13 (1.8–4.0)	Bahr and Lanier 1981
Gulf coast of southern Florida	2–3.5 (0.6–1.1)	Stout 1984
Gulf coast, western Florida to Mississippi	less than 2 (0.6)	Stout 1984
Mississippi River delta of Louisiana	1 (0.3)	Gosselink 1984
Southern California	3.5 (1.1)	Zedler 1982
Pacific Northwest	up to 10 (3.0)	Seliskar and Gallagher 1983

Differences between the temperature of freshwaters and saltwaters entering estuaries cause similar fluctuations in temperature in the areas of mixing. Because of its much smaller volume, water in coastal streams and rivers responds much more rapidly to daily and seasonal changes in air temperature than does water along the coast. Generally, upper waters of an estuary will be cooler in the winter and warmer in the summer than waters near the estuary's mouth (Reid and Wood 1976).

Thus tidal activity creates a dynamic system that is constantly fluctuating, causing the development of several different habitat types arranged sequentially from the inland edge of the estuary to its mouth. Estuarine habitats are typically classified according to physical and chemical conditions, and to the dominant vegetation that occurs in areas with particular sets of those conditions. Major habitat types include the high salt marsh, intertidal (between the high and low tide marks) mud flats, low salt marsh, and deeper channels and areas that are generally inundated with water.

High Salt Marsh. This portion of the estuary is not flooded much of the time but is affected by high waters of the tidal cycle. The boundaries of the high salt marsh are most frequently defined by the distribution of vegetation in this habitat rather than by environmental conditions established by tides. Upland from the high marsh may be a band of freshwater wetland vegetation, such as rushes from the genus *Juncus.* This vegetation separates the high salt marsh from typical upland tree and shrub vegetation that is intolerant of saltwater (Nixon 1982). The high salt marsh is dominated by salt-marsh hay (*Spartina patens*), spike or salt grass (*Distichlis spicata*), or other salt-tolerant species (Nixon 1982; Seliskar and Gallagher 1983). Bertness (1991) found that the distribution of high salt marsh vegetation is determined by interspecific competition. Salt-marsh hay and spike grass cannot compete with rushes in the freshwater wetland margin of the estuary. Thus these plants are restricted to more saline conditions of the high salt marsh. In Bertness's study, both high marsh plants exhibited maximum growth rates when transplanted into freshwater wetland margins of estuaries where rushes were eliminated. In addition, the distribution of spike grass is determined by the presence or absence of salt-marsh hay, as the former species cannot compete with the latter.

Substrates in high marsh areas consist of fine silt and clay particles, with some organic detritus. This habitat typically has tidal creeks winding and branching through it, and it is dotted with numerous pond holes, or pannes—shallow depressions of the marsh area lacking vegetation and often holding small, permanent pools of water. Rising tides move through tidal creeks, finally wetting the land between the channels and flooding the pannes. Fishes are largely absent from the high marsh except during high tides. However, killifishes, sticklebacks, and other small-bodied species that can tolerate substantial changes in salinity within the daily tidal cycle may be residents of some pannes.

Tidal Flats. Tidal flats are unvegetated sand or mud substrates that occur in estuaries between the high and low tide marks (Peterson and Peterson 1979). Tidal flats are often bordered by salt marshes landward and by deep channels or eelgrass subtidal habitats seaward (Whitlach 1982).

Mixtures of sand and clay particles make up the sediments of tidal flats, with sand dominating where tidal currents are greater, usually toward the mouth of the estuary. Tidal flat sediments contain relatively high levels of organic detritus that is brought in and deposited by tides washing over and receding from the salt marsh and tidal creeks (Welsh 1980) and from eelgrass beds of subtidal habitats—habitats below the low tide line that are permanently inundated (Whitlach 1982). Tidal flats are well known for their highly productive food webs.

A variety of benthic epifauna—fauna living on the surface of the substrate— are present on tidal flats, including gastropods such as periwinkles and mud snails, blue crabs, fiddler crabs and other macrocrustaceans, whelks, and bivalves (Peterson and Peterson 1979; Whitlach 1982). The blue mussel often dominates the epifauna on the Pacific coast, whereas the American oyster does so in many south Atlantic and Gulf coast flats (Peterson and Peterson 1979). Expansive oyster reefs occur in the intertidal flats of many estuaries from southern North Carolina to mid-Florida (Bahr and Lanier 1981). The density of infauna (organisms living in the substrates) of bare mud flats can be substantially lower than in substrates covered with seagrasses (Peterson and Peterson 1979), but numerous polychaete worms, bivalves such as the quahog, soft-shell clam and hard-shell clam, amphipods, and other invertebrates are common.

Because they are productive feeding areas, tidal flats teem with fishes during high tide. Small-bodied species that gather food in the water column and bottom-feeding fishes such as flatfishes and skates are particularly dependent on tidal flats for food (Whitlach 1982). Due to their complex food webs, tidal flats are noted for converting plant production into animal biomass, as opposed to the more landward marsh habitats that are seen largely as exporters of plant materials to flats and other adjacent waters.

Regularly Flooded or Low Salt Marshes. Low salt marshes are flooded by all tides under normal conditions (Teal 1986). Smooth cordgrass (*Spartina alterniflora*) usually dominates the vegetation of the low salt marsh of Atlantic and Gulf coast estuaries and open coastal marshes, whereas species such as cordgrass (*Spartina foliosa*), pickleweed (*Salicornioa virginica*), and Lyngbye's sedge (*Carex lungbyei*) do so on the Pacific coast (Zedler 1982; Seliskar and Gallagher 1983; Gosselink 1984; Teal 1986). Low salt marshes house many fishes, including small-bodied residents such as mummichogs, killifishes, sticklebacks, northern anchovy, and Atlantic silversides. Such taxa congregate in tidal creek channels during low tides and spread throughout the low marsh as the tide rises. Other, larger-bodied species move into and out of this habitat according to water levels during rising or dropping tides.

In an estuary, most of the primary production that occurs in the low salt marsh enters the food webs as detritus supporting decomposer organisms. Only about 10 percent of the smooth cordgrass production in the low salt marsh is grazed on directly by animals (Seliskar and Gallagher 1983).

Subtidal Flats and Seagrasses.   The lower (seaward) reaches of estuaries typically have deeper channels than are seen landward. Larger systems have extensive areas that are subtidal and sufficiently deep to exclude emergent vegetation such as smooth cordgrass. The substrates of this portion of an estuary may be unvegetated or may possess submergent rooted vascular plant taxa that collectively are referred to as seagrasses. Subtidal habitats may constitute a substantial proportion of the total estuarine area. About 84 percent of the surface area of Texas estuaries is subtidal, with over 80 percent of that area composed of bare substrate and about 20 percent covered with seagrass (Armstrong 1987). Two major types of seagrass meadows occur within coastal waters of the United States: eelgrass communities (*Zostera* spp.) on the Atlantic and Pacific coasts and turtle grass communities (*Thalassia* and *Halophila* spp.) of the Gulf of Mexico and the Atlantic coast from Mid-Florida southward.

Planktonic algae are the major primary producers in the open waters of the subtidal area of estuaries (Armstrong 1987). A variety of benthic invertebrates inhabit these areas, including polychaete worms, clams and other mollusks, and crabs and smaller crustaceans. Many fish species feed on abundant invertebrates and smaller fishes of this subtidal habitat.

Seagrasses contribute substantially to primary production of subtidal areas. Light penetration into the water column determines the depth of water in which seagrass beds can grow. Eelgrass meadows are limited to waters less than 7 feet deep in turbid coastal plains estuaries such as those in North Carolina, but they can occur in waters greater than 30 feet deep in less turbid estuaries of New England (Thayer et al. 1984). Seagrass meadows help stabilize subtidal substrates, reducing the abrasive and eroding effect of tidal currents. Such habitats also provide nursery habitat to young fishes, protecting them from predators and providing a plentiful food supply. Because the young of some fish species, and all ages of many small-bodied species, utilize seagrass so extensively, large predators tend to feed around the fringes of seagrass meadows (Thayer et al. 1984). The density of fishes in south Florida's estuaries are much higher in seagrass meadows than any other habitats (Zieman 1982).

Due to the high levels of primary production by seagrasses and the algae that live on them, seagrass meadows play an important role in the food webs of estuaries and adjacent coastal waters. These habitats are responsible for the export of plant materials to the coast via tidal currents, and the export of other portions of the food web by mobile predators that feed around seagrass and exit the estuary for open water. Seagrass meadows also provide important nursery and foraging habitats in open coastal areas for fishes that reside in adjacent coral reefs or other open-water habitats of the inner continental shelf (Zieman 1982; Jaap 1984).

Oyster Reefs of the Southern Atlantic Coast.   Although seagrass is present in most regions of the coastal continental Unites States, seagrass meadows are absent from the south Atlantic coast from southern North Carolina to mid-Florida. In this region, large reefs of the American oyster provide intertidal shelter and feeding opportunities for fish communities that are similar to those offered by subtidal seagrass meadows in other regions (note that many oyster reefs in North Carolina are subtidal; see Peterson and Peterson 1979). Oyster reefs stabilize substrates of estuarine flats and provide stable islands of hard, highly structured habitat in otherwise mud-bottomed, open areas of estuaries. These reefs provide shelter sites to a diverse community of invertebrates and small-bodied fishes, attracting large-bodied fish predators that feed on these forage taxa (Bahr and Lanier 1981).

The Mississippi River Delta Plain

The Mississippi River Delta of Louisiana, the largest continuous wetland system of the United States, has nearly 1.8 million acres of marsh habitat. Much of this area is composed of brackish water and saline marsh habitats (Gosselink 1984). Unlike estuarine systems, which are partially enclosed basins, the salt marshes of the Mississippi Delta are exposed to the open coastline. The delta has been built through the deposition of sediments by the Mississippi River. Deposition during this century has been markedly reduced because of levees and dikes that no longer allow floodwaters to flush over the marshes of the delta; the rate of deposition is now only about 60 percent of that characteristic prior to regulation of flow in the lower river (Gosselink 1984; see Chapter 9 for a discussion of wetland loss in this delta).

Zonation of vegetation of delta marshes is similar to that of estuaries. Salt-marsh hay dominates the brackish water marsh, which is the equivalent of the high salt marsh of estuaries, whereas smooth cordgrass dominates the saline marsh, the equivalent of the low salt marsh of estuaries. As in estuarine salt marshes, food webs are based on the decomposition of plant detritus. Decomposing bacteria are fed upon by small invertebrates that form the food base of larger invertebrates and small fishes (Gosselink 1984). The delta marshes are extremely important as nursery and feeding grounds to nearly all of the fishes and shellfishes that support important commercial fisheries of the region.

Habitats of the Coastline

Sandy Shorelines.   Tidal action on open shoreline areas influences the types of biotic communities, and of fisheries resources, that inhabit inshore habitats. Many shoreline areas are dynamic habitats, with sandy sediments constantly being deposited, eroded, and shifted by the actions of tides and wind-driven waves. Sediments of the intertidal zone are particularly unstable; particles are constantly being lifted, carried, and deposited by turbulent waters. Because of seasonal changes in the force of shoreline waves (the changes are due to the seasonal frequency and force of storms that drive waves onto the shoreline), the shape and configuration of the submerged portion of the beach area may change greatly throughout the year and between years. Because of the unstable nature of sand sediments, few attached plants will be found living on submerged sand substrates, although waves will wash in large amounts of algal masses or seagrasses "uprooted" from somewhat deeper waters.

Intertidal and subtidal areas serve as active feeding sites for fishes because the turbulent waters and shifting sediments stir up small invertebrates upon which many fishes feed. In turn, larger predators pursue the abundant smaller fishes. Tidal rips can be particularly active feeding areas for shoreline fishes. A rip occurs where rising or dropping tides flow essentially unidirectionally through a physically constricted area, such as the narrow opening to an estuary, or low areas in elongate submerged sand or gravel bars that lie just offshore, parallel to the shoreline (Davis 1987). In the latter instance, the tide flows along the shore until it empties through the openings in the bar. Tides may flow similarly along certain jetties or other artificial shoreline structures. In such areas, the forceful movement of water actively stirs sediments and creates an active feeding site for fishes.

Rocky Shorelines.   Rocky shorelines provide a more stable substrate than do sand beaches. Thus these habitats contain a broad array of attached algae, including large green, red, and brown algae that form mats on the rocks and boulders of the intertidal and subtidal zone. The intertidal zone possesses a series of algal assemblages (groups of specific taxa), organized in vertical groupings between the high and low tide marks. Algal taxa that live in the intertidal zone are each adapted to survive particular levels of exposure to air during the tidal cycle. Thus they are vertically segregated along the boulders and rocks of this zone (Reid and Wood 1976). Other algal forms sensitive to exposure to air are abundant below the low tide line in the subtidal zone. A variety of mollusks and other invertebrates inhabit the intertidal and subtidal rocky shorelines, and numerous fish taxa feed within the intertidal or feed and reside within subtidal rocky habitats.

Reefs of Southeastern United States.   Although much of the natural structure that provides shelter and nursery areas for fishes is associated with estuarine or shoreline areas, vertical structure is found in more open coastal waters. For example, rock outcroppings, coral reefs, and artificial structures (see Chapter 10) constitute nearly 30 percent of the continental shelf between Cape Fear, North Carolina, and Cape Canaveral, Florida (Plan Development Team 1990). These structures provide a critical habitat for a variety of fish taxa that support important recreational and commercial fisheries in southeastern United States.

Reef fishes utilize a variety of areas with abundant shelter sites, or hiding places, including oyster reefs, seagrass meadows, rock and boulder shorelines, coral reefs, and artificial structures including sunken ships, waste concrete rubble, mounds of tires, and other materials (see Chapter 10). In the United States, coral reefs are limited to the eastern coast of Florida and Hawaii and the South Pacific Territories (Jaap 1984). Coral reefs house extremely diverse fish communities, including recreationally and commercially important taxa such as grunts, snappers, and groupers. Shellfishes such as the spiny lobster also support important commercial fisheries (Jaap 1984). Because many of the fish taxa use reefs for shelter and feed in adjacent sand flats or seagrass meadows, fishes constitute a major proportion of the animal biomass in coral reef habitats. Although a wide variety of species may be present on coral reefs, a small number usually make up the majority of the recreational or commercial harvest (Jaap 1984).

Kelp Forests of the Pacific Coast. Kelp forests also support unique fish communities. Kelp forests occur where dense stands of attached algae essentially fill the water column from the substrate to the surface (Figure 2.6). Although kelp forests are widespread in the southern hemisphere, north of the equator the taxa of algae that create this type of habitat occur only from Alaska to Baja California, Mexico (Foster and Schiel 1985). Within a region of the west coast, particular species of algae are restricted to particular depth zones. For example, the brown alga *Macrocystis pyrifera*, which forms the giant kelp forests of southern California, is usually found in waters between 4.6 and 20 meters (15 and 65 feet) deep, although it may occur in waters as deep as 30 meters (100 feet). Kelp forests are highly productive systems. Kelp itself has been harvested, first for potash to make gunpowder during World War II and more recently to manufacture algin, an emulsifying agent used in foods and pharmaceuticals. Abalone, sea urchins, spiny lobster, and fishes such as sea basses and rockfishes are recreationally and commercially harvested from kelp forests (Foster and Schiel 1985). Kelp forests provide the greatest proportion of harvests of some of these taxa. As much as 90 percent of the "rockbass" (kelp bass and sand bass) harvested in southern California are taken from giant kelp forests (Foster and Schiel 1985).

Figure 2.6 A coastal California kelp forest. (Photograph courtesy of Gregory Ochocki; Photo Researchers, Inc.)

Continental Shelf

The greatest proportion of fisheries resources harvested within territorial boundaries of the United States and on a worldwide basis comes from continental shelves (Davis 1987; NOAA 1991). Continental shelves are submerged extensions of continental masses, extending seaward from the coastline to the 200-meter-depth (approximately 660-feet) contour. The width of the shelf area varies according to the region of the country; it is greatly expanded in areas of the Atlantic coastline (particularly New England and the fishing banks of the Atlantic Maritime Provinces of Canada), the eastern Gulf of Mexico, and Alaska (particularly the Eastern Bering Sea; see Figure 2.7). Along west coast states other than Alaska the shelf is narrow and steeply sloping (Reid and Wood 1976; Ingmanson and Wallace 1979). In some regions of shelf expansion, the outer shelf displays a series of offshore elevations, called banks, which are great expanses of shallow water that are separated from each other and the mainland of the continent by broad, deep channels (Figure 2.8).

Continental shelf waters are shallow enough that they are thoroughly mixed from surface to bottom for much or all of each year. This mixing, caused by tidal currents and storm-driven waves, carries nutrients from bottom sediments up into the

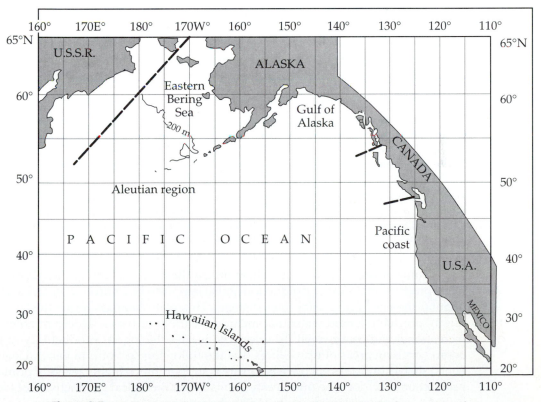

Figure 2.7   The continental shelf off Alaska, marked by the 200-meter depth contour, is as large an area as much of the western United States. (Reprinted from Loh-Lee Low (editor), 1991. Status of living marine resources of Alaska as assessed in 1991. U.S. Dept. of Commerce, NOAA Tech. Memo, NMFS F/WC-211, 95 p.)

Figure 2.8 Georges Bank and the Scotian Shelf banks of the New England and southern Atlantic Maritime Province region of North America. (Reprinted by permission of *Recreational Fisheries of Coastal New England*, by Michael R. Ross. With contributions by Robert C. Biagi [Amherst: University of Massachusetts Press, 1991], copyright © 1991 by the University of Massachusetts Press.)

water column, where planktonic algae utilize them to sustain high levels of primary production. Primary production is also supported by nutrients entering continental shelves from coastal river systems. This plant biomass forms the basis of highly productive food webs that include a variety of fishes widely pursued in recreational and commercial fisheries. The highest biological productivity in the ocean occurs in areas of upwellings (see the following section). In general, continental shelves and adjacent open ocean areas support much higher levels of primary production than does the mid-ocean (Davis 1987). Biological productivity of large expanses of the open ocean is limited by the extremely low concentrations of soluble nutrients (particularly phosphorus and nitrogen) that occur in the upper water column. The distance from continental coastlines and the lack of mixing of the water column from surface to bottom limit the concentrations of these critical nutrients.

Thus marine fisheries harvests are greatest on continental shelves not only because of their closer proximity to human populations but also because of the greater productivity of the resources themselves in these regions.

Ocean Currents.   Surface currents of the ocean result from friction caused by the circulation of the atmosphere over the water's surface (Davis 1987). In the northern hemisphere, major subtropical currents circle the Pacific and Atlantic oceans in a clockwise direction. Tropical gyres—circular current patterns resulting from surface winds, rotation of the earth, and the Coriolis effect—circulate in a counterclockwise direction (Figure 2.9). On the Atlantic coast of the United States the Gulf Stream flows in a northeasterly direction. On the Pacific coast the West Pacific Current splits as it approaches the U.S. coastline, flowing southward along southern California (the California Current) and northward to the Alaskan coast (the Alaskan Current). The divergence of surface waters on the Pacific coast creates a large area of upwelling—the rising of water from ocean depths to the surface. Upwellings of the world's oceans transport high concentrations of nutrients from bottom sediments to the surface, creating high levels of primary production. Highly productive pelagic fish populations, such as the Pacific sardine on the west coast of the United States and the Peruvian anchoveta, have supported expansive fisheries in upwelling regions of oceans.

Within regions, current patterns can be very complex. Complex patterns result from the combined effects of mixing or divergence of major currents, topographic features of the coastline, and the influence of freshwater input from major river systems (for example, currents of the Northwest Atlantic coastline; see Figure 2.10).

Temperature Patterns.   Much of the ocean is cold. Less than 10 percent of the ocean is warmer than 10°C and less than 25 percent is warmer than 4°C (Davis 1987). This is largely due to the huge volume of water that is below the thermocline. Surface water temperatures generally correspond to latitudinal position (Davis 1987), although major currents alter this basic pattern. The Gulf Stream and the Alaskan Current carry warm water much further north on the Atlantic and Pacific coasts, respectively, than would otherwise occur. In addition, the upwelling of deep waters that occurs in the mid-Pacific coast of the United States produces coldwater conditions in warmer months of the year. For example, August surface water temperatures in southern California are colder than those in the much more northerly waters of Nova Scotia on the Atlantic coast of Canada; this latter region is swept by the warm waters of the Gulf Stream (Davis 1987). Similarly, in New England the surface temperatures of the Gulf of Maine (which receives waters from the Labrador Current and the Gulf of St. Lawrence to the north) are considerably colder than adjacent Georges Bank, whose outer reaches are influenced by the Gulf Stream. Seasonal and yearly temperature regimes influence the distribution and extent of migration distances of oceanic fishes.

Salinity.   The salinity of open ocean surface waters averages 35 parts per thousand, or 35 grams of salt ions per kilogram of water, and does not vary beyond 33 and 37 parts per thousand (Davis 1987). The flow of freshwater from rivers affects salinity not only of water along the coastal shoreline but waters well onto the continental

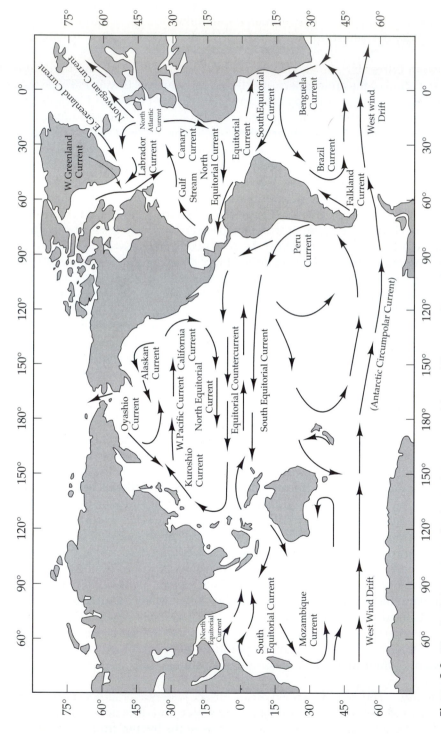

Figure 2.9 The circulation pattern of major surface currents of the world's oceans. (From: Stowe, *Ocean Science*. Copyright © 1979. Reprinted by permission of John Wiley & Sons, Inc.)

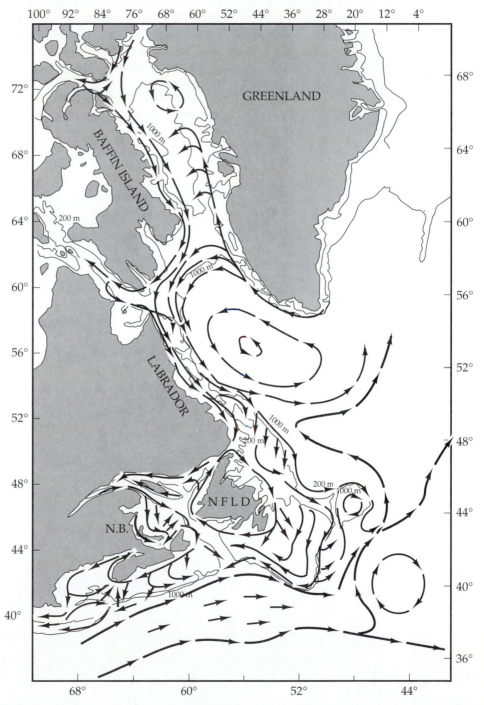

Figure 2.10 Surface currents of the Northwest Atlantic coastline. (From W. B. Scott and M. G. Scott, *Atlantic Fishes of Canada*. Copyright © W. B. Scott and M. G. Scott, 1988. Reprinted by permission.)

shelf. Salinity increases steadily from the coastline seaward until it reaches standard seawater concentrations. For example, freshwater from the Gulf of St. Lawrence and rivers of New England mixes with saltwater and flows southwestward along the New England continental shelf. This input of freshwater keeps salinity below 32.5 parts per thousand throughout most of the continental shelf of that region (Budgen et al. 1982). Salinities along much of the rest of the shelf area of the 48 conterminous states are somewhat higher, whereas those of the Alaskan shelf areas are similar to those of New England (Ingmanson and Wallace 1979).

3

Life Cycles, Population Processes, and Community Interactions

With nearly 25,000 living species fully described and perhaps several thousand more yet to be named, fishes are the most diverse group of vertebrates (Moyle and Cech 1996). Although the oceans cover about 70 percent and freshwater systems only 1 percent of the earth's surface, more than 40 percent of the described fish species live in freshwater systems, and an additional 1 percent spend part of their life in freshwater and part in saltwater; the rest live in saltwater throughout their lives (Moyle and Cech 1996). Because of their evolutionary diversity and of the wide variety of aquatic habitats they inhabit, fishes exhibit an array of life-cycle patterns that allow survival under the set of conditions characteristic of the environment around them.

The fisheries manager must understand how basic components of a fish's life cycle and its interactions with other members of biotic communities function to predict how fishing and other human activities might affect specific fish populations. This chapter focuses on aspects of fish ecology that most centrally influence conservation and management strategies, including the features of the life cycle that determine the reproductive capacity of fishes, the potential of fish populations to produce biomass that is available for harvest, and interspecific interactions that determine the role a particular species fills in fish communities. The chapter also examines how fisheries agencies use such information to develop effective fisheries conservation and management programs.

HOW ADAPTATIONS ARISE

Species are typically divided into geographically separated units called populations. Populations are groups of organisms that reproduce freely among themselves but are largely isolated, normally due to geographic separation, from breeding with individuals of the same species in other populations. Evolution is the result of interaction between a gene pool—the total genetic makeup of all members of a population—and the environment in which the population exists. Genetic makeup varies among individuals within populations due to mutation and other genetic changes as well as the recombination of gene pairs that occurs when two individuals reproduce sexually. If genetic differences among individuals produce traits that influence the relative number of surviving offspring the individuals produce, the frequency of occurrence of such traits within the population will change with time. Genetic traits that improve the average rate of offspring survival will increase in frequency in the gene pool, whereas the frequency of traits that lessen the rate of offspring survival will decline. Note the importance of the term *surviving offspring:* Adaptations arise when a genetic trait is passed from one generation to the next, gradually increasing in frequency because individuals carrying the trait possess some survival or reproductive advantage over those that do not. If traits produce such advantages in reproductive success persistently from generation to generation, the genetic material that produced the trait should ultimately become the dominant form in the population. Natural selection is the process whereby advantageous traits increase in frequency within a gene pool. Thus adaptations arise in populations because some individuals contribute more to the population's future gene pools than do others not possessing the same genetic traits. In order to best understand how traits might be adaptive, they will be presented and discussed using this evolutionary perspective.

LIFE HISTORY CHARACTERISTICS AND REPRODUCTION

Fecundity

Fecundity is typically defined as the number of eggs produced by a female (Moyle and Cech 1996). The number of offspring produced by a male fish is related to the number of females with whom he mates but is largely independent of the number of gametes (sperm) he produces. However, the number of offspring produced by a female is obviously determined by, and limited to, the number of eggs she produces. Similarly, the number of offspring produced by a population is related to total egg production of females within the population. Fishes display extreme variation in female fecundity, both among individuals within species and among species.

Variability of Fecundity Within Populations. Within a population of fishes, fecundity is directly related to body size of females. The difference in the number of eggs produced by small and large females of the same species is related to the maximum fecundity typical of that species. Thus the number of eggs produced by female Atlantic cod, a species exhibiting relatively high levels of fecundity, ranges from about 200,000 to 12 million eggs, whereas egg production in the smallmouth bass ranges from about 2,000 to 20,000 eggs (Carlander 1977; Scott and Scott 1988).

At any time in its life, a fish has specific levels of energy reserves that are distributed among all of its current needs, such as the need to avoid predators, forage for food, grow, reproduce, and retain physiological balance, or homeostasis, that is necessary to maintain health. In general, the amount of energy reserves available to meet these needs is directly related to body size of the fish. Thus, as body size increases, so does the amount of energy available for reproduction in general and for egg production in particular. For example, the ovary weight of Pacific herring, which is largely a function of the number of fully developed eggs that are in the ovary, is directly related to body size (Ware 1985).

Some theorists (such as Schaffer 1974) have proposed that the percentage of total available energy that is expended for reproduction may increase with age in species with prolonged adult life spans. In part due to high energy demands, reproduction confers a risk of death to an individual. As the percentage of total available energy that is expended for reproduction increases, the likelihood that an individual will survive to reproduce again decreases. At some maximum level of energy expenditure, death is essentially inevitable (Stearns 1992). Reproduction confers both benefit and risk to an individual: The benefit is the production of offspring, and the risk is death that prevents any future production of offspring. The relative importance of possible loss of future offspring production decreases as an adult ages, because more and more of its possible lifetime reproductive effort has already been accomplished. Thus, in some long-lived fish species, the substantial difference between the relatively modest number of eggs produced by young females and the much greater number produced by older females may in part be the result of natural selection favoring modest energy expenditures at young adult ages and higher energy expenditures at older ages.

High levels of energy are allocated to reproduction by fishes that spawn once in their lives and die, perhaps because energy reserves to support recovery from the physiological stress caused by reproduction are not required. For example, all American shad migrating into rivers of the south Atlantic coast of the United States die after spawning; in northern rivers of New England and Canada a high percentage of adults of the same species migrate back to the sea after spawning and survive to spawn in subsequent years. A southern female shad, which does not reserve energy to survive, produces several times as many eggs as the average female in more northerly populations, even though southern shad mature at a younger age and smaller size than northern shad (Leggett and Carscadden 1978).

The Relationship Between Fecundity, Egg Size, and Larvae Size.   In general, fecundity of a species is inversely related to the size of eggs that the species produces (Mitton and Lewis 1989; Elgar 1990). Elgar (1990) compared the relationship between egg size and fecundity among 42 species of fishes from 26 different taxonomic families. After removing any effect that different body sizes among the species might have on fecundity and egg size, he found a statistically significant inverse correlation between these two parameters. The size of an egg is a general measure of the amount of yolk reserves available for embryological development: The larger the egg, the greater is the amount of energy reserves and the larger the larvae once yolk reserves are fully resorbed after hatching.

Table 3.1 presents the relationship among fecundity, egg size, and larval size at hatching for some moderate- to large-bodied species of marine and freshwater fishes. Species included represent a variety of different families, fecundities, and egg and larval sizes. Even though this table contains species that attain greatly different maximum sizes as adults, the basic relationship between fecundity, egg size, and size that offspring have attained by the time they first need to gather their own food resources is generally apparent.

The most fecund marine species, such as members of the cod and tuna families, flatfishes, and other taxa, produce very small, buoyant eggs that are spawned, or broadcast, into the open water column. Small and relatively poorly developed larvae hatch from these eggs, often after only several days of development. For example, cod eggs are only about 1.6 millimeters in diameter, and larvae are 3.3 to 5.7 millimeters long at hatching (Scott and Scott 1988; Table 3.1). Mortality rates of these eggs and larvae are high. Thus broadcast spawners allocate minimum amounts of energy into each small egg and produce very large numbers of offspring that offset the high rates of mortality the offspring suffer. Extremely high fecundities have been measured among the broadcast spawners; the female ocean sunfish may produce as many as 300 million eggs per year (Hart 1973).

Many species exhibit markedly lower fecundities than oceanic broadcast spawners. Species that produce large eggs typically exhibit longer periods of embryological development and larger and relatively better-developed larvae than

TABLE 3.1   Maximum fecundity, egg diameter, and length of offspring at hatching or birth of selected species of North American freshwater and marine fish.[a]

Species	Fecundity	Egg Size (mm)	Offspring Length (mm)
Bluefin tuna	60,000,000	1.2	2.4
Atlantic cod	12,000,000	1.6	5.7
Winter flounder	3,330,000	0.8	3.5
Walleye	612,000	1.5–2.	6–8.9
Northern pike	595,000	3	7–9.5
American shad Connecticut R.	321,000	1.8	10.0
Largemouth bass	80,000	1.5–1.7	3–5.4
Atlantic wolffish	40,000	6	18.0
Arctic char	5,000	4–5	25 at emergence from gravel[b]
Blue shark	135		440
Spiny dogfish	15	45	300

[a]Only medium- or large-bodied species are included to reduce any effect that body size might have on the relationship between fecundity and the size of eggs and larvae.

[b]Like most members of the Salmonidae, Arctic char fry remain in the gravel of their nest for prolonged periods of time after hatching, using large yolk reserves during this extended period of growth and development before they leave the protection of the gravel.

those with small eggs. For example, when young salmonids absorb their yolk reserves and leave the gravel nest in which they were spawned and hatched, they are well-developed juvenile fishes. Thus, by the time they are first exposed to predators and the problems of foraging for food, their advanced stage of development and size increases the probability of survival over that typical of broadcast spawning species. Ovoviviparous sharks that incubate fertilized eggs within the female's reproductive tract display extremes in egg size and in fecundity. The eggs of these species contain enough yolk reserves to allow extensive periods of development while incubating. Such sharks give birth to very well-developed, large offspring as much as a year or more after the eggs were initially fertilized. Because of their large size, newborn sharks have relatively high survival rates.

Parental Care

Fecundity and Parental Care.   Fecundity is also clearly related to the level of parental care provided to eggs and larvae. Baylis (1981) broadly defined parental care as any reproductive specialization that increases offspring survival rates. This includes not only direct behavioral care but also physiological and morphological adaptations that provide offspring with an advantage, such as traits that allow internal fertilization and incubation of eggs and direct nourishment of developing embryos. Generally, the greater the level of parental care provided offspring, the lower is the fecundity exhibited by fish species.

Broadcast Spawning—Minimal Parental Care.   Broadcast spawning of buoyant eggs represents the least specialized mode of reproduction in fishes. Eggs are broadcast into the water column where they are fertilized, afterward passively drifting near the surface with prevailing currents. Buoyant eggs are common among marine fish species but are atypical of freshwater taxa. Once newly hatched oceanic larvae deplete their yolk reserves, they must feed on plankton near the water's surface in waters of considerable depth. Buoyancy keeps eggs within the upper portion of the water column where feeding will need to occur at some point after hatching. In freshwater, estuarine, and some coastal marine systems the chemical and physical conditions, availability of shelter from predators, and the quality of food resources for larval fishes may vary considerably over relatively small distances. In such systems, buoyant eggs and drifting larvae could easily be carried from favorable to unfavorable habitats by currents and wave patterns. Apparently, for most species that spawn in freshwater systems, natural selection has favored demersal (sinking) eggs that adhere to substrate or stationary objects, thus staying in areas offering the best conditions for survival (Baylis 1981).

A great many species of fishes select the type of habitat or substrate on or in which eggs are deposited and protect or tend offspring. Fish ecologists have developed several schemes for classifying such patterns of parental care (as in Nikolsky 1963; Balon 1975).

Egg Scatterers.   Many fish species release adhesive, demersal eggs over specific types of substrate. The selection of appropriate substrates is critical for offspring survival. Because egg scatterers do not modify the surfaces on which they lay eggs, they are usually separated from those that do, the nest builders (see the section on nest builders in this chapter). Egg scatterers are often classified according to the types of surfaces on which they scatter eggs (as in Balon 1975).

The sand lance scatters eggs over sandy substrates in open continental shelf waters (Scott and Scott 1988), and the winter flounder, the only species of Northwest Atlantic flatfish that lays demersal, adhesive eggs, typically spawns over sandy substrates in estuaries (Buckley 1989). A variety of freshwater species scatter eggs over coarse substrates. The lake trout scatters eggs over boulder and rocky substrates of lakes or rivers. Sauger and walleye scatter eggs over gravel and boulder shoals of lakes or rapids of rivers (Scott and Crossman 1973).

Other taxa scatter adhesive eggs over vegetation. Although the anadromous rainbow smelt scatters eggs on a variety of substrates in coastal streams including gravel, rocks, and boulders or aquatic vegetation, survival rates of eggs are significantly higher on vegetation than on other substrate types (Sutter 1980). The estuarine spawning Atlantic silverside spawns on mats of filamentous algae in the intertidal zone (Conover and Kynard 1984). Adhesion to algal mats may have the double advantage of hiding eggs from predators when they are submerged and preventing them from desiccating when exposed to air during low tides (the mats of algae tend to remain moist throughout such periods). Members of the pike family *Esocidae* scatter eggs over dense aquatic vegetation or temporarily flooded herbaceous terrestrial vegetation of the shores of lakes or the floodplains of rivers (Scott and Crossman 1973).

Shelter Spawners.   Some fish taxa seek a specific type of shelter in which to lay eggs. This shelter, which may be living or nonliving, typically serves to reduce the level of predation on eggs or provides suitable physical and chemical conditions for optimal embryological development.

Female yellow perch drape long gelatinous strands of eggs over aquatic vegetation or logs (Smith 1985). Some fishes use living invertebrates as substrates for egg deposition; for example, the sea raven and longhorn sculpin (family *Cottidae*) lay their eggs at the base of or among the branches of living sponges (Scott and Scott 1988), Pacific snailfishes of the genus *Careproctus* lay their eggs in the gill cavities of bivalve mollusks, and damselfishes lay theirs among the tentacles of sea anemones (Peden and Corbett 1973; Balon 1975).

A variety of species lay their eggs in crevices or cavities of hard substrates, including species of darters; freshwater sculpins of the genus *Cottus;* species of North American catfishes; and coastal species·such as the shorthorn sculpin, the Atlantic wolffish, and the ocean pout (Scott and Crossman 1973; Scott and Scott 1988; Ross 1991).

Nest Builders.   Many species modify specific substrate types before or during egg deposition. This nest-building behavior may serve to cover and hide eggs from predators or improve the microhabitat conditions that might affect egg survival.

Many freshwater and anadromous fishes construct gravel nests in streams. The action of moving gravel and rocks dislodges fine silt particles into the current, which carries such materials away from the nest. Removal of silt particles allows excellent percolation of water through the coarse gravel, which provides optimum aeration and removal of waste products from the buried egg clusters. Eggs that settle into the gravel also are protected from egg predators.

The male sea lamprey constructs a shallow pit of gravel and small rocks by pushing materials around with his mouth (Scott and Scott 1988). Males of numerous stream-dwelling minnow species such as the creek chub and fallfish construct gravel nests in riffle areas by excavating and moving gravel with their mouths (Figure 3.1; Ross 1977; Ross and Reed 1978). Many species in the sunfish and salmon and trout families (*Centrarchidae* and *Salmonidae,* respectively) excavate nest depressions by fanning the tail over substrate or by lying on one side and vigorously flexing the tail and caudal peduncle upward. Male sunfishes and black basses excavate circular pit nests in this manner (Figure 3.1). Female salmon and trout construct nests called *redds* by excavating elongate pits; after spawning, these fishes fill the pits with a low-lying ridge of gravel by using the same motion (Chapman 1988). Some coastal fishes, such as the fourhorn sculpin, sweep similar depressions in soft-bottom muds for spawning (Scott and Scott 1988).

Some fishes construct simple or elaborate nests from vegetative materials. The male garibaldi, a West Coast species, tends a nest consisting of a living patch of a particular form of red algae. The nesting male clips sections of the algal mass with his mouth until the algae is shaped into a specific size and form (Clarke 1970). Male sticklebacks (family *Gasterosteidae*) construct elaborate vegetative nests from fragments of aquatic vegetation. The material is glued together by a secretion of the kidneys (Courtenay 1985). Completed nests are hollow compartments with two openings. Particular species construct characteristic shapes (tunnel, spherical, cuplike) either on substrates or attached to vegetation, sticks, or other projections from the substrate (Balon 1975).

Guarding.   Many nest-building species actively guard spawning sites from egg predators and from conspecifics that may attempt to spawn on the same site. The male guards the nest site in the majority of nest-guarding species. Theoretically,

A   B

Figure 3.1   (*A*) A fallfish nest in the Deerfield River, Massachusetts. (*B*) A bluegill nesting area from Lake Caroline, Virginia. The fallfish nest is above the water line because of the regulation of river flow that occurred at upstream dams after the nest was constructed. Sunfish nesting areas often have numerous nest pits crowded very closely together. (Photographs courtesy of Mark Bain.)

males may benefit more from the evolution of nest-guarding behavior than would females of the same species (Baylis 1981). If a male guards a nest to which multiple females are attracted and prevents other males from mating with females at that nest, guarding behavior not only protects offspring but also potentially increases the number of offspring produced by the guarding male. In many species, males that construct and successfully defend a nest spawn more frequently than males that do not (as in Ross 1977; Gross and Charnov 1980). If a female guards a nest, she can only increase the rate of offspring survival because the number of offspring she produces is limited to her fecundity, not to the number of males with which she mates.

Brooding.   Brooding (also called bearing), the carrying of offspring during embryological development or longer, is the most specialized mode of parental care. Brooders may be divided into two major groups: external brooders—those that carry fertilized eggs outside of the female's reproductive tract, and internal brooders—those that carry developing embryos within ovaries, oviducts, or uteri.

External bearers have evolved a variety of ways to carry young. They may carry them attached to the body's surface, as in some species of South American catfishes; in specialized pouches, as with seahorses and pipefishes; in the mouth, as with some sea catfishes and cichlids; and in gill chambers, as with North American cave fishes (Lagler et al. 1977; Moyle and Cech 1996).

Eggs of internal brooders, including the elasmobranchs (sharks, rays, and skates) and several evolutionary lines of bony fishes (such as the live-bearing family *Poeciliidae*), are fertilized within the reproductive tract of the female. Most species of internal brooders give birth to morphologically well-developed young. Many internal brooders are ovoviviparous, but some are viviparous; embryos of the latter species are nourished by fluids secreted by the female or, in some taxa, by placental capillary beds (Moyle and Cech 1996). Regardless of the mode of offspring nourishment, most livebearers give birth to well-developed and large offspring, significantly reducing the mortality that these young fishes would otherwise suffer during egg and early larval stages.

Fishes expend substantial energy when reproducing. Some species commit minimal energy to the individual egg but produce an enormous number of offspring. A variety of modes of parental care have evolved that commit a greater amount of energy to each of a smaller number of offspring and in doing so improve the probability that a particular offspring will survive. No reproductive strategy seems to be clearly superior. Differences in fecundity and energy allocated to individual offspring are alternative specializations that evolved due to the pressures of offspring survival and the success of an individual's reproductive effort, and due to constraints that influence the direction that adaptation may take.

Egg and Larval Mortality

Fish eggs and larvae suffer the greatest rates of death of any stage of a fish's life cycle. Although juvenile and adult death rates of some large-bodied species may average about 5 percent to 10 percent annually (Woodhead 1979), and mortality of smaller-bodied species may be much higher, eggs and larvae of these same species may suffer similar levels of mortality on a daily or weekly basis. Between 90 and 99 percent of rainbow smelt eggs die before hatching (Sutter 1980). About 70 to 85 percent of larval American shad in the Connecticut River die 4 to 9 days after hatching (Crecco et al. 1983). The daily mortality rate of newly hatched capelin larvae may exceed 50 percent (Wootton 1990). Winter flounder and striped bass larvae suffer a 1.2 percent and a 1.5 percent average daily death rate, respectively (Pearcy 1962; Dey 1981). A daily egg and larval death rate of 2 to 10 percent may be typical of broadcast spawners (Pitcher and Hart 1982). In contrast, more than 90 percent of the brown trout eggs buried in redds may survive the 3-month incubation period to hatching (Wootton 1990).

Mortality of an individual larva may be caused by physical and chemical conditions or by biological circumstances such as starvation and predation. Larval mortality rates in populations are generally a result of not one but several interacting factors (Pitcher and Hart 1982).

Environmental Causes of Mortality. Physical conditions, such as water temperature, rate of flow of river or ocean currents, or water levels at time of spawning, can cause extreme fluctuations in larval survival from year to year. Survival of North Sea plaice larvae is greatly affected by water temperatures and current speeds. Eggs and larvae produced on spawning grounds of the Flemish Bight are carried by surface ocean currents toward the Frisian Islands, where the larvae typically metamorphose into bottom-dwelling juveniles. In years when water temperatures are low or current speeds are high, eggs and larvae may be swept far past nursery grounds before they are ready to metamorphose and become demersal. Conversely, if water temperatures are high or current speeds low, larvae metamorphose and settle to the bottom before reaching nursery habitats (Pitcher and Hart 1982). The annual survival rate of larval Pacific hake is affected by drifting rates in a manner similar to that of the North Sea plaice (Bailey 1981). Ocean current patterns are believed to be responsible for annual fluctuations in offspring survival of the bluefish, the Atlantic menhaden, and many other open-ocean spawning species (Iselin 1955; Norcross et al. 1974; Nelson et al. 1977).

Variability of environmental conditions also causes high annual variability in egg and larval survival of fishes that are not broadcast spawners. The level and frequency of inshore winds affect the survival rates of larval capelins (Frank and Leggett 1981; Leggett et al. 1984). Wind-induced wave action is required to free capelin larvae from the sand and pebble substrates of the beaches where they are spawned. If time intervals between onshore winds exceed several days, larvae remain buried in beach substrates too long, resulting in total yolk resorption and rapid deterioration of their physical condition.

The average river discharge or volume of water passing downstream per given amount of time, water temperature, and total monthly precipitation are all significantly correlated to survival of larval American shad (Crecco and Savoy 1984). During years of high June rainfall, high river flows and resulting low water temperatures not only adversely affect rates of egg and larval development but also reduce the density of zooplankton, which serve as the primary food of American shad larvae. Thus unfavorable feeding conditions result in high larval mortality.

Walleye year class strength—the number of individuals added to a population as a result of a specific year's reproductive effort—is directly correlated to water levels in some lakes. Higher water levels increase the expanse of available, high-quality spawning habitat (Kallemeyn 1987). Kallemeyn (1987) also found a positive correlation between reproductive success of walleye and yellow perch and air temperature prior to and during the spawning season: Years of higher and more stable air temperatures produced higher survival rates for eggs and larvae. Reproductive success of largemouth bass is higher in years when temperature and water levels are stable than when they fluctuate (Carlander 1977). Egg mortality of brown trout in streams increases during high-water periods when turbulent water dislodges eggs from the redd (Wootton 1990).

Thus mortality rates of early life stages not only are high but vary yearly due to a variety of environmental factors. Although physical parameters of the environment are clearly correlated with mortality, in many instances these factors merely affect eggs and larvae in some way that weakens their condition. In such situations, starvation or predation may be the ultimate cause of death.

Biological Causes of Egg and Larval Mortality. Predation has been identified as a major cause of larval fish mortality (as in Moller 1984). Predator-caused mortality can be particularly severe in pelagic ocean communities (Rothschild 1986). Invertebrate plankters such as copepods and chaetognaths (Pitcher and Hart 1982) may impose the greatest predation mortality on fish eggs and larvae in oceanic communities. Small fishes may also eat substantial numbers of eggs and larvae including members of their own species; nearly 30 percent of the natural mortality of northern anchovy eggs is due to cannibalism by anchovy (Rothschild 1986). Fishes can be a major source of predation mortality in freshwater systems. For example, high densities of bluegills can cause high mortality of largemouth bass eggs and larvae (Carlander 1977).

The impact of predation on larval fishes may vary according to other conditions affecting the larvae. Slow growth caused by low water temperature or poor feeding conditions can increase their vulnerability to predation. Many predators, particularly invertebrates, feed only on smaller fish larvae. As the larvae grow, they become too large to be captured and eaten effectively by specific predator taxa. The longer the time period before larvae grow larger than the vulnerable size range for particular predators, the longer they will be vulnerable to those predators. Thus poor feeding or growth conditions can increase rates of mortality caused by predation.

Field collections and laboratory experiments provide evidence that poor feeding conditions directly cause mortality of larval fishes (Rothschild 1986). Larval survival may be strongly dependent on food availability after total yolk absorption (Pitcher and Hart 1982). This time has been called the critical period (Rothschild 1986) because most fishes that cannot gather some minimal level of food die, even if they ultimately encounter suitable densities of food before they starve. For example, Laurence (1974) starved newly hatched haddock larvae for various numbers of days, after which they were provided with high densities of zooplankton food. Over 50 percent of larvae deprived of food for 6 days or less after hatching survived. However, larvae starved for 8 to 14 days ultimately died, even though some larvae had actively fed after being provided with food. Experiments with the anchovy produced similar results (Lasker et al. 1970). Although some authors have adhered to the critical period hypothesis, others have suggested that larval survival occurs at a somewhat constant or constantly increasing rate and is not dependent on one short time period immediately after yolk absorption (Rothschild 1986).

Some laboratory experiments have indicated that fish larvae are unable to feed effectively when presented with densities of food items that simulate the average plankton densities found in the ocean (Pitcher and Hart 1982). Hunter (1972) and others have invoked the "plankton patch hypothesis" to explain how planktonic fish larvae find suitable densities of food to survive in the ocean. Currents and wind-driven waves tend to concentrate plankton into high densities in localized areas. These patches are separated by oceanic reaches of very low plankton concentrations. Thus localized patches have markedly higher densities than the average for a large oceanic region. Although some biologists feel that the disparity between the densities of food that are necessary to prevent starvation and the average densities of food in the ocean may be due to inaccuracy of data collection techniques, some studies indicate that fish larvae might be able to survive only if they encounter patches of food (Werner and Blaxter 1980).

In theory, encountering plankton patches should not be difficult for marine pelagic fish larvae. There is a strong relationship between the time of spawning of many ocean fishes and the time of peak plankton production (Pitcher and Hart 1982). Because larvae are planktonic, ocean currents should concentrate them with their planktonic food. However, weather conditions in a given year may not allow production of abundant plankton food resources, or wind-driven surface waves produced by storms might disrupt plankton concentrations (Rothschild 1986). Thus abiotic conditions may strongly influence the severity of starvation-induced mortality in a given year.

Overwinter Survival and Year Class Strength.   Year class strength of some freshwater fishes may be strongly influenced by the rate of first-year growth and the severity of winter mortality that occurs during the first year of life. Late-season spawning caused by weather conditions or slow growth after hatching due to poor feeding conditions will produce small young-of-the-year largemouth bass that enter their first winter with limited fat reserves. Survival during the first winter of life can be size dependent, with smaller fishes experiencing much higher death rates than larger ones (Miranda and Hubbard 1994). Length-dependent winter mortality of young-of-the-year largemouth bass has been found by some researchers but not others (Toneys and Coble 1979; Kohler et al. 1993; Miranda and Hubbard 1994). Researchers have found evidence of size-dependent first-winter survival of rainbow trout and walleye (Toneys and Coble 1979; Smith and Griffith 1994). A general inverse relationship between first-year growth and first-winter mortality may exist for many temperate-region freshwater species. Thus, even during years when high densities of eggs and surviving larvae are produced, the ultimate abundance of that year class of fishes may be significantly lower than would be predicted based on reproductive activity.

Early Life Stage Mortality.   Mortality in the early life stages has many causes. If food resources are limited, larvae or young-of-the-year juveniles may die of starvation, or their slow growth and poor physical condition will increase their vulnerability to predation or other mortality factors. These factors result not only in extremely high but also in extremely variable and unpredictable death rates. Theorists have proposed that the highly variable and unpredictable nature of larval mortality has been a strong factor in the evolution of the reproductive cycles of fishes (as will be described in the section on reproductive schedule).

Body Growth

Fish growth is indeterminate and highly plastic. That is, a fish can grow throughout its life cycle, and the rate at which it grows is highly variable. A variety of biotic and abiotic factors influence the rate of growth experienced by fishes.

Growth According to Life Stage.   Fish growth is most rapid during early stages of the life cycle. As a fish reaches sexual maturity, energy is diverted to reproductive activities such as gamete production, migration, and spawning. The amount of energy contributed to reproduction can be great. Thus the growth rate of a fish slows markedly at sexual maturity due to the reduction in energy reserves that are available for growth. Fishes that approach maximum age typically exhibit additional decreases in growth due to the combined needs of reproduction and increasing maintenance requirements of foraging, shelter seeking, and maintaining physiological homeostasis; total maintenance energy requirements generally increase with age and body size. Although the increase in weight or length gained during a time period relative to the size of the fish at the beginning of that period is greatest in larval and early juvenile life stages, the actual length or weight gained is small because the fish is small. Thus, if the size of an individual from a long-lived species is compared with its age, the relationship forms a sigmoid (S-shaped) curve (Figure 3.2). Sexual maturity typically occurs at the point where the increase in size from one age to the next begins to decrease, and the ultimate flattening of the curve is due to the combined energy costs of reproduction and maintenance.

Growth Variation Among Individuals.   Rates of growth typically vary among individuals of a given age within a population. Differing capacities to grow are likely carried throughout life. Thus, although 1-year-old fishes of a long-lived

Figure 3.2   The general relationship between body length and age of a fish.

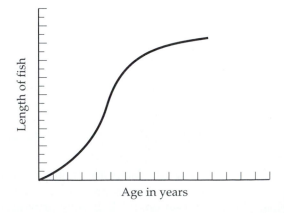

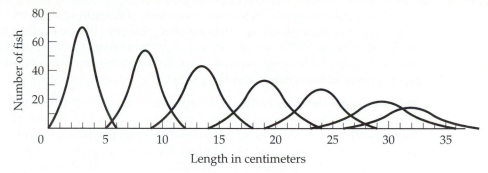

Figure 3.3 Theoretical frequency distribution of the lengths at age of fishes sampled from the same population. The differences among sizes of fishes of the same age tend to increase with age, causing greater overlap among the sizes of the faster-growing fish of one age and slower-growing fish of older ages.

species may not vary much in size, differences in size between fast- and slow-growing individuals accumulate with age. Ultimately, some fast-growing individuals may be larger than older but slower-growing ones (Figure 3.3).

Growth rates may also vary significantly between males and females within a species. The sex that exhibits larger average sizes at a given age often achieves greater growth at least in part by deferring maturity (that is, by reaching sexual maturity at a later age than the other sex). Such individuals spend a prolonged period of life as fast-growing juveniles, thus reaching larger sizes before growth is slowed due to the energy demands of reproduction. The sex that matures at an older age may also realize a longer average life span and reach a larger maximum size than members of the other sex. Thus all large striped bass are commonly called cows, because it is well known that they are females. (The reproductive advantages of deferred maturity are discussed in the section on age at sexual maturity.)

Environmental Factors and Growth. Experimental work has clearly demonstrated that water temperature affects the efficiency and rate at which food is eaten and converted into energy for growth. Generally, the warmer the average water temperature, the greater is the rate of intake and digestion of food, as long as the temperature falls within an optimum range for that species (Brodeur and Pearcy 1987). Changes in the pattern of daily intake of food are followed by corresponding changes in the growth rate of fishes (Cochran and Adelman 1982). In temperate regions of the world, many fishes tend to grow more rapidly during warm seasons than cold. Differences in growth between cold and warm seasons are so great that hard tissues, such as bones, otoliths, and scales, exhibit yearly growth rings that reflect the effect of temperature-controlled metabolic rates on the density and composition of such tissues (Jearld 1983).

The same relationship between temperature and growth can produce measurably different average growth rates among populations of the same species residing in different geographic regions. Likewise, changes in average annual temperatures can produce highly variable growth rates of fishes from year to year within populations, as shown for walleye (Carlander and Whitney 1961) and Atlantic herring (Anthony and Fogarty 1985).

Other environmental factors stimulate annual variability in growth rates within populations. Water-level fluctuations affect the growth rate of the walleye (Carlander and Payne 1977). Salinity affects growth rates of coastal species due to the energetic costs of adjusting internal physiology to withstand the stresses of fluctuating salinity levels (Pitcher and Hart 1982).

Density-Dependent Growth.   Intraspecific competition for food resources is generally believed to have a major influence upon individual growth rates. Theoretically, an increase in abundance of the fish population should result in a decrease in the growth rate of individual fishes, because the food ration available to each fish would be smaller. Conversely, if a population's abundance declines, individual growth rates should increase due to a higher available food ration.

Early experimental work (such as Swingle and Smith 1942) established that a relationship clearly exists between population abundance and individual growth, if all other variables that might affect growth, such as fluctuations in water temperature, are held constant. Indeed, natural populations that suffer a major, rapid decline in abundance have displayed subsequent increases in individual body growth (as reviewed by Ross and Almeida 1986).

Some field studies have found a correlation between changes in abundance and the rate of growth, whereas others have not. For example, annual fluctuations in environmental factors such as water temperature and water levels have influenced growth rates of walleye to a greater degree than concurrent changes in population abundance (Carlander and Whitney 1961; Carlander and Payne 1977). Results of studies investigating the relationship between population density and growth have been so inconsistent that fisheries scientists have concluded the following:

1. An inverse relationship between abundance and growth has been clearly demonstrated (Beverton and Holt 1957).

2. Abundance consistently affects growth in natural populations, but often the relationship is not discovered because of the difficulty in gathering accurate, precise measures of abundance (Backiel and LeCren 1967).

3. Abundance is not "generally and systematically related to growth" (Weatherley 1972).

Recent analyses of the Atlantic herring (Anthony and Fogarty 1985), silver hake (Ross and Almeida 1986), and haddock and yellowtail flounder (Ross and Nelson 1992) clearly identify the strong correlation between abundance and growth in populations of these species during series of years when abundance is relatively high. These studies also demonstrated that abundance had less effect on growth rates during prolonged periods of low population abundances (as seen in Figure 3.4 for silver hake). Apparently, when population density is low, such as during times of intensive human harvest, food resources are not limited in availability to the relatively small number of fishes that are feeding on them. However, when population density and competition for food are high, yearly fluctuations in the population's abundance will change the average ration available to each fish, thus affecting its annual growth.

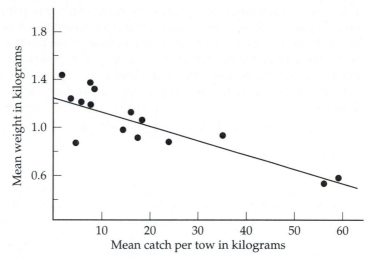

Figure 3.4   The relationship between population density and body weight at age of Georges Bank haddock. Density (or abundance) is an index measured as the average weight of haddock captured per net tow during bottom trawl research cruises of the Northeast Fisheries Science Center (NMFS). Regression analyses compare the average weight of age-2 fishes captured during each year's cruises to the average biomass of the stock. Because weight at age is a result of growth conditions to which fishes have been exposed in the past, average weight of age-2 haddock is compared with the catch per tow for the population 12 to 24 months prior to the time that individuals were captured and weighed (the time period that demonstrated the best correlation between density and weight for age-2 fishes). Regression line $r^2 = 0.73$. (Reprinted by permission of *Recreational Fisheries of Coastal New England,* by Michael R. Ross. With contributions by Robert C. Biagi [Amherst: University of Massachusetts Press, 1991], copyright © 1991 by the University of Massachusetts Press.)

This hypothesis provides an explanation for the absence of patterns of density-dependent growth in some field studies. Many studies focus on economically or recreationally important fish species that are being maintained over long periods of time at low population densities because of high rates of fishing mortality. One should not expect to find a strong correlation between population density and growth under these conditions. Anthony and Fogarty (1985) and Ross and Nelson (1992) found that, although density influenced growth strongly during time periods of high population density, temperature had a greater influence than density within the same populations during times when density was relatively low.

Age at Sexual Maturity

Fishes reach sexual maturity at different ages. Individuals of some species may mature by the end of their first year of life, whereas those of other species mature after a prolonged juvenile life stage; for example, the Atlantic sturgeon may not spawn until about 28 years of age (Scott and Scott 1988).

Age and Size at Maturity.   Within a species, age at sexual maturity often is correlated with growth rate (as in Jonsson et al. 1984). An acceleration in the rate of growth and a subsequent lower age at maturity are typical life history responses to a decline in population abundance (as reviewed for herring in Ware 1985). The growth rate of North Sea Atlantic herring increased by 25 percent and the age of maturity declined by 2 years during a period when the stock was undergoing a decline in abundance due to a growing fishery (Murphy 1977). Body size at maturity remained constant owing to the accelerated rates of growth. Such observations have led many to conclude that maturity is size dependent (as in Alm 1959) and related to the body size that is required to provide sufficient energy reserves to reproduce successfully and, for individuals of many species, to survive the spawning season (Trippel and Harvey 1989; Trippel 1993). Roff (1982) suggested that maturity is most size dependent in large-bodied, older-maturing species. Constraints such as the energy and time necessary to develop eggs fully may require small-bodied, early-maturing species to reach a certain age before spawning regardless of their juvenile growth rates. For many species, age at sexual maturity may vary greatly between geographic areas, based at least in part on the prevailing growth conditions.

The Relationship Between Mortality and Age at Maturity.   Mortality is believed to influence the evolution of age at maturity. Species that have consistently high juvenile death rates generally mature at relatively young ages. Thus the scup, with annual mortality rates for juvenile and adult fishes of up to 80 percent, spawns at 2 years of age (Finkelstein 1969, 1971). The spiny dogfish, a small-bodied shark species noted by Scott and Scott (1988) to have few natural predators, may not spawn until 12 years of age.

Characteristics such as age at maturity should evolve in a manner that provides the greatest probability that an individual will achieve some maximum rate of offspring production for that species. If juvenile mortality is consistently high, an individual has little likelihood of surviving an extended period of life as a juvenile to mature at a relatively old age; the risk of death is far greater than any benefits associated with spawning for the first time at an older age and larger size. Thus sexual maturity at an early age might evolve (Stearns 1992).

In some species, the rate of juvenile mortality may be relatively modest, whereas the rate of adult mortality may be high due to stresses caused by reproduction. If death caused by the stress of reproduction is a great risk to future production of offspring, then older ages at maturity may evolve (Stearns 1992). By extending the period of life before sexual maturity, fishes can reach significantly greater sizes before they reproduce. Greater energy stores should allow an individual to produce a greater number of offspring. Therefore, the fish is risking spawning-induced death at a time when it can produce markedly greater numbers of offspring than it might had it matured at a younger age and smaller size.

Later age at maturity might also allow greater opportunity to survive the stresses connected with reproduction (production of gametes, migration, spawning, nest construction, offspring protection), because larger, older fishes presumably have greater energy reserves. This would allow a greater opportunity to live to spawn during some subsequent spawning season (Roff 1982). Atlantic salmon that must migrate over substantially greater distances in rivers to reach spawning grounds mature at an older age than those with shorter freshwater migrations, presumably because of the greater energy cost associated with extensive freshwater migrations (Schaffer and Elson 1975).

Offspring Production Rates and Maturity.   If reaching sexual maturity at a relatively old age carries the joint advantages of increased offspring production and postspawning survival, is there an advantage to spawning when young? Theoretically, the decrease in the number of offspring produced due to spawning at a young age and relatively small size may be offset by the reduction in time necessary to produce a new generation of descendants (Stearns 1992). An individual can produce more descendants during a given number of years if a younger age at maturity allows a greater number of generations to be produced during that time period. Figure 3.5 illustrates the advantage of decreasing age at maturity, even at the expense of reduced fecundity. Of course, such a simplistic representation does not consider such factors as surviving to reproduce in subsequent seasons, rates of mortality of both offspring and parent, the possible effects of growth after reproduction on future fecundity, and other conditions that may influence the probable offspring production rate of an individual. However, it does demonstrate that increasing fecundity and lifetime offspring production by increasing the age at maturity is not necessarily advantageous.

The relative importance of risks and benefits associated with maturing at a specific age obviously varies according to the conditions to which fishes are exposed. For example, Schaffer and Elson (1975) proposed that the possible loss of future reproduction that is risked whenever Atlantic salmon migrate from the sea to freshwater to spawn influences the relationship between rate of growth and age at maturity exhibited by this species. Fast-growing Atlantic salmon tend to mature at an older age and larger body size than do slow-growing ones (Schaffer and Elson 1975). This appears to contradict the general hypothesis that many fishes tend to mature upon reaching a specific size, with faster-growing fishes reaching that size at

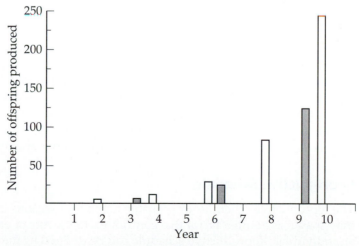

Figure 3.5   The hypothetical number of descendants produced within a 10-year time span by a female that matures at 2 years of age and produces three female offspring, and a female that matures at 3 years of age and produces five female offspring. This simplistic graph assumes that all females die after reproducing and all offspring live to reproduce once. The female that matures at 2 years of age is represented by the light bars, and the female that matures at 3 years of age by the dark bars. (Redrawn by permission of *Recreational Fisheries of Coastal New England*, by Michael R. Ross. With contributions by Robert C. Biagi [Amherst: University of Massachusetts Press, 1991], copyright © 1991 by the University of Massachusetts Press.)

a younger age than slower-growing fishes. However, Schaffer and Elson argue that the risk to slow-growing Atlantic salmon of remaining at sea for long periods of time before maturing was greater than the benefits gained from extending the juvenile life stage. Because such fishes are slow-growing, extra time spent as juveniles would not substantially increase the size, thus the fecundity, of females at maturity. Extended time as an ocean-dwelling juvenile increases the time period of vulnerability to predation before such fishes mature and spawn. Thus, if a slow-growing fish stays at sea, it risks death without receiving much of a benefit of additional egg production when it matures. Alternatively, the gain in fecundity accomplished by fast-growing fishes with a prolonged juvenile life stage may be a greater benefit than the extended period of juvenile vulnerability to oceanic predation is a risk, because the additional time spent as a juvenile would substantially increase body size and energy reserves available for reproduction. Thus slow-growing salmon may mature after as little as 1 year at sea, whereas fast-growing ones may remain at sea for as long as an average of 2.6 years before maturing. Schaffer and Elson (1975) found a statistically significant direct correlation between age at first spawning and juvenile growth rates for salmon from 14 river systems in the Atlantic Canadian provinces.

Deferred Maturity.   Age at maturity may consistently differ between sexes within populations of fishes. Females of many species reach sexual maturity at a significantly older age than males. Such instances of female deferred maturity are common among the flatfishes (Roff 1982) and many other broadcast-spawning or egg-scattering fishes. Female deferred maturity is thought to be adaptive due to the increased fecundity and higher probability of surviving reproduction that may be associated with increased body size. Because sperm production is not constrained by body size, males that mature at a younger age than females have the advantage of reduced generation time but experience no disadvantage in gamete production.

Male deferred maturity has evolved in some species that exhibit intensive male social hierarchies as a means of competing for mates and spawning territories. In such species, body size of a male is a critical element in determining his capacity to defend a nesting site or to mate as often as possible. Small males in such species typically fertilize a limited number of eggs (as in Gross and Charnov 1980; Ross 1983) and yet still face the risk of mortality associated with reproduction. The reproductive advantage associated with large body size may have been a selective force contributing to the evolution of male deferred maturity.

Reproductive Schedule

The reproductive schedule, or probable number of seasons in which an individual may spawn, is another factor that contributes to the rate of offspring production of an individual. Reproductive schedule can be defined as the number of spawning seasons in which a fish participates modified by the probability that it will survive to participate in each successive spawning season. An individual's probability of surviving between any two consecutive seasons of reproduction is based on the types and intensities of mortality factors to which it is exposed.

Semelparity and Iteroparity. Many short-lived and some relatively long-lived fishes, such as the American eel and species of Pacific salmon, have evolved a semelparous life cycle—adults invariably die after their first spawning season. However, other species are iteroparous—individuals may survive to spawn in two or more seasons. As a group, iteroparous fishes form a continuum of species, from those that exhibit such high adult mortality rates that few individuals survive to spawn in more than one year to those in which the average individual may survive to reproduce in multiple spawning seasons.

Intuitively, iteroparity would appear to be more advantageous than semelparity due to the greater number of offspring a fish might produce over a prolonged adult life. However, individuals from iteroparous species may die from the physiological stresses of reproduction; from increased predation pressure during spawning; or from infection, disease, or predation while in postreproductive weakened health. Energy reserves needed to resist stress and possible death obviously cannot be put into reproduction. There are several ways that iteroparous fishes might conserve energy to increase the chance of surviving through multiple spawning seasons:

1.   Energy may be conserved by reducing egg production.
2.   Maturity may occur at an older age, providing greater energy reserves when reproduction is first attempted.
3.   Individuals may not spawn every year after reaching maturity in order to grow and accumulate energy reserves between consecutive spawning periods.

Alternatively, semelparous fishes may be able to contribute substantially greater proportions of available energy reserves to offspring production because energy need not be reserved for survival. Thus, although semelparous fishes spawn only once, they may produce relatively large numbers of offspring from that spawning effort.

Some spawning populations of the American shad are semelparous, whereas others are iteroparous (Leggett and Carscadden 1978). The shad is an anadromous homing species; individuals typically return from the ocean to spawn in the river in which they hatched. Shad spawning in river systems in the southeastern United States are semelparous, whereas increasing percentages of adults spawn at least twice in their lives as one proceeds northward along the Atlantic coast to Canada. Female shad spawning in the St. Johns River in Florida mature at a younger age and smaller size but produce about three times as many eggs as do newly matured females in iteroparous populations in the Miramichi River of New Brunswick (Leggett and Carscadden 1978). Indeed, the average semelparous Florida female produces nearly twice as many eggs in her lifetime as the average iteroparous female of the Miramichi. Regardless of where they spawn, American shad are widely migratory along coastal North America during the oceanic, growth phase of their lives, generally experiencing the same array of environmental factors that may affect their growth rates. Yet southern shad grow significantly more slowly than northern ones. Glebe and Leggett (1981) found that St. Johns River (Florida) females use 70 to 80 percent of available energy reserves to migrate into freshwater and spawn; Connecticut River females use from 40 to 60 percent of available energy reserves to migrate,

spawn, and return to the sea. These researchers estimated that Florida females allocate about 1.6 times more energy to egg production than Connecticut River females.
Thus southern shad contribute greater energy reserves to reproduction and less to
growth and postreproductive survival than do northern ones.

The semelparous life cycle produces a greater average number of offspring for
female shad than does iteroparity. Additionally, the time it takes for a semelparous female to produce her next generation is shorter, further increasing her rate of production of descendants over that possible for an iteroparous female. Semelparity seems so
advantageous for the American shad that one might question why this way of life
does not dominate all shad populations. Leggett and Carscadden (1978) suggest that
differences in the severity of egg and larval mortality account in part for the reproductive schedules that have evolved for different shad populations. Rivers in the southeastern United States are relatively stable environments, with conditions that might
affect survival of young shad, particularly water temperature, varying relatively little
from year to year (Leggett and Carscadden 1978). Alternatively, northern rivers display greater year-to-year variability in environmental conditions, with annual fluctuations in temperature occurring in a random, unpredictable manner. Because southern
shad spawn in conditions that are favorable for consistent yearly larval survival, maximum rates of offspring production have evolved, even though the adult does not survive to reproduce more than once. However, shad genetically "tuned" to spawn once
and die in northern rivers would often contribute little or nothing to future gene pools,
because the mortality of all (or nearly all) offspring in the years when conditions are
unsuitable for survival would cause failure of an individual's entire lifetime reproductive effort. Under such circumstances, a life history should evolve that allows more
than one opportunity to spawn, even if iteroparity requires that the rate of offspring
production be reduced from that possible for individuals that spawn once and die.
Spreading the reproductive effort over at least 2 years would increase the likelihood
that some offspring would be spawned under conditions suitable for survival.

Several theorists (Murphy 1968; Charnov and Schaffer 1973; Schaffer 1974)
have hypothesized that iteroparity and prolonged adult life spans should evolve
when the success of reproduction in a given season is highly unpredictable. Animals
that spread their reproductive effort through time ultimately achieve some level of
reproductive success, whereas those that put all of their reproductive output into
one effort may be spectacularly successful occasionally, but most often they will
fail to produce any surviving offspring. The differences in life history among populations of the American shad provide support for this hypothesis.

Spawning Less Frequently than Every Year.   Most long-lived temperate-region
oviparous (egg-laying) fishes are believed to spawn annually after reaching sexual
maturity. However, some fishes (such as the Atlantic salmon, Arctic char, whitefish,
striped bass, species of sturgeon, and white sucker) are known to have nonannual
spawning cycles. Even though some portion of the population reproduces yearly,
many individuals spawn less frequently (Jackson and Tiller 1952; Roussou 1957;
Dadswell 1979; Morin et al. 1982; Quinn and Ross 1985; Dutil 1986). In some species,
the interval between consecutive spawning seasons of individuals can be quite
long. For example, individual female shortnose sturgeon may spawn no more
frequently than every 8 years after maturity (Taubert 1980). The spawning frequency
of males is generally greater than females in nonannual species.

Bull and Shine (1979) suggested that lifetime offspring production might be increased by nonannual spawning cycles if reproduction itself significantly increases adult mortality. If present reproductive activity serves as the greatest risk to future offspring production of an individual, then the most effective reproductive schedule might be one that includes prolonged periods of growth and energy storage between spawning seasons. By spawning on a nonannual basis, a female may markedly increase the energy stores that may be allocated to egg production and to postspawning survival. Dutil (1986) demonstrated that Arctic char require more than one year to replenish their depleted energy reserves after spawning. After spawning, fishes will not reproduce again until they have recovered some threshold level of energy reserves. Similarly, Trippel (1993) demonstrated that the occurrence of nonannual spawning in white sucker and lake trout is related to "poor nourishment." Atlantic salmon that conduct extensive freshwater migrations remain at sea at least 2 years between their first and second spawning seasons, whereas most or all that spawn in shorter river systems and survive will spawn again the following year. Schaffer and Elson (1975) related these differences to the energy reserves required to survive freshwater migrations.

The rate of offspring production is substantially lower for nonannual than for annual individuals. Thus, if nonannual spawning is adaptive, the benefits of increasing egg production and the number of reproductive seasons must outweigh the slow reproductive schedule in order for nonannual spawning to have evolved. Because a relatively small number of species are known to display this type of life history, these conditions may not be met often.

Iteroparity in the Presence of High Offspring Survival.   Iteroparity has evolved in many oviparous fishes whose offspring mortality rates are high and vary unpredictably from year to year. Prolonged adult life stages and iteroparity have also evolved in the elasmobranchs even though offspring mortality rates are believed to be low and less susceptible to annual variability caused by unpredictable conditions. In elasmobranchs, prolonged periods of incubation within the reproductive tract of a female allow offspring to be very large and highly developed at birth. Thus postbirth mortality is believed to be very low compared with oviparous fishes. However, high offspring survival rates in this instance are coupled with extremely low offspring production rates. Female elasmobranchs produce very few offspring during a reproductive cycle. Thus iteroparity in elasmobranchs is associated with high rates of offspring survival but low rates of offspring production. In such a life history pattern, an individual's contribution to future gene pools accumulates through a prolonged life span by the relatively high rate of survival of very small numbers of offspring.

POPULATION GROWTH

The rate of growth exhibited by a fish population is the result of additions to the population through birth and growth and subtractions from the population through death. In theory, a stable population whose additions through birth and growth equal the subtractions caused by death has reached the carrying capacity of the system in which it lives. In other words, it is existing at the maximum biomass level that the system can support.

The carrying capacity of a system is based on the availability of resources such as food that are critical to survival of a species. Any population whose biomass is below carrying capacity of a particular system will tend to achieve higher gains than losses, and thus its biomass will increase toward maximum levels that the system can support. The rate at which biomass of a population increases toward carrying capacity (that is, the difference between gains to the population versus losses from it per unit of time) is believed to be based on its current abundance or biomass, as shown by the logistic growth model in Figure 3.6. At very low densities, the rate of growth is small due to its extremely limited abundance. At very high densities, the rate is low because the system cannot support much of an increase in abundance because of growing competition. According to the logistic growth model, the rate of growth is potentially greatest at mid-density levels. Logistic theory has been used to predict the maximum harvest that any density of a population can sustain while still replacing the lost portion of the population by subsequent reproduction and biomass production (see the section on surplus production in this chapter).

INTERACTIONS WITHIN FISH COMMUNITIES

Fisheries scientists are also interested in the nature of the effects that species have on each other within fish communities. In theory, removal of a portion of a fish population through fishing or management actions such as the introduction of a new species may change the predator–prey and competitive relationships of resident species within fish communities. The ramifications of depleting a fish stock or of introducing a new species may be far-reaching, affecting a variety of other species within the same fish community.

Predator–Prey Interactions

Early fisheries biologists thought that predators were capable of decimating populations upon which they preyed. Baird (1873) credited the decline of a variety of coastal New England fish species in the mid-1800s to predation by the bluefish (see Chapter 6). Although this species has a reputation as a voracious predator, it hardly was capable of causing simultaneous regional declines of numerous fish stocks with which it had coexisted previous to that time.

Figure 3.6   The logistic growth curve demonstrating the theoretical increase in abundance of a newly introduced population that expands toward carrying capacity.

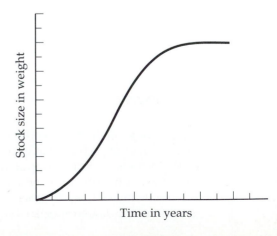

This early philosophy of the detrimental effects of predators on fish communities persisted well into the 20th century. Interestingly, this thinking seemed to trouble the fisheries manager only when the prey species in a predator–prey relationship was a prized fishery; fish predators were taking individuals that humans might otherwise catch. However, when the predator was the prized member of a predator–prey interaction, its value as a gamefish was probably enhanced by its reputation as a predator. Predator control programs conducted by fisheries management agencies have paralleled similar programs conducted by wildlife agencies ("vermin" control programs aimed at eliminating terrestrial predators to "protect" game wildlife species). Such fisheries and wildlife efforts shared a common trait: Humans were unwilling to recognize that the effects they created through harvest or habitat deterioration and loss were centrally responsible for the decline of species that they proposed to protect.

By the latter half of the 20th century, fisheries scientists were developing a better understanding of the manner in which predator and prey interact. Efficiency of food gathering by predators corresponded closely with availability of prey, with prey abundance serving as a major factor in determining availability. Predator–prey interactions between the predaceous walleye and the yellow perch provide an excellent example of interactions between a single predator and a preferred prey species. When walleye populations increase in abundance, their feeding activity may lead to a population decline of the perch. However, as this prey species declines in abundance, the efficiency with which the walleye can harvest it also declines. As a result, walleye switch to other prey (including younger, smaller walleye), which allows a gradual recovery of the yellow perch population (Forney 1977). Such cyclic changes in population abundances of predator–prey complexes have been identified elsewhere in fish communities. For example, the number of newly matured fishes that are added each year to spawning stocks of the Pacific cod in Hecate Strait, British Columbia, are positively correlated with the abundance of a major food species, the herring, whereas additions to spawning stocks of herring are inversely related to the intensity of cod predation. Thus population densities of these two species cycle up and down through time, probably due at least in part to the nature of their predator–prey interaction (Walters et al. 1986).

In theory, switching from one prey to another occurs when the cost of energy expended to capture prey begins to have a negative impact on the energy gained from eating it. Such switching does not always provide fully suitable feeding conditions for the predator. Magnusson and Palsson (1989) demonstrated that Icelandic cod consume their major prey item, the capelin, at a rate that is positively correlated with the capelin's abundance. When the capelin stock declines, cod consumption of capelin decreases accordingly. Additionally, total consumption by cod of all foods declines; thus this predator cannot fully compensate for the loss of a major prey by feeding on other species. Although predators certainly influence the status of prey populations, the availability and abundance of food determine the foraging success and ultimately affect the population status of the predator species.

Competition

Because many species of fishes are generalist feeders eating a wide variety of prey species, early researchers concluded that interspecific competition was intense between most or all taxa coexisting in a fish community (as in Hartley 1948; Larkin 1956). Although counter to general ecological theory, which held that species coexist only if they fill separated, largely nonoverlapping niches, such philosophy guided the development and expansion of fish extermination programs by public agencies through much of the 20th century (see the section on controlling undesirable fish species in Chapter 10).

Keast (1965, 1966) demonstrated that a large portion of the feeding overlap found by earlier researchers was probably an artifact of somewhat crude analytical techniques. Identification of food items found in the stomachs of fishes was usually very general, with foods being categorized no more precisely than "unidentified fish remains, insects, crustaceans," and others. Keast's more precise taxonomic identification indicated that the diet of any species in the communities that he studied differed somewhat from other coexisting taxa. In some instances, species that did eat many of the same types of food foraged in different habitat or microhabitat types. Thus they would not be competing directly for the same resources. Further, many morphological adaptations, such as body form, mouth size, and size and shape of teeth, evolved as a means of allowing fishes to become niche specialists (Keast and Webb 1966), gathering particular food resources efficiently while reducing interspecific competition to a level that permits coexistence with other species. Because individuals within the same species are most similar in morphology and behavior, intraspecific competition should be more intensive than interspecific competition among coexisting species. Zaret and Rand (1971) demonstrated that broad feeding overlap often occurs among species at times of the year when particular resources are rapidly increasing in abundance. In this instance, overlap was not indicative of competition because the food resources were temporarily so abundant that they were not limited in availability to the taxa feeding on them; a resource must be limited to the degree that the demand of all users cannot be met in order for competition to exist. Zaret and Rand concluded that overlap can be indicative of the absence, rather than the presence, of competition.

Werner and Hall (1976, 1977) conducted controlled experiments to determine the effects of interspecific competition in pond communities supporting three species of sunfishes from the genus *Lepomis,* the green sunfish, bluegill, and pumpkinseed. They found that the bluegill and pumpkinseed exhibited somewhat different patterns of habitat use and foraging behavior, depending on whether the other two species were present or absent. Werner and Hall believed that this change in foraging was the result of direct competition for food and shelter resources. As a result, interacting species exhibited a reduction in growth rate and reproductive rate and an increase in mortality. Thus these researchers demonstrated that, although many species may be morphologically and behaviorally specialized enough to coexist, potential pressure of interspecific competition can affect population characteristics of interacting species.

THE MANAGER'S USE OF LIFE HISTORY
AND COMMUNITY ECOLOGY INFORMATION

Understanding the growth, mortality, and reproductive dynamics of fish populations is critical to the development of effective fisheries conservation and management strategies. Analyses of egg and larval mortality and of growth can provide an indication of the suitability of specific habitats for the introduction of particular fish taxa or of the impact of programs focused on mitigating habitat deterioration (see Chapter 9) or improving habitat conditions (see Chapter 10).

Similarly, an understanding of species interactions can guide stocking programs. The dynamics of fish community interactions often have been disrupted by the stocking of new species into basins, resulting in the decline or collapse of native fish stocks due to predation or competition imposed by the newly introduced species (see Chapter 8). Accumulated experience with the negative impact of introduced species on native fish communities has led fisheries agencies to reduce significantly the number of species introductions, and in some instances to attempt to reduce or eliminate populations resulting from past introductions (as in recovery programs for some endangered species; see Chapter 11). Many recent introductions of taxa into reservoir systems have been guided by careful selection of species that will provide desirable predator–prey interactions or prevent the negative impacts on growth and mortality caused by interspecific competition.

Life history and community interaction data are also critical for understanding the impacts of fishing harvest on fish populations and for establishing harvest regulations designed to maintain viable, productive fisheries.

The Impact of Fishing on Life History and Production Characteristics

To maintain viable fish populations and the fisheries associated with them, the portion of a population that is harvested must be replaced through reproduction and biomass production. The process of adding new individuals through reproduction and growth is often called *recruitment*. When referring to the recruitment potential of a population, fisheries scientists most often are describing the production of offspring that survive long enough either to reach sexual maturity and reproduce— recruitment to the spawning portion of a population—or reach a size that is harvestable—recruitment to a fishery. Managing harvest cannot be successful without an understanding of the capacity of a stock to replace lost numbers and biomass through recruitment.

The history of fisheries management is fraught with examples of the inability of fisheries agencies to match harvest levels with the potential of populations to withstand depletion caused by fishing. The replacement potential of many fish populations should be considered quite limited regardless of life history patterns characteristic of those taxa. For example, fish species exhibiting high fecundities of females may not be resistant to high levels of fishing mortality because high levels of recruitment may occur only once every several years. Even when striped bass stocks are abundant, high numbers of surviving juvenile fishes are produced on the average only once every several years (Figure 3.7; Boreman and Austin 1985). Years of phenomenal reproductive success separated by several years of modest to little success is typical of many egg-laying fish species. Many species that exhibit consistently high rates of offspring survival, such as the livebearing elasmobranchs, display slow population growth rates due to late age at maturity and low fecundities. The reproductive potential of populations of fishes that spawn nonannually is quite limited because only a percentage of adult females produce eggs in a given year. The lake

Figure 3.7 Juvenile striped bass abundance indices, expressed as percentages of the maximum value in the time series, for the Roanoke River–Albemarle Sound (North Carolina). Yearly abundance indices were based on the average number of fish caught per effort during fisheries agency survey sampling. Note the varying time intervals between years of high recruitment of juvenile fishes into the population. (From: J. Boreman and H. M. Austin, Production and harvest of anadromous striped bass stocks along the Atlantic coast. *Transactions of the American Fisheries Society*, 114: 3–7. Copyright 1985. Reprinted by permission of The American Fisheries Society.)

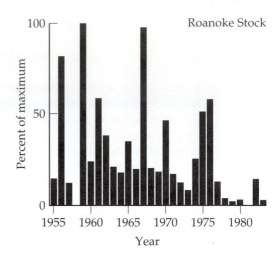

sturgeon was one of the first species to suffer fishing-related declines throughout the Great Lakes (Hartman 1972; Smith 1972). Sturgeon stocks typically exhibited declines shortly after they began to support directed fisheries. Deterioration of spawning habitat probably contributed to these precipitous declines, but overexploitation was a major factor (Hartman 1972). The late age at maturity (about 20 years; Hartman 1972) and nonannual spawning schedule made sturgeon vulnerable to overexploitation in spite of the relatively high fecundity of individual females.

Harvest may affect hermaphroditic species of fishes differently than it does the majority of gonochoristic species (taxa in which individuals spend their adult life as either male or female). Huntsman and Schaaf (1994) modeled the impact of fishing mortality on the graysby, a protogynous grouper species. Nearly all fishes in protogynous hermaphroditic species reproduce as females when they reach sexual maturity. Some individuals that survive long enough to reach large body sizes physiologically and behaviorally switch sexes, reproducing as males for the rest of their lives. Thus nearly all of the largest fishes in graysby stocks are males, and the abundance of males within stocks is limited. Because of the difference in sizes between the sexes, harvest, which is most intensively focused on the largest fishes in many fisheries, would remove males from a graysby stock in disproportionately high numbers. The removal of males from stocks of protogynous hermaphroditic species stimulates the process of sex reversal, in which some females that are among the oldest and largest fishes present become functional males. If males were removed from a stock prior to the spawning season with enough time for the largest, oldest females to switch sexes, the effect of harvest on reproduction would be similar to the effect of harvest on a typical gonochoristic species: Sex change would cause a decline in the number of mature females, which would cause a decline in egg production. However, if males were intensively harvested immediately before and during the spawning season, sex change may not occur rapidly enough to fully replace lost males (Plan Development Team 1990). In this instance, the stock would exhibit a reduction in egg production due to the loss of larger females to harvest and an additional loss of offspring production due to an insufficient number of males to fertilize all available eggs. Populations exhibiting such uncompensated protogyny will suffer more precipitous declines in abundance than gonochoristic species that have similar life history characteristics and that endure similar levels of fishing mortality (Huntsman and Schaaf 1994).

Compensatory Mechanisms and the Impacts of Harvest.   Figure 3.8 illustrates the theoretical potential of populations maintained at different densities to replace individuals and biomass removed by harvesting (based on the growth potential of a population predicted by the logistic growth equation). The rate of population growth is greatest in populations being maintained at some mid-level of abundance, in part because of the change in life history characteristics brought about by the reduction in abundance normally caused by fishing. In moderately harvested fish populations, reductions in abundance initiate an acceleration in body growth of fishes, which in turn may cause a decline in the average age at maturity. For

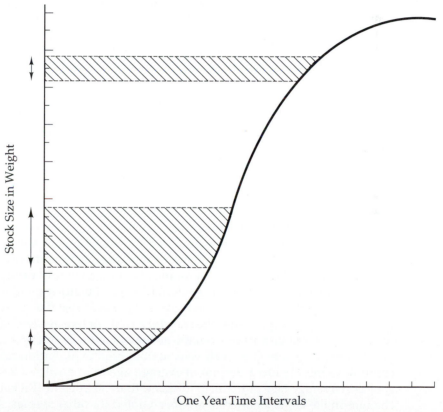

One Year Time Intervals

Figure 3.8   The capacity of a population to replace individuals removed by harvest. The hatched areas and corresponding arrows illustrate the increase in biomass that would occur in a one year time interval at low, medium, and high stock sizes. The potential increase that would occur over a one-year time span can be removed from a population during the same year without causing the population to decline. Note that a greater biomass can be replaced after removal if the population is maintained at some mid-level of biomass than if it is maintained at a low or high biomass level.

example, Atlantic herring grew faster and matured at a younger age as a result of fishing mortality (see the section on age at maturity in this chapter). Earlier age at maturity increases the rate of offspring production and also reduces the time that young fishes are subjected to potential death through harvest before they have had a chance to reproduce.

However, acceleration in rates of growth and reduction in age at maturity may not compensate fully for high rates of fishing mortality. High fishing mortality can reduce the production potential of a fish population in two ways (Cushing 1977):

1. Growth overfishing occurs when fishing mortality upon young, small fishes is too high, resulting in a mean size of fish in the harvest that is significantly smaller than the size that would allow maximum harvest of biomass from a stock.

2. Recruitment overfishing occurs when harvest reduces the adult stock to a level that causes a decline in reproductive success and recruitment.

These two processes are not mutually exclusive. Stocks in which the average size of individual fishes has declined significantly due to fishing mortality will exhibit significant decreases in total egg production because younger, smaller females typically produce markedly fewer eggs than older, larger ones. Also, if the average age of a stock decreases, and the stock includes fewer age groups of young adults, then the stock is more vulnerable to collapse in the event of several consecutive poor years of recruitment. The average time between years of potentially high recruitment may be longer than the average life expectancy of most adults in the heavily fished stock. Rago and Goodyear (1987) demonstrated that the likelihood of producing periodic strong year classes of the striped bass decreases substantially when intensifying fishing pressure reduces a spawning stock to relatively few ages of small, young adults.

Although shifts in average age at maturity initially occur because of increased growth rates that are stimulated by reductions in abundance, younger age at maturity can be genetically selected by high fishing mortality imposed on fish stocks through multiple generations. Fishing mortality produces genetic selection against traits such as large size, older age at maturity, and reproduction by older individuals in heavily fished stocks. Because most fishes are harvested before they become large, genes that produce traits of late age and large size at maturity will tend to disappear from the gene pool of these stocks through extended years of exploitation. Over a 60-year period, high exploitation rates imposed on chinook salmon led to a decline in average age at maturity of 2 years and a decrease in average size at maturity of more than 50 percent (Ricker 1981). Similar changes in life history patterns have been noted for other species, such as the vermilion snapper and gag grouper (Plan Development Team 1990). A major decrease in size at maturity coupled with a reduction in the adult life span would significantly reduce egg production of the average female and total egg production of the population.

The Impact of Harvest on Fish Community Dynamics

Overexploitation of fish stocks often modifies reproduction, growth, and mortality characteristics in ways that can strongly change the nature of predator–prey and competitive interactions in which these species are involved. Such changes can complicate the manager's capacity to determine the impact of fishing on the dynamics of fish stocks and to realize predicted responses of those stocks to specific levels of harvest.

The Georges Bank Groundfish Example. Heavy exploitation has caused the decline of most traditionally important commercial species in the Georges Bank demersal (benthic) fish community. As stocks of haddock, Atlantic cod, pollock, yellowtail flounder, winter flounder, summer flounder, and others declined from the 1960s to the 1990s, the spiny dogfish and several species of skates gradually increased. These elasmobranchs, which constituted about 21 percent of the total weight of demersal fishes in research vessel catches in the 1960s, constituted about 75 percent by 1989 (Murawski and Idoine 1992). Murawski and Idoine demonstrated that, although the species makeup of the research vessel catches had changed markedly over the years, the size composition of the catches had not. Apparently, the food resources of the Georges Bank region support an average total biomass of fishes, and the fish community will display a certain distribution of body sizes of fishes regardless of the species that are most abundant. Murawski and Idoine suggested that the change in abundances may be a replacement due to elasmobranchs taking advantage of food resources that had previously been shared with the more abundant gadids and flounders; dietary overlaps between flounder-skates and between cod-dogfish are typically broad (Grosslein et al. 1980).

This shift in abundance from traditional groundfish species to elasmobranchs occurred gradually. As late as 1980 Grosslein et al. published detailed analyses indicating that the decline in abundance of traditionally harvested species had neither stimulated a shift in the species composition of that fish community nor been followed by a large increase in abundance of any largely unexploited species. It is not surprising that Grosslein and colleagues had not identified this change during their study. The reproductive rates of the spiny dogfish and skates are quite low owing to late age at maturity and extremely low fecundities. Thus population expansion would not occur rapidly even if food resources and other needs became markedly more available because of the decline in the other members of the demersal community.

Fisheries scientists are concerned about this replacement in the Georges Bank demersal fish community. If elasmobranch abundance has stabilized at the markedly higher level of the 1980s, recovery of the traditionally important groundfish species may be quite difficult to achieve, even if fishing pressure imposed on them is sufficiently controlled. Fogarty et al. (1992) demonstrated that the level of recruitment for Georges Bank haddock from any given population size between 1968 and 1988 is substantially reduced from levels typical of earlier years (Figure 3.9). The increase in predator biomass (mainly the spiny dogfish, several ray species, and the Atlantic mackerel) during that time may have increased predation rates on pre-recruit life stages of haddock, consistently depressing recruitment below the levels that might be expected for specific abundances of spawning adults.

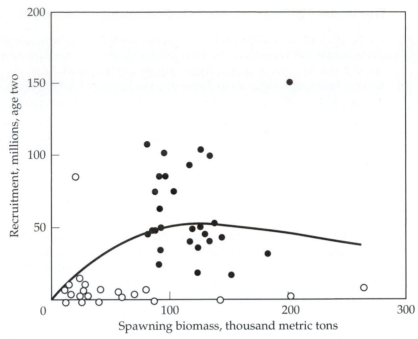

Figure 3.9 The relationship between spawning stock size and recruitment for Georges Bank haddock. Solid circles represent years between 1932 and 1967, and open circles represent years between 1968 and 1988. As with many stock–recruitment relationships, data points vary widely around the curve used to describe the relationship. However, note that most levels of recruitment after 1968 were consistently lower than those predicted by the stock–recruitment curve, whereas recruitment levels prior to 1968 were more evenly divided above and below the predicted curve. (Redrawn by permission of Fogarty et al. Copyright 1992 American Chemical Society.)

Managing Harvest by Using Life History and Production Information

Protecting Fishes from Harvest Until They Have Spawned. Basic life history parameters are used to develop fishery regulations that restrict harvest. For example, size limit regulations have often been established to provide fishes the opportunity to spawn at least once before they are legally harvested (see Chapter 7). Size-at-maturity information was used to establish harvest restrictions that produced a significant increase in the abundance of the Atlantic coast striped bass stock after two decades of persistent decline. The rapid recovery of that stock in the late 1980s and early 1990s was facilitated by size limits that prevented harvest of the relatively abundant 1982 year class. Size limits protected that year class of striped bass throughout the 1980s, until females had matured and produced a strong year class of offspring in 1989 (see Chapters 7 and 12).

Although useful, allowing fishes to spawn once before harvest may not be sufficient to protect the sustainability of a fish stock, particularly given the uncertainty of reproductive success and recruitment in a given calendar year. Size at maturity is used directly to establish harvest regulations when more detailed analyses of fish population productivity and harvest mortality cannot be conducted. For example, statewide size limit regulations for species of stream-dwelling trout or lake-dwelling largemouth bass that are based on size-at-maturity information often have been established because agencies may not be able to complete detailed analyses of the production potential of each population and the intensity of fishing pressure imposed on each population for a large number of small, separated populations (see Chapter 7). Agencies use life history and fisheries data for more complex analyses when single stocks and associated fisheries cover large geographic regions, such as coastal areas or the Great Lakes.

Predicting Recruitment.   In the mid-20th century, fisheries scientists developed analytical techniques to predict the level of recruitment that would occur when specific biomasses or numbers of adult fishes, called the "spawning stock," reproduced. Such predictions were considered central to fisheries conservation efforts because they would allow populations to be maintained at levels capable of producing sufficient numbers of recruits to replace individuals lost to harvest.

Two basic types of stock–recruitment models were developed to describe the relationship between the size of a spawning stock and the expected rate of recruitment. The Ricker model predicted that recruitment would peak at some mid-level of stock abundance and would decline at high stock abundances due to density-dependent mortality rates. This model was based on the hypothesis that offspring mortality rates should be high when the number of young-of-the-year fishes is high because of competition and depressed growth among the young fishes or because of high levels of cannibalism imposed on them by the abundant adult stock. In contrast, the Beverton-Holt model predicted that recruitment would increase with increasing stock size, but at a constantly smaller rate because of some maximum abundance of offspring that food resources can support (Ricker 1975). The shape of the curves that describe these two theoretical stock–recruitment relationships are shown in Figure 3.10. Unfortunately, in nature stock and recruitment often are not well correlated (Rothschild 1986). Figure 3.11 presents two stock recruitment relationships calculated from data gathered over a series of years that compare the abundance of recruits to the abundance of the stock that had spawned them. The level of recruitment in a given year can be substantially higher or lower than that predicted by the stock–recruitment curves, as demonstrated by the data points on the graph. The lack of precision in predicting levels of recruitment is not surprising considering the unpredictable levels of mortality suffered year to year by eggs, larvae, and young-of-the-year fishes. In addition, substantial error can be associated with measuring both stock size and abundance of offspring. Although stock–recruitment theory would seem to provide the fisheries conservation agency with a powerful tool for regulating harvest, the imprecision of stock–recruitment analyses has led fisheries agencies in the United States to reject this method as a sole means of establishing harvest restrictions (Fletcher and Deriso 1988).

Surplus Production.   As can be seen in Figure 3.8, populations maintained at a density of approximately half the maximum possible for that system should achieve the highest rate of replacement per year of harvested fishes. This replacement, which is called the yield or surplus production of the stock, represents the surplus biomass that can be removed by fishing and fully replaced by reproduction and growth by the following year. Surplus production models are based on the premise that, on the average, fish will produce more offspring than are necessary to replenish a stock. These models were developed to predict the stock size and level of fishing effort that would provide the maximum annual yield of a fish stock that can be sustained through time.

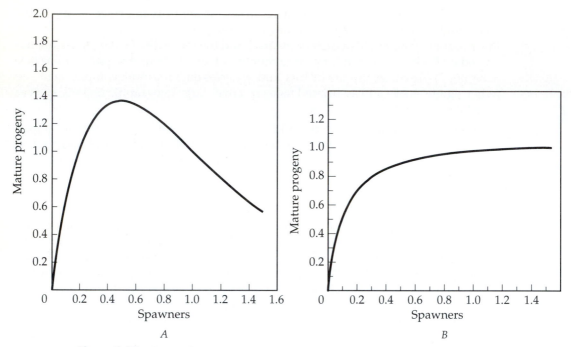

Figure 3.10   Curves demonstrating the general stock/recruitment relationships of Ricker (*A*) and Beverton-Holt (*B*). (From: W. E. Ricker, "Computation and Interpretation of Biological Statistics of Fish Populations." Fisheries Research Board of Canada. Bulletin 191. Copyright 1975. Reprinted by permission.)

The instantaneous rate of surplus production of a stock is directly proportional to the current biomass of the stock and to the difference between the current biomass and the maximum biomass the area will support. Mathematically, this can be expressed as follows:

$$\frac{dB}{dt} = \frac{kB(B_\infty - B)}{B_\infty}$$

where: B =   stock size
B_∞ =   maximum stock size
k =   intrinsic rate of increase of a stock (equal to the instantaneous growth rate of a stock at densities near zero)
t =   time, usually in years (Ricker 1975)

Thus the basic relationship between stock size and surplus production forms a parabolic curve (Figure 3.12; Van den Avyle 1993). The surplus production model allows calculation of yield only when fishing effort is relatively constant (Tyler and Gallucci 1980). The catchability of fishes within a stock, that is, the probability that a fish will be caught with a specific unit or level of effort, is assumed to remain constant. Thus any specific level of fishing effort will remove a set percentage of the stock being fished regardless of the stock's abundance. Based on this assumption, yield will follow the same general relationship with a measure of fishing effort as it does with direct measures of stock size. Although surplus production theory is

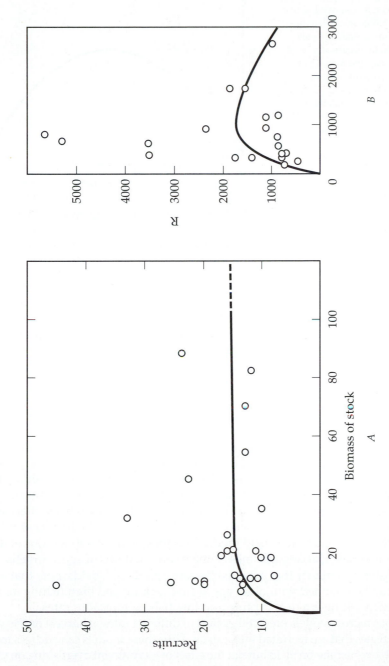

Figure 3.11 Two stock/recruitment relationships. *A* is a Beverton-Holt curve demonstrating the relationship between the number of recruits and the biomass of the adult stock of North Sea plaice. *B* is a Ricker curve of the biomass of recruits compared to that of the spawning stock of chum salmon. Note the divergence of data points from the stock/recruitment curves, indicating a relatively poor fit of the data to the curve. Such relationships may not allow a precise prediction of the recruitment level that will be achieved by a specific size of spawning stock. (From: W. E. Ricker, "Computation and Interpretation of Biological Statistics of Fish Populations." Fisheries Research Board of Canada. Bulletin 191. Copyright 1975. Reprinted by permission.)

Figure 3.12 Surplus production curve, illustrating the proportion of a fish stock that can be removed without causing a change in stock abundance. Maximum sustainable yield (Y_{max}) theoretically occurs at a stock size that is one-half the maximum (B_∞) that a habitat will support. (From M. J. Van Den Avyle, "Dynamics of Exploited Fish Populations." Pages 105–125 in C. C. Kohler and W. A. Hubert, editors. Inland fisheries management in North America. Reprinted by permission of the American Fisheries Society.)

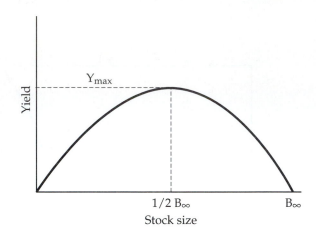

based on population characteristics of birth (or recruitment), growth, and mortality rates, surplus production analyses may be conducted using annual effort and catch statistics from commercial fisheries that are readily available for many fisheries (Van den Avyle 1993; as in Figure 3.13).

Yield-per-Recruit Models.   These models, also called dynamic pool models (Van den Avyle 1993), are based on the premise that yield is equal to the integration of instantaneous rate of fishing mortality (F), the number (N_t) and average weight (W_t) of fishes alive at each age between the age at which fish are vulnerable to capture (t_c), and the maximum age in the fishery (t_l):

$$Y = \int_{t_c}^{t_l} FN_t W_t \, dt \text{ (Van den Avyle 1993)}$$

Once a cohort of fishes reaches the age at which it is first harvested, those that do not die each year from natural causes or harvest will grow to some average larger size. Thus the yield or harvested biomass is determined by the intensity of fishing mortality at each given age and the survival and growth that occurs between consecutive ages. Yield-per-recruit analyses are conducted to determine what yields can be accomplished by specific combinations of age at first harvest and the rate of fishing mortality. Figure 3.14 illustrates yield predictions for specific combinations of these two fishery variables. The isopleth lines connect combinations of age at first harvest and rate of fishing mortality that produce the same predicted yields; note that identical yields can be obtained with older age at first capture and high fishing mortality rates as with younger age at capture and low fishing mortality rates.

　　Yield-per-recruit models predict the effect of different rates of harvest on the yield gained from fishes that are recruited into a stock. Thus these models can address the effect of growth overfishing on yield but not the effect of recruitment overfishing on yield. Spawning stock biomass per recruit (SSB/R) analysis, which is an extension of yield-per-recruit methods, can be used to evaluate the effects of age at first capture and the rate of fishing mortality on the reproductive potential of the population (Gabriel et al. 1989). Spawning stock biomass is calculated by multiplying the number of fishes alive

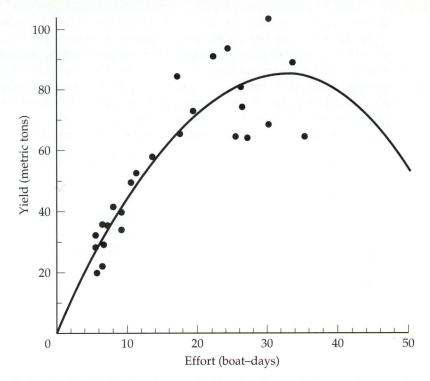

Figure 3.13 Catch and effort data from the Pacific yellowfin tuna fishery fitted to a surplus production curve. (Redrawn by permission of Pella and Tomlinson 1969. Inter-American Tropical Tuna Commission Bulletin 13(3): 421–458.)

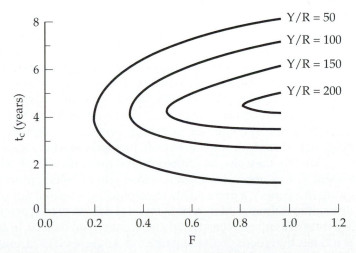

Figure 3.14 Yield predictions from a dynamic pool yield model for a variety of combinations of age at first vulnerability to capture t_c and instantaneous fishing mortality (F). Isopleth lines connect various combinations of t_c and F that produce the same yield per recruit. (From M. J. Van Den Avyle, "Dynamics of Exploited Fish Populations." Pages 105–125 in C. C. Kohler and W. A. Hubert, editors. Inland fisheries management in North America. Reprinted by permission of the American Fisheries Society.)

at each age by the fraction that are mature at that age and by the average weight of individuals of that age. The maximum spawning stock biomass per recruit occurs in populations that suffer no fishing mortality. Thus any combination of instantaneous fishing mortality and age of fish at first entry into the fishery reduces the SSB/R to some percentage of the maximum. As with yield analyses, low levels of fishing mortality applied at a young age at first vulnerability to capture can have a similar effect on spawning stock biomass as higher levels imposed at an older age.

Eggs per recruit (EPR) is another variation of the yield per recruit model. The eggs per recruit calculates the change in maximum lifetime fecundity of the average female that will occur with changes in the rate of fishing mortality or age at first vulnerability to harvest. This method has also been used to estimate the increase in fishing mortality that a population might be able to endure while remaining at a specific abundance if survival of offspring in the first year of life is significantly improved by habitat cleanup programs (Boreman 1991). Species with late age at maturity, or the greatest difference between age at maturity and the age when first vulnerable to the fishery, display the greatest reduction in EPR as fishing mortality increases.

Virtual Population Analysis.   Another group of harvest models, called cohort models, calculate the declining abundance of year classes as they age through a fishery. Virtual population analysis, one form of cohort models, uses estimates of natural mortality, estimates of fishing mortality for the oldest age taken in a fishery in the most recent year, and annual catch statistics for each age of fish within a stock to develop estimates of annual fishing mortality for each age of fishes in the fishery and estimates of total stock abundance. Once VPA analysis is completed, different effort and catch scenarios can be applied to age-specific abundance estimates to predict the effect of different levels of fishing mortality in the coming year on stock abundance.

Effectiveness of Modeling Techniques.   All of the above modeling techniques are somewhat constrained by their inability to describe precisely the effect of harvest mortality upon fish stocks. The mathematical basis for specific models may require assumptions that do not accurately describe fish–human interactions in specific fisheries. For example, the Beverton and Holt yield-per-recruit model assumes that the rate of fishing mortality is equal for all ages of fishes that are vulnerable to the fishery (Van den Avyle 1993). When effort is used in surplus production calculations, changes in catchability can bias the calculated relationship between effort and yield. A variety of factors can change the catchability of fishes within a fishery, including changes in the average size of fishing vessels, changes in gear that improve the efficiency of capture, and changes in methods used to locate fishes.

Models may also be limited due to the difficulty of calculating certain parameters. For example, the instantaneous rate of fishing mortality for the oldest age group in a fishery, a prerequisite to conducting a virtual population analysis, is very difficult to estimate. Also, certain types of data may lack precision, such as catch-at-age data in a mixed recreational and commercial fishery. In addition, the traditional models that focus on the dynamics of single species cannot predict the impact to a stock of predator–prey and competitive interactions that are modified by harvest.

Although analytical techniques may have constraints, when stocks and the fisheries associated with them collapse, it is often due to the management system's ignoring calculated predictions of overharvest rather than to a complete inability to predict the impact in the first place.

4

The Human Factor in Fisheries Conservation and Management

We need to conserve and manage fisheries resources primarily because we harvest them, or because our activities change their surroundings in ways that are detrimental to them. Indeed, most conservation or management programs do not manage fisheries resources directly; they regulate the manner in which people interact with fisheries resources. Such programs may include the following:

1. Controlling the nature and extent of harvest by people who fish (see Chapter 7).

2. Providing advice to agencies concerning land- and water-use issues that are related to protection of the quality of aquatic habitats (see Chapter 9). Note that fisheries agencies have only limited responsibility to protect habitat quality. However, numerous federal and state laws require those agencies that do have the responsibility to receive and consider advice on habitat quality and the impact of habitat alterations on fisheries resources (see Chapter 6).

3. Rearing and stocking fish to enhance depleted fish populations or simply to provide fishes for people to catch (see Chapter 8).

4. Manipulating fish communities or aquatic habitats to provide greater availability of fish species that people consider important (see Chapter 10).

In all instances, resource agencies are either addressing human effects on fisheries resources or changing the availability of resources to meet the perceived needs and interests of people.

Wild and stocked fisheries resources living in public waters normally are the property of the general public and are conserved and managed by federal and state agencies or other legislatively created commissions or councils on the public's behalf (some resources in public waters are reared by aquacultural enterprises; these "farmed" resources may be privately owned). Historically, most funds used to conserve and manage fisheries resources have come from the following sources:

1. The sale of recreational fishing licenses and excise taxes or duties collected from the sale of recreational fishing equipment, boats, and fuels (see Chapter 6)
2. General revenues from state and federal taxation

The first source comes from one segment of the general public—anglers. Thus many agency programs focus on recreationally important resources in a manner that addresses the interests of anglers. Agencies design other conservation and management efforts to broadly address the interest of the general public. These interests can be varied; for example, the major federal law that establishes guidelines for fisheries conservation in federally managed coastal waters, the Magnuson Fishery Conservation and Management Act, requires that fish harvests be regulated in a manner that provides the optimum benefit to society ("optimum yield"; see Chapter 6). Such benefits include not only the maintenance of healthy fisheries resources but also the economic and social benefits that derive from harvest of these resources. Other legal mandates such as the Endangered Species Act (ESA) and the Marine Mammal Protection Act (see Chapter 6) have required full protection of some living resources in spite of the economic consequences of such actions.

Generally, laws define the responsibility of government to act in the best interest of society. When a law is passed concerning natural resource conservation, goals are listed that describe government's responsibility, and the agency, council, or commission that must work to meet those goals is identified. Goals listed in such legislation often are very general, such as management for optimum yield in the Magnuson Act. Once a law is passed, agencies or other identified groups must develop programs with well-defined objectives that help meet the goals of the law. Obviously, at any time the types of objectives that are developed in such programs will be influenced by the personal philosophies and values of people administering the management system and of the general public. State and federal administrations will also vary in the emphasis they place on carrying out the goals listed in specific laws. For example, since passage of the ESA, the number of new species listed and the number of new conservation plans developed in any given year have changed significantly as federal administrations have changed. During the first year of the Reagan administration, no new species were included on the list of federally endangered species even though the number of species in line for review and possible listing was increasing (see Chapter 11).

Sociological and economic considerations often constrain the types of fisheries conservation programs that may be developed. For example, numerous Pacific coast salmon stocks are severely depleted due in part to changes in freshwater systems that they inhabit. Timber harvesting, damming for irrigation and production of electricity, development, and other land- and water-use practices are major factors leading to the depletion of these resources (see Chapters 9 and 12). These activities are important to the economic health of the Pacific coast region. Thus fisheries agencies cannot simply call for the elimination of these land- and water-use practices. Instead, conservation efforts have had to work within the constraints created by a public that demands that these activities continue. The axiom that fisheries managers deal with people, not fishes (paraphrased from Orbach 1980), describes the fisheries conservation and management arena well.

This chapter describes some characteristics of people that use fisheries resources and human factors related to those uses that must be considered when developing conservation and management strategies. The types of social conflicts that can arise when resources are intensively fished or when aquatic systems in which they live deteriorate as fish habitats due to human activities are also reviewed.

TYPES OF FISHERIES RESOURCE USERS

People use a fishery resource in one of three ways: They eat it, sell it, or play with it (Orbach 1980; Royce 1987). Subsistence harvesters fish in nearby waters to eat their catch or to barter it in local communities; fishing puts food on their tables. Fishing also can be an important expression of their cultural heritage. Commercial harvesters fish for a living, and most depend on an array of support businesses such as suppliers of equipment, fuel or boating supplies, and repair facilities. Thus not only do they earn a living, but their fishing activity spreads money throughout local and regional economies. Recreational users fish for a variety of reasons, including aesthetic ones. The landing and eating of fishes represents only part of their reason for fishing; many spend money simply to have an enjoyable outdoor experience. Some recreational users are nonconsumptive. Whale-watcher enthusiasts may value their interactions with marine mammals as much as anglers do their interactions with sport- or gamefishes.

The categories listed above do not fully separate users into specific groups with different, nonoverlapping interests and motivations. Subsistence users, who fish largely to eat what they catch, may also supplement their income from local sale of their catch. The importance of that supplementation may lead individuals to upgrade their boats, equipment, or methods to become more efficient harvesters. Thus they can make decisions relating to their level and type of fishing activity that are similar to those made by commercial harvesters.

Commercial users, whose standard of living depends on selling fish, may find as much pleasure and fulfillment in fishing as do anglers who fish solely for recreational enjoyment. Commercial users often choose to fish in part because it offers them a lifestyle that they desire rather than simply because it is a way to make money. Harvesters may be "lured to the sea by the independence and freedom of a fishing life, where the rewards come from hard work" (Canfield and Garber 1994). Indeed, when many fish stocks become depleted from overfishing, harvesters may continue to fish even though their income declines significantly (as described in Lantz 1994).

On the surface, an enjoyable outing of fishing might seem less critical to a recreational angler's well-being than a successful harvest is to the subsistence or commercial user. Although the angler may be fishing for pleasure, the quality of the fishing experience may be as meaningful to his or her well-being as it is to the commercial harvester; the goal may be aesthetic, but achieving it is still important. As with commercial harvesters, the money that anglers spend stimulates local and regional economies where they fish (Royce 1987). Some anglers may fish in part for economic gain. The most successful tournament anglers may supplement their yearly income more than the part-time commercial harvester whose major earnings come from other sources.

Regardless of user group, people fish in part due to cultural, ethnic, community, or group traditions and to individual experiences. Thus, when agencies manage resources, they are really dealing with a variety of people whose motivations for fishing are influenced by numerous sociological and economic settings.

Subsistence Users

Subsistence activities are usually confined to a community or small geographic region. Although some portion of the harvest may be bartered or sold locally, true subsistence users eat most of their catch. In the United States, subsistence users may depend on traditional ways of harvest (Figure 4.1), although inefficient methods

Figure 4.1 Native American subsistence harvesters, dipnetting salmon on the Celilo Falls, Willamette River (Oregon side), circa 1950. (Photograph courtesy of Oregon Historical Society.)

generally have been replaced with more effective ones. For example, in Alaska Native American subsistence harvest of migrating salmon may be conducted with modern gill nets rather than traditional, less efficient fish traps (Muth et al. 1987).

On a worldwide basis, subsistence users harvest substantial amounts of fisheries resources. In industrialized nations, natural resource policy makers have generally discounted the importance of subsistence use of fisheries resources. However, in recent years it has become increasingly evident that rural subpopulations rely on consumption of fish, wildlife, and plant resources even in technologically advanced areas of industrialized nations (Muth et al. 1987). For example, more than 75 percent of the farm families in northern Florida "raise their own meat and/or hunt and/or fish"; these activities provide the average farm family with more than 50 percent of the meat and fish that are consumed (Gladwin and Butler 1982).

Because of legally binding treaties signed with the U.S. government, many Native American tribes have the right to subsistence fishing on, and in some instances outside, tribal lands (Lamb and Coughlan 1993). These rights supersede the right of state and federal agencies to determine how all resources in public waters will be harvested. Examples of such court-upheld Native American subsistence fishing rights are presented in Chapter 6. The federal Alaska National Interest Lands Conservation Act in 1980 established the right of both Native American and non–Native American rural residents of Alaska to subsistence harvests of fisheries and wildlife resources (Muth in press). This legislation, specifically designed to protect the cultures and lifestyles of rural resource users, favors subsistence use in certain areas of the state over commercial and recreational uses if the abundance of particular resources is insufficient to meet all user needs.

Recreational Users

The roots of angling are at least several thousand years old. Ancient drawings depict Egyptian rulers engaged in fishing (Royce 1987). One can assume that rulers of that culture likely were not directly engaged in subsistence or commercial harvest but rather fished for enjoyment. Recreational fishing continued to be largely restricted to the aristocracy and the rich well into the 19th century. In the monarchic governments of Europe, royalty and the rich controlled use of and access to fisheries resources (Nielsen 1993). Once the United States became a sovereign nation, its fisheries resources became a publicly owned resource whose use would be overseen by state or federal government (Nielsen 1993). However, recreational fishing did not become popular in the United States until the advent of affordable equipment and, more particularly, leisure time for the working classes. With nearly explosive growth in participation throughout much of the 20th century, recreational fishing has become one of the most popular outdoor leisure-time activities in the United States (Figure 4.2). More than 36 million anglers fished a total of 511 million days in the United States in 1991 (USDI and USDC 1993). About 85 percent of the total fishing days were spent in freshwater systems.

Angling Motivations and Satisfaction Goals.   Numerous motivations and goals lead people to fish recreationally. Anglers may seek:

Figure 4.2 The quality of an angling trip may be determined by a variety of factors, including the natural setting and the social experience gained while fishing. (Photograph by Clyde H. Smith, courtesy of Peter Arnold, Inc.)

1. Temporary escape from the daily routine and the tension of work
2. Time spent in clean, natural surroundings
3. Companionship of sharing an experience with family and friends
4. Excitement or adventure in new (to the individual) waters
5. Meeting the challenge of trying to catch fishes
6. Catching and eating fishes

Recreational fishing may provide additional benefits in specific settings. For example, some anglers pursue the challenge of competing for prizes, money, or the prestige of winning trophies in angling tournaments (Royce 1987). Tournament anglers consider catch-related motives such as size of fish, species of fish, and the experience of the catch itself more important to the quality of their overall angling experience than do anglers who do not fish in tournaments (Loomis and Ditton 1987).

Interestingly, studies have shown that the quality of a fishing experience to many nontournament anglers can be less related to actually catching and eating fishes than to noncatch factors such as experiencing nature in a pristine setting (as in Moeller and Engelken 1972; Kennedy and Brown 1976; Dawson and Wilkins 1981; Holland and Ditton 1992; Fedler and Ditton 1994). However, many anglers do consider catching and eating fishes important to the success of fishing trips. Some anglers do not consider the challenge of catching a wild fish in a natural setting a mandatory component of a successful fishing experience; they may value catching a fish produced by nature no more than a fish raised in a hatchery. Thus, whereas some anglers expect fisheries agencies to provide them with the opportunity to pursue wild resources in pristine habitats, others may expect agency hatcheries to stock sufficiently large numbers of fishes to provide them with a "guaranteed" catch whenever they fish. Anglers who fish on charter and party boats often relate a trip's success to their catch (Graefe and Fedler 1986; Fedler and Ditton 1994). Analyses of anglers in Wisconsin indicated that the number and type of fish caught were key variables determining the quality of a fishing experience. For these anglers success during an angling trip was achieved by "filling the box" (carrying the daily catch limit in the boat's live box; Stoffle et al. 1987).

There is considerable variability among motivations and goals of anglers within and among specific recreational fisheries (Loomis and Ditton 1987; Fedler and Ditton 1994). Ditton et al. (1992) demonstrated that goals leading to satisfaction with an angling experience were related to the level of fishing activity that specific anglers exhibit. Anglers who fish often, place greater emphasis on catching big, distinctive, or trophy fishes—so-called rare-event fishing experiences—than do anglers fishing less frequently. Anglers who fish frequently also consider aspects of a specific fishing trip unrelated to catching fishes, such as experiencing natural surroundings with family or friends, more important than do low-activity anglers. Catching fishes is relatively more important to low-activity anglers than is experiencing nonspecific elements of a fishing trip (Ditton et al. 1992).

Thus fisheries agencies are dealing with a wide variety of human clients when managing recreational fisheries. There is no such thing as the average angler. Not only will motivation and satisfaction goals differ among fisheries, but specific resources are being used by humans who have a variety of interests and expectations related to their fishing activity.

Recent Trends of Recreational Fishing in the United States.   Although the number of anglers in the United States has grown substantially through the latter half of the 20th century, the rate of increase declined from 3 percent per year from 1955 to 1980 to 2 percent per year during the 1980s. The yearly sale of fishing licenses declined in 11 states during the 1980s (SFI 1993). In addition, in recent years the average number of days spent fishing per year by license holders declined (SFI 1990). Reduced numbers of license holders spending less time fishing can reduce revenues that state agencies use to conduct their fisheries programs. When facing declining revenues, states must consider raising the cost of fishing licenses, which can cause a decline in license sales if marginal anglers decide that fishing is not worth the extra cost (SFI 1991). The average age of those who fish increased after 1984, indicating that declining numbers of young people are being exposed to fishing as a form of recreation. Studies have shown that most people who fish as adults were first exposed to angling before they were 9 years old (SFI 1990).

Thus recent trends indicate that the general public is becoming more urbanized and less attached to fishing and to nature in general. However, the stewardship responsibility of fisheries conservation agencies has not been made easier by a decrease in the growth of recreational fishing activity in the United States. Many of the conservation and management problems faced by state agencies have been caused not simply by too many anglers fishing for too few fishes but by a multitude of issues, including various causes of habitat deterioration, that affect the condition of wild fisheries resources nationwide.

Commercial Harvesters

As with the recreational user, there is no such thing as a typical commercial harvester. Participants in commercial fisheries vary widely in cultural background, reasons for fishing, means of fishing (including types of gear, equipment, and vessels), and means of monetary support for their activities. This variability occurs not only regionally and among different types of fisheries but even from port to port in specific regions of the country and within specific fisheries (Doeringer et al. 1986). All commercial harvesters have one thing in common: They fish to make money. Beyond that, their defining characteristics are very diverse.

Why Do Commercial Harvesters Fish?   People often fish for a living because of family and community tradition. More than 69 percent of the boat captains and crew members working on vessels in New Bedford, Massachusetts, in the late 1970s had fathers who had fished for a living; 57 percent worked on a vessel with at least one relative (Doeringer et al. 1986). Crew members are commonly related to each other, to the boat captains, or to boat owners in some fisheries (Hall-Arber 1993). Parents or other close relatives often loan money to young Maine lobstermen so that they can fish (Acheson 1988). Many harvesters in other fisheries have similar intergenerational ties to fishing as an occupation.

Cultural ties are also strong among harvesters within a community, although these ties vary greatly from site to site. For example, people of Sicilian descent hold about 50 percent of all fishing jobs in Gloucester, Massachusetts. Many of these harvesters have descended from families that fished in Italy and continued this occupation after emigrating to the United States. More crew members of vessels out of New Bedford are of Portuguese descent than of any other nationality, whereas nearly 40 percent of the boat captains are "Yankees" (descended from multiple generations of New Englanders with English ancestry) or Norwegian (Doeringer et al. 1986). Most Texas shrimpers are Mexican, Texan-born Anglo-Saxon, or Mexican American. Louisiana Cajuns and resettled Vietnamese entered these fisheries after World War II and in the late 1970s and early 1980s, respectively (Maril 1983).

Regardless of their cultural background, many harvesters fish in part because of the independence it offers. The family-owned boat operator has an opportunity to be his or her own boss; the success of a boat is wholly attributable to the decisions and skills of the boat captain. Although not everyone becomes a captain or owner of a fishing vessel, many believe that they ultimately will reach that status (Hall-Arber 1993). Harvesters also appreciate the peaceful setting offered by fishing, as well as the opportunity to work outdoors or, more specifically, at sea (Doeringer et al. 1986; Gatewood and McCay 1990). Some feel that the opportunity to earn money is good (Doeringer et al. 1986), whereas others feel that alternative occupations are limited due to the lack of formal training needed to follow other career directions (Maril 1983). Although they appreciate the independence that fishing offers, many harvesters believe that spending substantial time away from home, family, and community is a major shortcoming of fishing (Doeringer et al. 1986; Gatewood and McCay 1990).

Harvesters often earn modest to low incomes that fluctuate depending on daily catch levels and constantly changing prices. However, harvesters are proud that their income is directly related to their own success. Even when their income is modest, they prefer the independence of being their own boss rather than being involved in other occupations where they would be dependent on employers who might be outside of their family or community circle.

What Constitutes a Fishing "Fleet"?   Although parent companies harvest, process, and market the catch in a few fisheries (such as the Pacific tuna fishing industry), in many others the harvester is separated from the rest of the marketing process (Figure 4.3). Captains, who often own their vessel, sell their catch at the pier to fish processors, who set the daily price for particular species according to their own marketing opportunities. Not only may a family fish together in these fisheries, but as a unit it must carry the entire cost of operating and maintaining a vessel; no parent company provides money for investment in vessels, equipment, or supplies.

It is difficult to describe the makeup of a fishing fleet in many large-scale commercial fisheries due to the variable characteristics of the participants. Even in fisheries where types of gear and vessels employed are similar among most participants, harvesters tend to view their own activities and those of the fishing community they live and work within as very separate from others involved in the fishery. For example, neither Louisiana Cajuns nor resettled Vietnamese were accepted well into the coastal Texas fishing communities when they entered the coastal shrimp fisheries of that state (Maril 1983). The fishing "fleets" of many coastal commercial fisheries are composed of family-owned vessels, operated within a coastal community functioning independently of other fishing communities of the region that participate in the same fisheries. Maine lobstermen form strong ties with others in their community involved in fishing but view harvesters even from adjacent communities as outsiders (Acheson 1988). Individuals or small groups of harvesters coming from the same community become extremely territorial toward lobster fishing grounds. Traditional fishing areas are considered the property of specific harvesters. When others attempt to fish within these areas, those perceiving that their territory has been invaded may respond by destroying fishing gear and equipment of the "interlopers" (Acheson 1988). When the New England groundfish fishery

Figure 4.3   Commercial trawlers at Woods Hole, Massachusetts. In many commercial fishing fleets, most vessels are family owned and operated. In such fisheries, families must carry the entire cost of operating and maintaining a vessel. (Photograph courtesy of Mike Ross.)

collapsed due to overfishing in the 1990s, some trawlers from Maine blamed the collapse on the big-vessel operators out of southern New England ports: "Southern New England boats have damaged the region's fisheries, threatening a way of life in Maine that dates back to Colonial times" (subheading of an article by A. Garber and C. Canfield in the *Maine Sunday Telegram,* September 18, 1994). An increase in the number of large vessels from southern New England during the 1980s did result in a substantial increase in fishing pressure in that fishery. However, such reasoning indicates that "others from the outside" were the sole cause of the overfishing that caused the collapse of the groundfish resource and its fishery (see Chapter 12).

Overcapitalization and Overfishing.   To conserve a harvested fishery resource successfully, agencies must restrict landings to a level that allows the abundance and availability of the resource to be sustained for future use. Harvesters, who are trying to earn a living from fishing, may have little concern for future benefits they might gain from abundant resources if they believe that their present income is significantly reduced by fishing restrictions. When a living is at stake, a commercial fisherman may be more interested in current income than in potential income of the future. In addition, harvesters normally try to maximize their catch in the belief that others will catch whatever fishes they do not. If they personally restrain their own fishing activities, the income they lose will simply be earned by others. Unless agencies restrict the number of participants by limited entry, or the maximum amount of a population that can be harvested with quotas (see Chapter 7), the number of participants and effort each harvester imparts in the fishery may grow until severe overfishing causes a decline in the abundance of the fish stock. Such a decline may be followed by economic collapse of the fishery that the stock supports.

Two events tend to lead to overfishing in commercial fisheries:

1. As a fishery develops, the promise of high profits causes overcapitalization of the fleet or fleets harvesting the resource. So many people invest money in boats and equipment that they all cannot earn a suitable profit unless the stock is overfished, causing a decline in its abundance. As the fish stock declines, catch levels of individual vessels decline.

2. Initially, as the catch of individual vessels goes down, an increase in its value may offset possible losses of income due to reduced catches. However, ultimately catches may decline so much that profits decline. To make a profit, boat owners must become more efficient, significantly increasing daily catch levels of the depleted resource to offset declining profits. Boat owners tend to invest in bigger vessels, better or bigger harvesting gear, and more efficient, costly electronic equipment to increase their catch. Thus they incur a bigger and bigger personal debt in order to increase profits at a time when maintaining catch levels becomes an increasingly more difficult challenge.

Thus, in many commercial fisheries, overinvestment occurs at the fleet level—too many vessels fish for a limited resource—and at the personal level—individual harvesters assume too large a debt in order to maintain profits in a declining fishery. This process leads to progressive overfishing, collapse of the resource, and decline in the profitability of fishing (Bell 1980). Royce (1987) refers to this progression of events as "irrational fishing." Once harvesters face increasingly lower profits and greater debts, government agencies often subsidize the industry, providing support for those harvesters who choose to upgrade their vessels and equipment in an attempt to continue to earn a living fishing for a depleted resource (Royce 1987).

Agencies may try to maintain a sustainable balance between harvest levels and the abundance of a fish stock by creating maximum allowable catch levels that cannot be exceeded. Fishing industries tend to vigorously oppose such restrictive regulations, because reductions in harvest will reduce the income of harvesters, processors, and marketers as well as the support services that supply the industry (see the section on economic influences upon fisheries conservation and management later in this chapter). Harvesters often view fishing regulations as the "last straw" that will drive them into bankruptcy and force them to leave the fishery and fishing as an occupation. Under circumstances described above, it is not surprising that many fishing industries of the United States, Canada, and western European coastal countries have overinvested in their activity, that their fishing efforts have not been effectively regulated to prevent overfishing, and that an increasing number are being subsidized in some manner by their governments even while resources continue to decline (Royce 1987).

ECONOMIC INFLUENCES ON FISHERIES CONSERVATION AND MANAGEMENT

Nationally, commercial fishing produces substantial income for harvesters. Although the total dockside value of commercial seafood landings (the earnings of harvesters when they bring their catch to the dock) was about $3.5 billion in 1993, it was $200 million less than in 1992 (NMFS 1994). The decline was due to depleted resources and restrictive regulations intended to support resource recovery. In spite of recent declines in dockside value, numerous people still benefit economically from commercial fishing. Harvesters depend on shipyards, repair facilities, equipment makers, supply stores, and other providers of services and supplies in order to fish. In many fisheries, harvesters sell their catch to processing companies, who produce filets or other products that require distribution and ultimate sale in retail markets. Processors, shippers, and marketers also pay for support services. Thus the economic activity generated by harvest of fisheries resources has a much greater impact than simply the value of the catch to the harvester.

Recreational harvests also feed substantial money into local and regional economies. Anglers spend money on boats, motors, electronic equipment, rods, reels, lures, baits, and other supplies to support their fishing activity. In addition, anglers may pay for services provided by marinas, gas stations, restaurants, hotels and motels, guides, and party- or charter-boat operators during fishing trips. Anglers spent about $24 billion to fish in 1991; about 75 percent of angler expenditures were associated with fishing trips to freshwater systems (USDI and USDC 1993).

The total economic impact of fishing is estimated by applying multipliers to angler expenditures or dockside values. These multipliers predict the total income realized as these monies move through various segments of local and regional economies (Weithman 1993). Money circulates within local and regional economies very differently in recreational and commercial fisheries. Because of this, it is not appropriate to compare directly the economic importance of money spent by anglers to income earned by commercial harvesters. However, commercial uses of fisheries resources are clearly linked to the economic well-being of many coastal communities, and recreational fishing affects local and regional economies in both coastal and inland areas of the United States. Thus economic considerations play a major role in the fisheries conservation and management decision-making process.

Interestingly, an increase in fishing activity can affect economic impacts within recreational and commercial fisheries differently. An increase in the number of participants in a commercial fishery increases the potential for overharvest: The resource collapses, the income of individual harvesters declines, and the economic impact of the fishery is reduced. If government decides to subsidize the economically ailing industry, the fishery becomes an economic drain rather than a stimulus (Royce 1987). An increase in the number of anglers participating in a recreational fishery can have the opposite effect, as long as anglers are willing to accept restrictive fishing regulations that protect the abundance and stability of the fishery resource even as the number of anglers increases. As participation in a recreational fishery increases, so does the total amount of money that enters local and regional economies due to angler expenditures. Thus growth of recreational fisheries stimulates local and regional economic activity (Royce 1987).

CONFLICTS IN THE FISHERIES CONSERVATION AND MANAGEMENT PROCESS

Most of the conservation and management programs developed by fisheries agencies focus on either (1) the allocation of resources to user groups whose total demand or potential to harvest specific resources would be greater than could be sustained unless their fishing is regulated, or 2) the protection of resources affected by land- and water-use activities that have altered the condition of aquatic habitats. These programs potentially conflict with the interest or activities of groups using either fisheries resources or the habitats in which the resources live. Much of the conservation process involves decision making and policy development in the face of conflicting human interests and goals.

Decision-Making Conflicts Created by Harvesting

Many fisheries resources are harvested by a variety of users with different interests, motivations, and goals associated with fishing. Whenever resources are not abundant enough to fully support the fishing activities of all groups, agencies must decide how to allocate the resources within and among the groups. Allocation often results in the perception that needs and interests of some users are better met than those of others. This section reviews how resource allocation affects the conflicting interests of different user groups.

Allocation Among Users Within Commercial Fisheries. Allocation of harvest within some fisheries can be complicated by the variety of users participating, and the variety of species that are simultaneously being harvested by certain types of gears. Demersal fisheries resources off the Alaskan coast support commercial harvesters using several different harvesting techniques and gears while fishing for particular species within the assemblage of bottom-dwelling taxa. Commercial fisheries associated with these resources include the following:

1. A Pacific halibut longline fishery
2. A king crab and tanner (snow) crab pot fishery
3. A groundfish fishery in which trawlers, longliners, and trap harvesters target a bottomfish assemblage including walleye pollock, flatfish species, Pacific cod, sablefish, and others

Maximum allowable harvest levels (quotas) are established for taxa in all three fisheries. Longliners and trawlers operating in the groundfish fishery incidentally capture halibut, crabs, and salmon, which often die due to landing and handling stress. In order to protect these resources from overharvest, the groundfish fishery is restricted to specific maximum levels of allowable bycatch of halibut and crabs. Once these levels of incidental catch are reached, the groundfish fishery is closed for the year (NOAA 1993). Limits for incidental catch have been set at such low levels in recent years that the groundfish fishery reaches those limits and is closed before harvest limits for the groundfishes themselves have been reached.

In 1992 the groundfish landings had a total dockside value of $658 million, substantially more than the value of Pacific halibut ($110 million) and crabs ($305 million) combined. Of the two major Alaskan fishing areas, crabs are harvested mostly from the Bering Sea and halibut from the Gulf of Alaska (NOAA 1993). In 1992 dockside value of groundfishes from the Bering Sea ($522 million) was substantially higher than that of crabs ($305 million), whereas in the Gulf of Alaska the value of groundfish ($136 million) was higher than that of halibut ($71 million). Groundfish harvesters have openly complained about fishery closures imposed on them to protect two other fisheries that produce less valuable landings.

In some fisheries, managers have considered allocating a guaranteed catch level to each individual harvester in the fishery (individual transferable quotas [ITQs]; see Chapter 7). The total of all individual quotas equals the catch level that a fish population can sustain without a major decline in abundance. Commercial harvesters have opposed individual quotas because such regulations do not allow individual harvesters to use their expertise and persistence to land significantly greater catches and earn significantly higher profits than others in the same fishery. Many harvesters feel that such management strategies intrude on their right to decide how much they will fish to earn a living.

Allocation Within Recreational Fisheries. Although allocation conflicts may not be as severe in recreational as in commercial fisheries, they do occur. In many recreational fisheries, allocation is focused on restricting the number or type of fishes that anglers can catch and keep. Daily limits on the number of a species that can be kept, or limits on the minimum size of a species that can be kept, spread harvest restrictions evenly among participants. Such regulations may be acceptable to anglers if they believe that all are sacrificing to ensure high-quality fishing opportunities in future years.

Some recreational fisheries restrict the use of certain types of gears while allowing the use of others (see Chapter 7), which in essence is a form of allocation among different user groups. For example, trout fisheries that are restricted to fly fishing only or largemouth bass fisheries that do not allow the use of live bait may be established to meet appropriate conservation goals, and not simply to favor one angling group over another. No-live-bait or fly-fishing-only regulations may be established in fisheries to increase the survival of fishes returned to the water after being caught. Fly-fishing-only regulations may also be used to reduce the total number of anglers fishing for a resource that is particularly sensitive to overfishing or to provide a small number of anglers with a particularly desirable angling experience. Low angler densities will enhance an atmosphere of being in a wild, pristine setting. Although these last examples may focus on appropriate management objectives, they also favor fly anglers over those using other equipment. Compliance with regulations is difficult to achieve if enough anglers feel that they are being excluded simply to favor others. Whenever such regulations are instituted, anglers are most apt to comply if they understand the conservation value of the restriction. Anglers may not comply if they feel that such restrictions reserve an angling experience for some but not others while having no benefit in protecting the quality of the fish stock being managed.

Allocations Between Recreational and Commercial Users.   Anglers in the United States have become effective lobbyists at the state and federal level, in part due to the organizational network they have created. Organizations such as the Izaac Walton League of America, Trout Unlimited, the Fly Fishing Federation, the Bass Anglers Sportsman Society, and the American Sportfishing Association influence the passage of legislation such as the Dingell-Johnson/Wallop-Breaux federal aid program (see Chapter 6) and the creation of fishery regulations by state and federal agencies. In part through this organizational network, and because they outnumber commercial harvesters by more than 300 to 1 (Royce 1987), anglers have exerted increasing influence on how fisheries resources are allocated to harvesters.

Most freshwater resources have been allocated to recreational uses for some time. On the other hand, commercial fishing interests often have held precedence over other user groups when coastal resource allocations were made. However, in recent years agencies and management councils have been reserving an increasing number of coastal resources for recreational use. Management plans established in the 1980s to stimulate recovery of the depleted Atlantic coast migratory striped bass stock included major reductions in recreational and commercial harvest (see Chapter 12). Since recovery of that stock, management strategies allocate most of the allowable harvest to recreational users. Southern Atlantic and Gulf of Mexico state agencies have reserved inshore harvest of red drum resources for recreational users. Perhaps the clearest example of the growing influence of anglers in coastal areas occurred in Florida in 1994. Angler groups managed to place a referendum on the state ballot that banned most commercial netting within state territorial waters. The referendum passed by a very wide margin. Although allocation between recreational and commercial users has typically been made by public agencies, in this instance the angling lobby went straight to the voting public to seek its goal.

Agencies may favor allocation to recreational rather than commercial users for reasons beyond simply the political power of angling lobbyists. Many resources may support greater economic activity when fished recreationally than commercially. Thus economic considerations may favor recreational use. In addition, anglers may be more willing than commercial harvesters to have their activities restricted to protect future fishing opportunities. If well-managed resources remain stable while supporting a recreational fishery, growth in recreational fishing activity may stimulate long-term growth in the economic impact of the activity. Alternatively, the management of commercial fisheries often has proven difficult. Decisions concerning resource use are confounded by the impact that restricting harvest will have on the income of harvesters and the industry in general. Management strategies that respond to economic concerns of commercial harvesters may risk the decline and collapse of the resources supporting the fishery. Achieving long-term optimal use of a resource is thus often easier to achieve in a recreational rather than a commercial setting.

Allocation Between Native Americans and Other Consumptive Users.   Historically, the allocation of fisheries resources to users was solely the province of federal agencies on federally controlled lands and waters and of state agencies on all other public waterways. By the mid-20th century, Native American tribal authorities were actively seeking access to fisheries resources guaranteed to them by federal treaties (see Chapter 6). Some of these rights promised access to fisheries resources held under the stewardship of state governments.

Treaty rights challenged a state's authority to be solely responsible for allocating public fisheries resources occurring within state boundaries. Federal court decisions generally have recognized tribal rights to manage sport and commercial harvests on reservations and have verified the responsibility of state government to provide access to specific resources off of reservation lands if guaranteed under treaty. Courts have interpreted the proportion of any designated allowable harvest that must be allocated to Native Americans and in some instances have required that public agencies provide tribal authorities a meaningful role in allocation decisions (see Chapter 6).

Due to these court rulings, agencies sometimes find themselves embroiled in controversy created by reaction of non–Native American users. For example, commercial shellfish harvesters and landowners threatened violence toward and discharged firearms over the heads of Native Americans after federal courts ruled that Native American tribes in Washington had the right to half the allowable harvests of shellfishes in the Puget Sound (*The New York Times*, p. A12, 27 January 1995; see Chapter 6). Although uncommon, violent reactions, including the threat of physical harm, occur when strict allocations or other harvest regulations are perceived as unfair by some user groups (see the section on conflicts created when endangered species are vulnerable to fishing gear in this chapter and the section called "New England Groundfisheries: Failure to Control Overfishing" in Chapter 12).

Controversy continues in some fisheries even after appropriate allocation has been established. For example, in August 1994 the Columbia River Inter-Tribal Fish Commission threatened to file a lawsuit to force a September salmon harvest in the Columbia River. Harvest had previously been closed because of the poor condition of the fall migrating salmon stocks. The tribal commission contended that the collapse of these resources was due to causes other than harvest pressure, which were not being adequately addressed to stimulate any stock recovery even if harvest were not allowed.

The Protection of the "Public Interest" by Private Conservation and Environmental Organizations.   Resource agencies are responsible for making decisions that are responsive to the best interest of the public. However, disagreement can arise concerning what actions constitute the society's best interest. Current conservation strategies directed toward New England's groundfish fishery illustrate such a conflict.

Since passage of the Magnuson Act, The New England Fishery Management Council has managed New England's groundfish resources in a manner that has been very responsive to the needs of the fishing industry. In its early attempts at management, the council allowed its own quotas to be exceeded annually by harvesters in order to prevent economic decline of the fishing industry. Ultimately it eliminated quotas altogether, attempting to control fishing through other methods while federal loan programs stimulated significant growth of the fishing fleet and of the harvesting capacity of individual vessels.

Regulations intended to protect the economic viability of the growing fishing fleet led to severe declines in the abundance of numerous groundfish stocks throughout New England's federally managed waters (see Chapter 12). In response to the worsening condition of the groundfish resources, the Conservation Law Foundation, a private environmental and conservation institute, brought federal lawsuit on behalf of the public against the New England Fishery Management Council and the National Marine Fisheries Service (NMFS) (Hall-Arber 1993). Although not responsible for developing management plans under the Magnuson Act, NMFS was included in the lawsuit because the foundation claimed that it and its parent agency, the U.S. Department of Commerce, had not aggressively advised the Management Council of the actions that needed to be taken to prevent further collapse of the groundfish resources. In settlement of this lawsuit, the council amended its management strategy to significantly reduce the effort and catch of the groundfish fishing fleet over a 5-year period (see Chapter 12).

Private environmental organizations have become increasingly involved in representing the public's interest by initiating lawsuits against management agencies. For example, in August 1994 several environmental organizations petitioned federal courts to halt all logging, mining, and other activities on specified National Forest Service lands in Idaho until suitable plans for protection of salmon habitats had been developed. In such instances, the public essentially is suing the government to meet its legal obligations as defined in federal or state laws.

Conflicts Created When Endangered Species Are Vulnerable to Fishing Gear.   Many types of gears are regulated to prevent or reduce the capture of nontargeted animals, including undersized, sublegal individuals of marketed species, and often nontargeted species. In some instances, such incidental catch consists of species that are protected under federal or state endangered species legislation. In these cases, management policy usually requires that harvesters modify their gear or collecting methods in specified ways to reduce the frequency of capture of protected species.

Turtle-excluding devices (TEDs; see Chapter 7) are required in the nets of all trawlers operating in the shrimp fisheries of the Gulf of Mexico and southern Atlantic states to reduce the capture and death of threatened and endangered sea turtles. Similarly, purse seiners in Pacific tuna fisheries are required to limit the incidental capture and death of dolphins to a small proportion of that which occurred before passage of the Marine Mammal Protection Act (see Chapter 7). Reduced mortality is achieved by using a process called *backing down* when retrieving seines in order to facilitate the escape of entrapped dolphins. In both instances, the industry initially fought such regulations, claiming that such gear or method modifications would significantly reduce its directed catch, lowering its profits. Shrimpers in the Gulf of Mexico reacted particularly angrily; many blockaded sea ports and loudly protested at public hearings, and some threats were aimed at fisheries agency personnel. As a means of resolving turtle capture, the federal government has funded a substantial research effort to develop devices that effectively minimize incidental capture of turtles while minimizing loss of shrimp in shrimp trawls. Objections from the tuna industry lessened once it became apparent that backing down the net would not cause high losses of captured tuna. Ultimate cooperation was also achieved in part because of threats by the environmental community to advertise boycotts of tuna products from companies that failed to comply with the program.

In fisheries that incidentally harvest endangered species, public agencies are not responsible for determining whether harvesting or protection should be emphasized if they conflict. Law clearly dictates that attention be focused on the protected taxa. Agencies do attempt to resolve such incidental capture issues in a manner that facilitates effective protection as well as effective harvesting.

Conflicts Created by Water- and Land-Use Practices

Water is a highly valued and intensively used resource. We have used our waterways to carry chemical wastes and sewage from urban centers. We have diverted water from river basins to irrigate farmlands or to meet the demands of water-limited urban centers, significantly reducing the quantity and quality of water in those basins. We have pumped water from underground aquifers, lowering water tables of soils and water levels of springs, ponds, and streams. We have dammed rivers to produce electricity, to control flooding, or to store water for local or regional uses. We have diked and channelized streams and rivers to reduce flooding, prevent erosion, and promote shipping. We have drained wetlands to create croplands or allow urban and suburban development. The western United States, where annual precipitation levels are modest, uses and reuses its sources of water with particular intensity.

All of these uses of surface and underground water support the economic well-being of communities and regions or provide what many people consider essential, everyday needs. Unfortunately, such water uses have the potential to significantly alter the condition of aquatic habitats that house our fisheries resources, causing major changes in the distribution and abundance of numerous fish populations. Additionally, land-use activities such as intensive farming, logging, grazing, mining of minerals, and urban and suburban development on lands adjacent to river and lake basins also modify habitat conditions in ways that are detrimental to native fisheries resources (potential mitigation of the above water- and land-use activities are discussed in Chapter 9). Although conservation strategies and policies are often focused on effective allocation of fisheries resources among various user groups, in many instances greater conflict arises in the attempt to protect habitat conditions necessary for healthy fisheries resources in the face of water- and land-use practices deemed essential by society.

Numerous laws establish the responsibility of government to consider the impact of land- and water-use activities on fisheries resources. Although many of these laws mandate consideration of impacts, they may not mandate specific mitigation of those impacts. When conflicts arise because human activities alter aquatic habitats to the detriment of fisheries resources, the priority placed on resource protection is somewhat open to interpretation by the executive branch of government, which must carry out the responsibilities defined in any law. For example, Section 404 of the Clean Water Act allows comment by the U.S. Fish and Wildlife Service, the National Marine Fisheries Service, and the Environmental Protection Agency before the U.S. Army Corps of Engineers can grant a permit for any dredge and fill operation within wetlands. The decision to grant a permit is based on which action provides the greatest public benefit, development or protection of the wetland. After receiving such recommendations, the Corps has often provided greater weight to benefits derived from development than benefits resulting from protection (Owen and Jacobs 1992; see Chapter 9). When multiple societal needs are met by land and water use within specific basins, such as those of the Columbia River of the western United States, even economically important fisheries resources such as the anadromous salmonids within that basin may not be considered as valuable to society as the benefits derived from the use of the basin itself (see Chapter 12). Ultimately, the current intent and interests of the society determine the outcome of fisheries resource–habitat deterioration conflicts.

Thus humans have created the need for fisheries conservation and management through their use of fisheries resources and their altering of aquatic habitats. They influence the direction that these activities take, either directly by working with lobbying agencies or other conservation and management entities, or indirectly by electing public officials with specific philosophies who pass or carry out laws. Most fisheries conservation and management actions developed by agencies and councils cannot be implemented without the society's formal input through some type of mandated public hearing process. In the management system created by the Magnuson Fishery Conservation and Management Act, harvesters become the decision makers as appointed members of the regional fishery management councils. In this instance, harvesters participate in determining what management strategies will guide the conservation of the resources they are harvesting. In such human-driven systems, it is not surprising that effective conservation and management efforts require understanding people as much as or more than they require understanding the condition of populations of fishes and other aquatic organisms.

5

Harvesting Methods and Fisheries Resources of the United States

The United States has one of the most diverse fauna of freshwater fishes and mussels of any temperate region of the world. Its vast freshwater, anadromous, and oceanic fisheries resources historically have supported extremely active commercial and recreational fisheries. Many of these resources have declined in abundance due to the exploitation that has been imposed on them and to changes in the condition of aquatic habitats caused by human land- and water-use activities. A great number of resources are depleted, in some instances to the point of extinction or near extinction, due to a wide variety of direct and indirect human influences.

This chapter provides an overview of harvest methods that have been used in various fisheries, and of the general types of fisheries resources that have been harvested or otherwise affected by humans.

HARVESTING METHODS IN FISHERIES

Most recreational fishing for fishes is accomplished by the use of hook and line or, in a few instances, gears such as dipnets—for example, in rainbow smelt fisheries. Recreational and commercial users of shellfishes may harvest nearshore mussels and clams with rakes or shovels or by hand, and species of crabs with small baited pots or traps. This section will not detail different hook and line or other methods of recreational users. Although traditional harvesting methods used by Native Americans or others in some subsistence fisheries (see Chapter 4), such as throw nets, fish wheels, and other gears, and some commercial methods such as harpooning are significant techniques in some fisheries, they will not be reviewed because they are not widely used. This section will focus on major gear types that are used in particularly large commercial fisheries or are used to harvest a wide variety of different fisheries resources.

Numerous types of gear and harvesting methods have been employed to harvest fisheries resources commercially. Many of these can land extremely large biomasses of fisheries resources within relatively limited amounts of time expended by participants in the fishery (for example, the 60-metric-ton catch of walleye pollock in one net haul of an ocean factory trawler shown in Figure 5.16). In addition, some types of gear harvest a wide variety of fishes and other resources that are not specifically being sought. For example, in the Gulf of Mexico shrimp fishery the biomass of organisms other than shrimp that are killed and discarded—the by-catch, or incidental catch—greatly exceeds the biomass of shrimp that the fishing effort is directed toward—the directed catch. Recreational fisheries can be managed with the expectation that a reasonable percentage of individuals that are returned to the water survive the capture experience. In many commercial fisheries, gears and methods do not allow this to occur. Very often, a captured fish is a dead fish whether it is kept or discarded. Thus the fisheries agency must understand the behavior of gears and the outcome of using particular methods in order to attempt to regulate the interaction between the resource and its user. Such characteristics of gears and methods can greatly influence the effectiveness of management strategies and regulations.

Hook and Line Gear

Hook and line capture is one of the oldest fishing techniques (Everhart et al. 1975). Longlines, also called trotlines or setlines, are left in the water and later retrieved (a passive capture method), whereas handlines, pole and lines, and trolling lines are fished actively under direct control of the harvester.

Longlines. Longline gear consists of baited hooks and leaders, also called gangions, that are attached in series to cables or ropes called mainlines that are left in the water to fish. Longlines used to harvest bottom fishes have an anchor, a buoy line, and float at each end of the mainline. Longlines used to harvest pelagic species—those that dwell in the water column rather than on the bottom—have buoy lines and floats, but these lines generally are not anchored (Everhart et al. 1975). The spacing of the leaders and hooks will vary according to species being sought. Ten-inch leaders may be spaced at 3- to 4-foot intervals in the Pacific sablefish fishery, whereas 3-foot-long leaders may be spaced as much as 30 feet apart in the Pacific halibut fishery (High 1989). The spacing of the gangions is related to the general distance that a baited hook must be from the nearest hooked fish to attract other fishes. Mainlines may stretch for several miles or more in length. After fishing, many longlines are retrieved by power winches. Some longline fisheries have become highly automated, with machines retrieving the mainline, removing fishes from hooks, removing old bait from hooks and then rebaiting them, and untangling twisted gangions.

Many types of marine pelagic and demersal bottom-dwelling fishes are taken in longline fisheries, including tunas, sharks, swordfish, sablefish, Pacific halibut, Pacific cod, reef fishes, and others (Joseph et al. 1988; High 1989; NOAA 1991).

Active Hook and Line Fishing.   Handlines are held and retrieved by hand, or they are wound on a hand-turned or power reel. Lines may possess a single baited hook or a series of hooks. Because it is labor intensive, this method is restricted to a relatively small number of valuable, specialty-market fishes such as grouper species in the Gulf of Mexico (High 1989). Pole and line fishing has been used extensively to harvest several tuna species commercially. When a school of tunas is sighted, live bait such as anchovies, sardines, or other small-bodied fishes are thrown overboard, a process called *chumming*, to attract the tuna toward the vessel. Once attracted, they are caught with pole and line using artificial lures with barbless hooks (Joseph et al. 1988).

Baited hooks and artificial lures imitating natural foods are also dragged through the water by moving vessels in a method called *trolling*. Vessels troll multiple mainlines, each one attached to a power or hand-wound reel. Each mainline may hold a single hook or a series of hooks at specific intervals along its length. King and Spanish mackerel of the Atlantic and Gulf coasts, chinook and coho salmon along the northwest Pacific coast, and albacore in the Pacific Ocean have been commercially harvested by trolling (Everhart et al. 1975; Joseph et al. 1988; High 1989; NOAA 1991).

Active Entrapment Gear

Active entrapment methods include gears that are towed or pulled through the water, encircling or trapping fishes or shellfishes during the towing process.

Trawls and Dredges.   Trawls are cone-shaped nets, closed at the posterior end, that are towed through the water by vessels (Figure 5.1). As the net is towed, fishes swimming in front of the trawl enter the open mouth and are collected in the bag at the rear end of the net, called the *codend*. Trawls can be fished along the substrate or in the mid-water column. There are two basic trawl designs, the beam trawl and the otter trawl. The mouth of a beam trawl is held open by a rigid frame or a pole attached to the headrope at the top of the net mouth. The footrope along the bottom of the net mouth is often weighted with a chain (Hayes 1983). The beam trawl is awkward to handle because of the rigid mouth of the net (Pitcher and Hart 1982). An otter trawl is held open by otterboards or doors attached to both sides of the net mouth. Ropes leading to the vessel are attached to the anterior edge of each door or to the surface of each door that is facing inward toward the middle of the net mouth. The net is attached to the surface of the boards facing outward. Thus, as the net is pulled forward through the water, the boards spread outward from each other due to water pushing against the inward-facing surface of each board. Floats are attached to the headrope and chains to the footrope in order to spread the net vertically. When bottom trawls are towed over somewhat rough substrates, large rollers may be attached to the footrope to prevent the net from becoming entangled

TRAWL NET

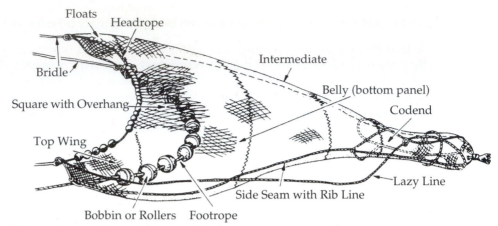

Figure 5.1   An otter trawl. (From Hayes, M. L., 1983. Active fish capture methods. Pages 123–145 in L. A. Nielsen and D. L. Johnson, editors. *Fisheries techniques*. Reprinted by permission of The American Fisheries Society.)

on rocks. The size of net used depends on the power of a vessel needed to tow the trawl through the water and to handle tow cables and otterboards (High 1989). The largest trawls may be as long as 100 meters with a mouth opening of 400 to 500 square meters (Pitcher and Hart 1982). A trawl normally is not towed at speeds fast enough to immediately overtake and capture most fishes in its path. Many fishes swim in advance of the trawl, avoiding its mouth until they tire and are gradually overtaken by the gear.

Trawls are used on both the Atlantic and Pacific coasts. A variety of continental shelf groundfish (bottom-dwelling) species are harvested in the northeast, including ocean perch or Acadian redfish, Atlantic cod, haddock and other members of the cod family (*Gadidae*), and yellowtail and winter flounder, American plaice, and other flatfishes. Gadids such as Pacific whiting, Alaskan walleye pollock, Pacific cod, yellowfin sole and other flatfishes, numerous rockfishes, and other groundfishes are harvested in trawl fisheries along the Pacific coast of the conterminous states and Alaska. Shrimp species are harvested by trawling in the Gulf of Mexico and the Gulf of Alaska and along much of the Atlantic coast of the United States. Pelagic species such as Atlantic and king mackerel, menhaden, long-finned and short-finned squid, drum and seatrout species, and butterfish are harvested by mid-water trawling.

Trawls typically capture a variety of organisms other than those that might be targeted for harvest in a fishery. This bycatch can include sublegal-sized fishes of species being sought or nontargeted species. The biomass of bycatch can exceed that of directed or targeted catch in some fisheries. Capture in trawls and handling on board vessels can be extremely stressful to bycatch. In many trawl fisheries, bycatch exhibits high rates of mortality even when returned to the water after the catch is sorted. The reduction of bycatch of finfishes or other taxa such as sea turtles has become a major priority in the management of many fisheries (see Chapter 7).

Bycatch of small fishes can be reduced in some trawl fisheries by the use of large mesh netting in the trawl's codend. The size of the openings in the codend mesh determines the minimum size of fish that will be retained in the trawl rather than passing through it. Small fishes pass through the mesh, whereas higher percentages of fishes of each larger size category are retained within the codend. Retention of small fishes may increase for any specific mesh size as a trawl fills while being fished, because the growing mass of trapped larger individuals may prevent smaller fishes that enter the codend from contacting and passing through the mesh openings. Special devices that deflect small fishes, such as bycatch reduction devices (BRDs), or sea turtles, such as turtle-excluding devices (TEDs), out of an opening anterior to the codend of a trawl are required in some shrimp trawl fisheries (see Chapter 7).

Dredges are rigid framed gears that are dragged over substrates to harvest bivalve shellfish such as scallops, oysters, and clams (Hayes 1983). A scallop dredge has a metal frame with a scraper bar across the bottom of the mouth and a collecting bag constructed of metal rings. The scraper bar or blade drags across the substrate as the dredge is pulled through the water, deflecting scallops into the collecting bag (Figure 5.2). The size of ring openings can function much in the same manner as mesh openings in trawl nets: Bivalves smaller than the ring openings may pass through the collecting bag, whereas larger ones are retained. An oyster dredge will have rakelike teeth on the scraper blade to dislodge the oysters from their attachment point on the substrate (Figure 5.2). Surf clam dredges may have a hydraulic device that washes sediment away and flushes the clams into the collecting basket. Whereas bottom trawls may be equipped with rollers to keep the footrope

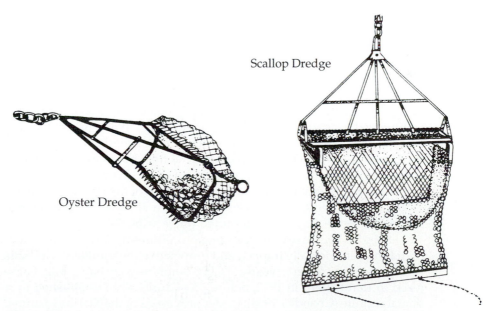

Figure 5.2 Scallop and oyster dredges. (From Hayes, M. L., 1983. Active fish capture methods. Pages 123–145 in L. A. Nielsen and D. L. Johnson, editors. *Fisheries techniques*. Reprinted by permission of The American Fisheries Society.)

from entangling upon objects on the bottom, dredges are constructed to scrape across and into substrates. Such harvesting devices can severely disrupt the surface substrate and the benthic fauna that live on or within it.

Numerous resources pursued in U.S. trawl and dredge fisheries are overfished, particularly principal bottomfishes of the Northeast (discussed elsewhere in this chapter) and sea scallops and other bivalves. Along with other regulations established to protect these depleted resources, groundfish trawl fisheries are often regulated with minimum-mesh size regulations that are intended to reduce the capture of undersized individuals and incidental capture of nontargeted species.

Seines. Seines are small-mesh nets used to encircle fish schools. The top line—called the *floatline,* or *corkline*—supporting the netting typically has floats, and the bottom line—or *leadline*—has weights to keep the netting spread vertically as it is pulled through the water (High 1989).

Beach or haul seines are used in shallow water to entrap fishes along the shoreline of coastal areas, lakes, or large rivers. Large beach seines are typically put into position by a boat, which encircles an area of water adjacent to the shore with the netting and places both ends at the shoreline. Once in place, a seine is hauled to shore by hand or by winches or other power equipment, with workers piling up the netting as both ends are drawn to shore. Striped bass, bluefish, Atlantic mackerel, weakfish, tautog, and other coastal species have been harvested with beach seines, as have freshwater fishes such as the common carp, freshwater drum, channel catfish, and other large lake and river species.

Purse seines are used in offshore deepwater fisheries. This gear consists of a long and relatively deep curtain of netting with a floatline and leadline, the latter having a series of rings attached along its length with a cable running through the rings. Once a school of fishes is found, the fishing vessel draws the seine around the school in a large circle while the free end of the seine is held in position by a small boat (skiff) or sea anchor. Once the seine encircles the school, the vessel retrieves both ends of the ring cable, closing the bottom of the net and ultimately pulling the leadline onto its deck (Figure 5.3). This traps the fishes within a "purse" of netting.

Purse seine fisheries harvest greater total biomasses of fishes than any other commercial gear used in the United States (High 1989). Before the collapse of Pacific sardine stocks in the early 1960s, purse seine fisheries for this species accounted for 25 percent of the entire U.S. commercial finfish harvest (NOAA 1993). Until recently, purse seine fisheries harvesting Gulf of Mexico and Atlantic coast menhaden stocks have accounted for as much as 40 percent of the total U.S. finfish catch (High 1989). Tunas, Pacific salmon species, northern anchovy, king mackerel, and other schooling species have also supported purse seine fisheries.

Standard purse seining operating procedures have caused high mortality rates of dolphins incidentally trapped and drowned in the southeastern Pacific yellowfin tuna purse seine fishery. Nearly 370,000 dolphins were either killed or seriously injured in that fishery in 1972, before procedures were established to reduce incidental capture (Chandler 1988). Under provisions of the Marine Mammal Protection Act, the number of dolphins that can be incidentally killed in the U.S. yellowfin tuna fishery each year is limited to a small fraction of the unregulated incidental take of years previous to regulation (see Chapter 6). A significant reduction in dolphin

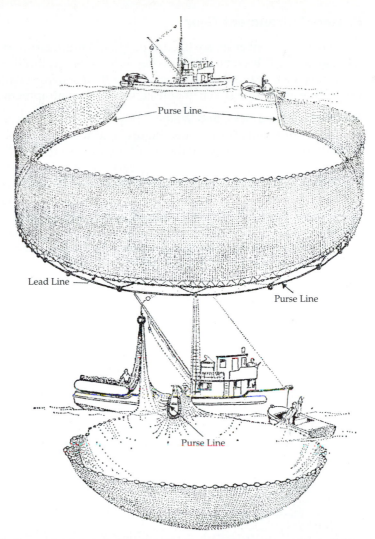

Figure 5.3 Purse seine being deployed and the purseline being drawn. (From: T. Bagenal, *Methods for Assessment of Fish Production in Fresh Waters*, 3/e. Copyright © 1978. Reprinted by permission of Blackwell Science, Ltd.)

mortality has been achieved by use of a procedure called *backing down*, which consists of lowering one edge of the float line below the water's surface after the seine has been pursed. This allows dolphins to escape while retaining most or all of the deeper swimming tuna. U.S. yellowfin purse seine vessels are also required to use a small mesh panel in the part of the net that the dolphins swim over, so that they will not entangle their snouts in the webbing and drown while trying to escape (Chandler 1988).

Many stocks that traditionally supported purse seine fisheries have collapsed or are seriously declining in abundance. Thus the use of purse seines to harvest some schooling pelagic species is regulated or prohibited.

Passive Entrapment Gear

Passive entrapment gears are fixed in position. Thus fishes and shellfishes must contact the gear and become entrapped in order to be captured. Fishes and shellfishes encounter such gear either because it lies in their standard path of movement or because they are attracted to bait housed within the entrapment area (Hubert 1983).

Trap Nets, Pound Nets, and Weirs.   Trap nets have an entrapment area consisting of a series of connected funnels (Figure 5.4); the final funnel is tied closed at its posterior end. The opening of the trap generally is anchored off the shoreline with its open end facing shoreward. Long barriers of netting (leads or wings) extend from the trap outward, often toward the shoreline. Fishes that encounter the netting follow it, eventually swimming into the entrapment area. Weirs and pounds are more permanent structures that function similarly to trap nets. Corralling areas and leads in herring weirs of coastal Maine and the Canadian Maritime Provinces consist of stakes driven into the substrate with brush woven through the stakes to create a barrier (Lagler 1978; High 1989).

Trap nets, pound nets, and weirs have been used to harvest salmon species on the Pacific coast; Atlantic herring, Atlantic mackerel, and other schooling species on the Atlantic coast; and species of whitefishes, ciscoes, and other Great Lakes fish-

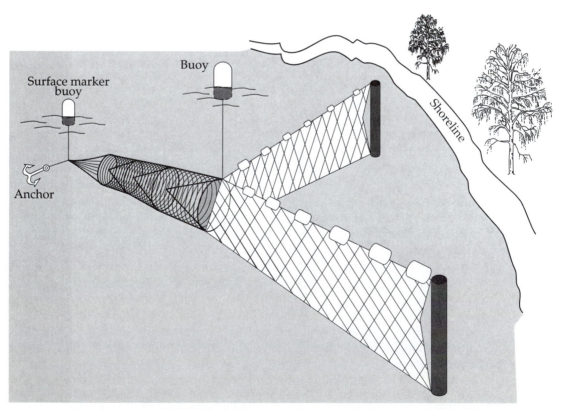

Figure 5.4   Trap net.

es. The potential impact of pound nets and weirs on coastal migratory species was recognized early by the first U.S. Fish Commissioner, Spencer Baird, who recommended prohibiting the use of these gears in New England coastal fisheries in the early 1870s due to the depleted condition of numerous fish stocks (see Chapter 6). The use of large Pacific salmon trap nets was prohibited in Alaska before the 1960s (High 1989). Such devices are now highly restricted or prohibited in many fisheries.

Pots and Traps.   Pots and traps are small, portable gears normally consisting of a rigid frame and mesh netting with constricted openings through which fishes or shellfishes pass (Figure 5.5). Devices designed to capture fishes are called *traps,* whereas those designed to capture invertebrates such as crabs or lobsters are called *pots.* Pots and traps are usually set individually, with an anchor to hold them in place and a float tied by line to mark their position. Some are attached in series along a long groundline, as in the sablefish trap fishery of the Pacific coast and the offshore lobster fishery of the Atlantic coast (High 1989). Although some species may enter pots or traps to gain shelter, most are attracted only if the devices are baited with food.

Many types of invertebrates are harvested with pots, including American lobster along much of the Atlantic coast of the United States and Canada, blue crab in the mid-Atlantic states, spiny lobster and stone crab from the southern Atlantic coast and the Gulf of Mexico, king and tanner crabs from Alaska, dungeness crab from California to Alaska, and spiny and slipper lobsters on the Hawaiian Islands. Trap fisheries have focused on species such as the sablefish in Alaskan waters, black sea bass in the mid-Atlantic, American eels in Atlantic coast rivers, and groupers in the south Atlantic states. Traps are now highly restricted in Alaskan sablefish fisheries (High 1989) and are prohibited in Florida grouper fisheries.

Pot fisheries have caused significant declines in some crab and lobster fisheries. American lobster and southeastern U.S. spiny lobster are considered overexploited. Population abundance of the Bering Sea Alaskan king crab stock has declined precipitously since the mid-1970s (NOAA 1993).

American Lobster Pot

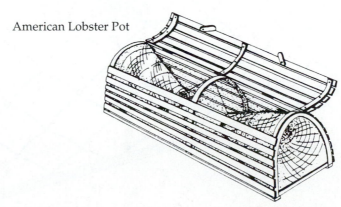

Figure 5.5   An American lobster pot. (From W. A. Hubert, 1983. Active fish capture methods. Pages 123–145 in L. A. Nielsen and D. L. Johnson, editors. *Fisheries techniques.* Reprinted by permission of The American Fisheries Society.)

Entanglement Gear

As with passive entrapment methods, entanglement gears must be encountered by fishes. Fishes encounter and are captured by these devices during normal daily or seasonal activity patterns or migrations.

Gill Nets.   The most commonly used entanglement gear, the gill net, is a long wall of large-mesh netting set in a straight line in the water column, with a floatline tied across the top and a leadline along the bottom of the wall of netting (Figure 5.6). Fishes may be captured in gill nets by being (Hubert 1983)

1.   Wedged—held by the mesh wrapped tightly around the body
2.   Gilled—held by mesh slipping under the opercles or gill covers
3.   Tangled—held by teeth, fin spines, or other protrusions without the head or body penetrating the mesh openings

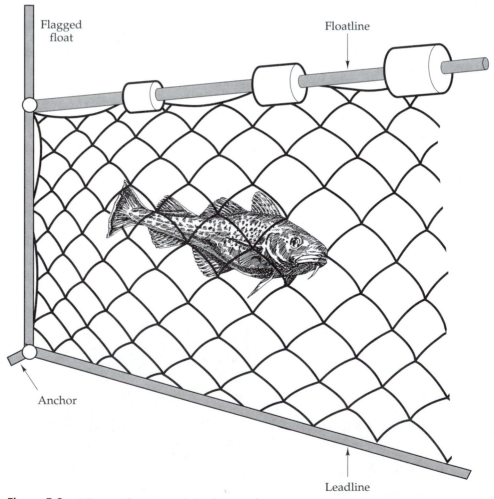

Figure 5.6   Gill net with an entangled Atlantic cod.

Gill nets can be anchored on or near the bottom or suspended in place anywhere within the water column of shallower systems by long anchor ropes attached to the ends. Drift gill nets, which are floated and weighted to remain spread but are not anchored to substrate, are used in rivers or in open ocean areas where anchoring is not possible.

Cotton and linen mesh webbing was used in early gill nets. However, more recently nylon and monofilament twine are most commonly used because they do not deteriorate as rapidly as cotton or linen. Monofilament nets are also markedly less visible to fishes than the other materials. Because wooden or plastic floats and lead weights tend to tangle in the net webbing when folded before or after fishing, foam-core floatlines and lead-core leadlines are commonly used to stretch the net vertically in the water column (Hubert 1983).

Gill nets are used to catch a variety of highly mobile fish species. However, specific gill nets efficiently capture only a narrow size range of fishes, based on the size of the openings of the mesh. For any given mesh size, fishes smaller than the size range of optimum capture efficiency may pass through the webbing, whereas fishes larger than the optimum size bump into the webbing without becoming entangled (Lagler 1978). As with trawls, gill-net size selectivity offers agencies the opportunity to reduce the capture of small fishes in gill-net fisheries by establishing regulations requiring minimum mesh size openings.

Capture efficiency also varies with time of day and length of time the net is fished in the water column. Efficiency may increase during the night, presumably because fishes are less likely to detect the net visually before contacting it. Efficiency also declines with increasing *soak time*—the time that the net is fished—because the frequency of detection and avoidance tends to increase as the net fills with entangled fishes (Hubert 1983).

Gill nets have been used to harvest a number of freshwater, anadromous, and marine fisheries resources. Their use has been restricted or prohibited in many fisheries, as in the Lake Erie walleye and yellow perch fisheries (see Chapter 12). Gill nets produce large biomasses of bycatch. Bycatch of fishes, sea birds, sea turtles, and marine mammals in the drift net fisheries of the Pacific Ocean, which employ gill nets that may stretch for miles, has caused particular concern in recent years (NOAA 1991). In addition, long-lasting monofilament materials, favored by harvesters for their endurance, cause high mortality of a variety of aquatic organisms when lost on fishing grounds. If not recovered from the water, such "ghost" nets can continue to entangle and kill fishes and other vertebrates as well as invertebrates for years (High 1989).

Trammel Nets.   A trammel net is composed of three walls of mesh netting lying parallel to each other and spread in the water column by floatlines and leadlines (Hubert 1983). The outer panels of netting are large-meshed, and the inner has small mesh. Fishes are captured in trammel nets when they strike the inner panel of small mesh netting and push it and themselves through an outer, large-mesh panel; this forms a pouch of gathered netting that entangles them. Although these nets are used in some fisheries, such as the California halibut fishery and formerly in the Alaska king crab fishery (High 1989), they are much less commonly employed than are gill nets. Because the middle panel of netting has small mesh openings, trammel nets capture a greater variety of sizes than do gill nets. Fishes and crustaceans can be difficult to remove from trammel nets. Thus fishing with this gear can be very labor intensive, and high mortality rates of bycatch can occur.

FRESHWATER FISHERIES RESOURCES

Although numerous freshwater resources have supported important commercial fisheries, many of these are no longer commercially viable, or commercial harvest is prohibited to reserve the resource for recreational uses, as with Lake Erie walleye in U.S. waters. On a regional and national basis, recreational fishing and harvest exceed commercial uses for most freshwater fisheries resources. In addition, it generates substantial economic activity: Anglers spent about $18 billion in 1991 to fish in freshwater systems (see Chapter 4). Thus most of this section will focus on recreational uses. Discussion will be limited to taxa that complete their life cycle in freshwaters. Fishes that spawn in freshwater but grow up in the ocean (anadromous species) or spawn in the ocean but spend their juvenile life stage in freshwater (catadromous species) will be considered later in this chapter.

Riverine Fisheries Resources

Riverine fisheries resources are frequently divided into two major categories according to annual temperature regimes to which they are exposed. Coldwater fishes cannot tolerate water temperatures above 22°C in the summer (Griffith 1993) and thus are restricted to coldwater habitats. Warmwater fishes live in systems that may cool toward the freezing point in the winter but regularly reach temperatures above 22° to 23°C in the summer.

Coldwater Streams.   Coldwater streams include all systems in which most recreationally important species are members of the trout family *Salmonidae* (Moyle 1993). Although a variety of habitat conditions can influence the presence of trout and nonsalmonid species in coldwater streams, for example, acidity and concentrations of dissolved oxygen (Moyle 1993), temperature is the primary factor dictating the geographic distribution of trout and nongame fishes typical of coldwater streams. Thus, in the continental United States, coldwater fish communities are limited to more northerly latitudes or to high altitudes. The southernmost distribution of the native range of eastern brook trout occurs in the Appalachian Mountains. Within a region, coldwater species tend to be more abundant in upstream areas than well down the river basin where temperature regimes will be warmer.

Production and standing crop levels of coldwater species in many soft-water or high-altitude streams are relatively low; thus they can be relatively susceptible to overfishing (Moyle 1993). Human activities within watersheds, such as logging, can cause significant deterioration in the quality of coldwater habitat requirements such as temperature or the presence of clean gravel spawning substrate. Such habitat deterioration not only might reduce the productivity of trout populations but can cause a local disappearance (see Chapter 9).

Distribution patterns of wild trout resources have been substantially modified by stocking activities of federal and state agencies in the last century. Wild populations of the European brown trout, first brought to North America in the late 1800s, are established in coldwater habitats of 40 states (Courtenay et al. 1984; see Chapter 8). The rainbow trout, native to much of the Pacific coast from Mexico to Alaska, is now established throughout the West, the upper midwestern states, the northeast and the Appalachian Mountains, south to the Carolinas, and northern

Georgia. The eastern brook trout, native to cold, small streams from eastern Canada to as far south as Georgia in the Appalachian Mountains and as far west as Minnesota and the Hudson Bay, has been widely established throughout the western
United States (Scott and Crossman 1973). Other western species, such as the cutthroat (Figure 5.7A), bull, golden, gila, and Apache trouts, are exhibiting reductions in distribution patterns or declines in abundance within their native ranges due
to the stocking of nonindigenous trouts (see Chapter 8) and habitat deterioration (see
Chapter 9). Some stocks are now extinct or endangered (see Chapter 11).

Although coldwater streams are classified according to the presence of trout
species, most of these systems contain species of nonsalmonids as well. Minnows
(family *Cyprinidae*), suckers (*Catostomidae*), and freshwater sculpins (*Cottidae*) are
common inhabitants of coldwater streams (Moyle 1993). Darters (*Percidae*) may inhabit coldwater streams in the eastern United States. Headwater trout streams house
few nonsalmonid species that are recreationally harvested. However, trout species
that tolerate summer water temperatures that may reach 20° to 23°C, such as the
brown trout, can be found in the same river reaches as recreationally important
nonsalmonid species such as the smallmouth bass.

Warmwater Rivers.   Warmwater rivers include all systems that warm sufficiently
to exclude salmonids from year-round residence or reproduction (Rabeni 1993);
summer temperatures are generally in excess of 23°C (Moyle 1993). Warmwater
fisheries predominate in recreational fishing activities in the majority of streams

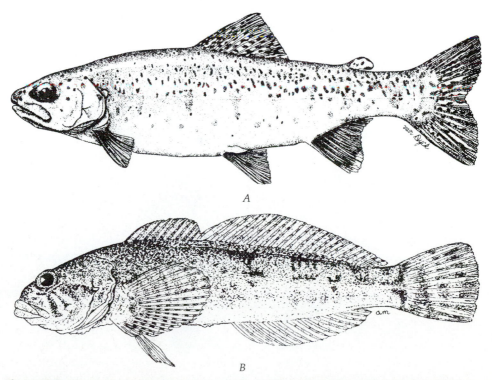

A

B

Figure 5.7   Two typical coldwater stream species. (*A*) The cutthroat trout. (*B*) A species of freshwater sculpin. (From: P. B. Moyle and J. J. Cech, Jr., *Fishes: An Introduction to Ichthyology*, 3/e 1996.
Reprinted by permission of Prentice Hall, Inc., Upper Saddle River, NJ.)

and rivers of more than half of the 48 conterminous states (Funk 1970). Warmwater systems vary greatly in size of river channel, volume of flow, gradient, depth, types of substrates, cover and vegetation, and other characteristics. Because of the diversity of habitat types, warmwater streams and rivers house a wide variety of fisheries resources (Figure 5.8). However, deforestation, agricultural and urban

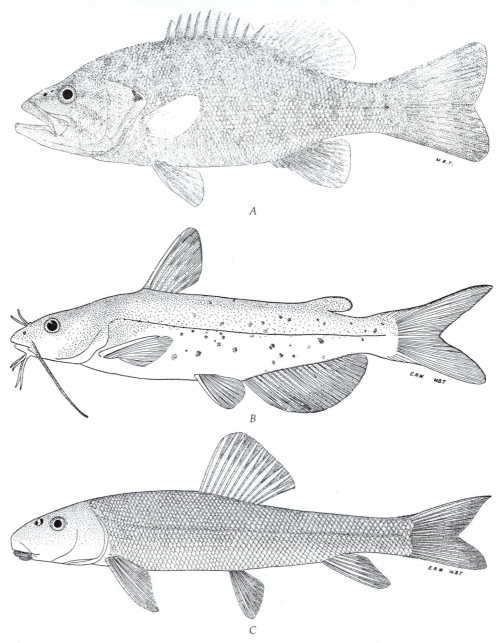

Figure 5.8 Some warmwater stream fishes. (*A*) Smallmouth bass. (*B*) Channel catfish. (*C*) White sucker. (Material from *The Fishes of Ohio*, Milton B. Trautman, is reprinted by permission. Copyright 1981 by The Ohio State University Press. All rights reserved.)

development, and industrial and domestic point source effluents have significantly altered many warmwater stream systems (see Chapter 9). Patterns of natural distribution of warmwater fish species, and the assemblage of fishes found in warmwater fish communities, have changed greatly from such human impacts.

Relatively few recreationally important fish species live in warmwater, headwater streams. However, these systems have many nongame species, including minnows, suckers, sculpins, darters, madtom catfishes (*Ictaluridae*), and lampreys (*Petromyzontidae*). Downriver from headwaters, a variety of species are harvested recreationally or commercially in streams and rivers.

The sunfish family *Centrarchidae* is well represented in such systems. Smallmouth bass or spotted bass may be abundant in mid-size streams with moderate gradients, whereas in larger, slow-flowing rivers with well-developed backwaters and within-channel aquatic rooted vegetation, largemouth bass and crappies may be abundant. Sunfish species may support productive "panfisheries" in these systems. The northern pike, muskellunge, and pickerel species; channel, white, blue, flathead and other catfishes; walleye; freshwater sturgeons; and other fishes support important fisheries in warmwater river systems.

Many aquatic resources that have been driven to extinction within the last century or are presently threatened with extinction are residents of warmwater streams and rivers, particularly east of the Mississippi River in the south and west of the continental divide formed by the Rocky Mountains (endangered taxa are reviewed in the section on endangered and threatened resources in this chapter).

Fisheries Resources of Lakes, Reservoirs, Ponds, and Pools

As with riverine habitats, fisheries resources within these standing water systems are often classified according to temperature regimes under which the resources persist, particularly in relation to salmonid (coldwater) and nonsalmonid fish communities. Lakes and reservoirs will be combined here and categorized according to the temperature regimes described by Moyle (1993): coldwater systems, coolwater systems, warmwater systems, and two-story systems. Rather than being clearly separate, permanent standing waters form a continuum of temperature conditions. Thus a species listed as common or dominant in one type of system may also be an integral part of the fish community in another. For example, the walleye, a common member of coolwater fish communities, may also be found in communities dominated by coldwater or warmwater fish assemblages, and lake trout may be present in coldwater, coolwater, and two-story lakes. True coldwater systems do not have warmwater species, nor do warmwater systems have coldwater species.

Coldwater Lakes and Reservoirs. These systems are typically found at higher latitudes. The most southerly occurring coldwater systems in the United States are restricted to mountain basins. Surface waters in the summer rarely warm to 25°C and typically remain much cooler than that. Most of these systems occur either in recently glaciated basins with large expanses of exposed bedrock or high in mountain ranges where the soils are thin. Thus relatively low concentrations of nutrients are washed into the water column. As a result, coldwater lakes exhibit low levels of biological productivity. The relatively few species of fishes that are typically present exhibit low standing crops and levels of biomass production (Moyle 1993).

Trout species, such as lake trout, brook trout, cutthroat trout, rainbow trout, and others, are common inhabitants of coldwater lakes, as are other salmonids such as the kokanee salmon (a landlocked form of the sockeye salmon); Arctic grayling; and whitefishes such as the lake whitefish, round whitefish, and mountain whitefish (Figure 5.9). The burbot (cod family), a coldwater predator found with these species, has supported modest recreational fisheries, although it is not highly revered in much of its range within Canada (Scott and Crossman 1973). As with coldwater streams, nongame fish taxa often include species of minnows, suckers, and sculpins.

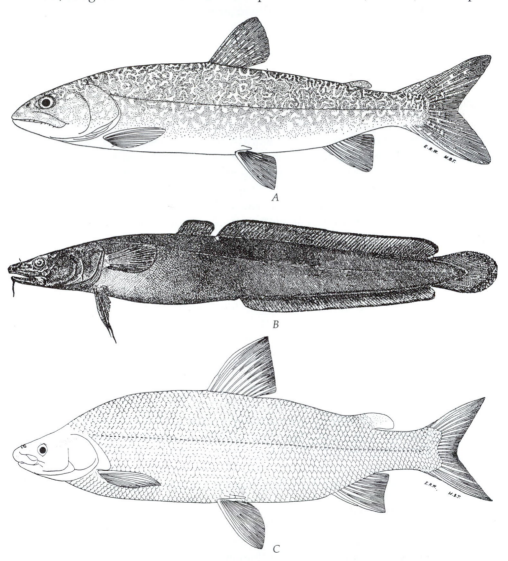

Figure 5.9 Some species of coldwater lake fishes. (*A*) Lake trout. (*B*) Burbot. (*C*) Lake whitefish. (*A* and *C*: Material from *The Fishes of Ohio*, Milton B. Trautman, is reprinted by permission. Copyright 1981 by The Ohio State University Press. All rights reserved; *B* from D. S. Jordan and B. W. Evermann, *The Fishes of North and Middle America: A descriptive catalogue*. U.S. Government Printing Office: Washington, D.C. 1900.)

Coolwater Systems.   Fish communities in coolwater lakes, which are transitional between coldwater salmonid lakes and warmwater lakes that are dominated by centrarchids (Trandahl 1978; Moyle 1993), usually occur in systems where surface waters do not exceed 20° to 25°C in summer months. Both salmonids and warmwater centrarchids, such as largemouth bass, crappies, and sunfish species, may be present in such systems, but a large proportion of the predator assemblage is composed of several or more of the following species: northern pike, muskellunge (muskie), walleye, sauger, and smallmouth bass (Figure 5.10). The yellow perch is often a major food or

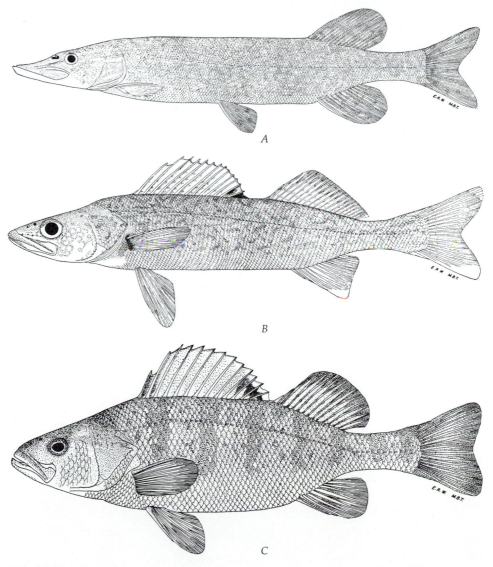

Figure 5.10   Some coolwater lake and reservoir species. (*A*) Northern pike. (*B*) Walleye. (*C*) Yellow perch. (Material from *The Fishes of Ohio*, Milton B. Trautman, is reprinted by permission. Copyright 1981 by The Ohio State University Press. All rights reserved.)

forage base for these predators. All of these species are the focus of important recreational fisheries and, in some instances, commercial fisheries, for example, walleye and yellow perch in the Great Lakes. The striped bass, an anadromous species whose natural distribution was limited to the Atlantic coast states, has been introduced widely into reservoir systems throughout many states. This pelagic species is often restricted to the thermocline during the summer in more southerly occurring reservoirs because of warm temperatures in surface and shallow inshore waters (see Chapter 8).

Coolwater systems are generally more productive than coldwater systems (Moyle 1993), and their fish communities are markedly more diverse. A variety of nongame species may inhabit coolwater systems.

Warmwater Lakes, Reservoirs, and Ponds.   The surface waters of warmwater systems commonly reach 25°C or higher in the summer. These systems are typically productive with diverse fish communities (Figure 5.11). Shallow waters may support expansive beds of aquatic vegetation that serve as optimal spawning, nursery, and feeding habitats for some warmwater species. However, reservoirs that exhibit major fluctuations in water level on an annual basis may be devoid of rooted aquatic

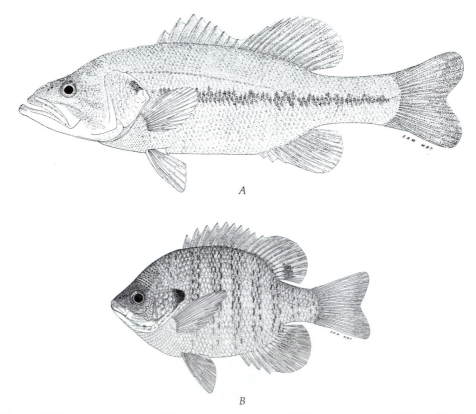

A

B

Figure 5.11   Some warmwater lake and reservoir species. (*A*) Largemouth bass. (*B*) Bluegill. (Material from *The Fishes of Ohio*, Milton B. Trautman, is reprinted by permission. Copyright 1981 by The Ohio State University Press. All rights reserved.)

plants. In many warmwater lakes and reservoirs, recreationally important species may include the largemouth bass; black and white crappie; bluegill, pumpkinseed, and other sunfishes; blue, channel, and species of bullhead catfishes; and white bass and yellow perch (Moyle 1993; Rohde et al. 1994). Darters, minnows, suckers, silversides, threadfin and gizzard shad, and other nonsport species are other common inhabitants of these systems (Figure 5.11). The common carp, a largely reviled introduction from Europe, can be abundant in warmwater systems. When abundant, this species can uproot aquatic vegetation and increase turbidity in the water column.

Small warmwater ponds differ from larger bodies of water in the number of species that are present, particularly newer ponds that are carefully stocked to develop recreational fishing opportunities. These ponds often house only one predator species, most often the largemouth bass, and no more than a few other species, including the bluegill sunfish or other sunfish species, golden shiners or other species of minnows to serve as forage for the bass population, and sometimes channel catfishes.

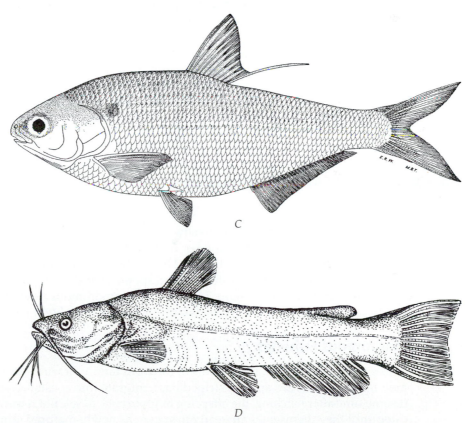

C

D

Figure 5.11 *continued.* Some warmwater lake and reservoir species. (*C*) Gizzard shad. (*D*) A bull-head catfish species. (*C*: Material from *The Fishes of Ohio*, Milton B. Trautman, is reprinted by permission. Copyright 1981 by The Ohio State University Press. All rights reserved; *D:* from P. B. Moyle and J. J. Cech, Jr., *Fishes: An Introduction to Ichthyology*, 3/e 1996. Reprinted by permission of Prentice Hall, Inc., Upper Saddle River, NJ.)

Two-Story Systems.   Two-story systems house coolwater and warmwater species in inshore waters and in surface waters above the thermocline in the limnetic zone, while coldwater species persist throughout summer months in the deep, cold waters of the hypolimnion. Well-developed two-story fishery resources are normally found only in large, deep lakes and reservoirs with a large volume of well-oxygenated hypolimnion water. Historically, the Great Lakes have supported a wide variety of recreational and commercial fisheries because these systems are suited for coldwater (lake trout, whitefishes, and ciscoes), coolwater (walleye, smallmouth bass, northern pike, and muskellunge), and warmwater (largemouth bass, crappies, channel catfish) species. Oxygen depletion caused by the decomposition of organic detritus can be particularly harmful to coldwater resources in two-story systems, because these fishes are restricted to the hypolimnion in the summer.

Desert Pools.   Numerous springs and seeps fed by underground aquifers occur in the desertlands of the arid West. Although these habitats contain only a minuscule proportion of the standing waters in the United States that support fishes, many of the nongame fishes that live in desert pools have received substantial national attention. A number of species of pupfish, poolfish, and springfish (family *Cyprinodontidae*) and minnows such as the Moapa dace are endemic to specific pools or outflows from them. Intensive water usage patterns of the West (see Chapter 9), particularly the withdrawal of groundwater for agricultural irrigation and other uses, have led to the deterioration or destruction of these habitats, which has threatened many of the species endemic to these habitats with extinction (see Chapter 11). These fishes were a focal point in the development of the nation's first endangered species legislation (see Chapter 6).

ANADROMOUS AND CATADROMOUS FISHERIES RESOURCES

Many fish species spawn in either freshwater or the ocean but spend much of their life in the other environment. A number of these taxa support extremely important recreational and commercial fisheries. This section briefly reviews anadromous species—those that spawn in freshwater and spend a substantial portion of their juvenile life stage in saltwater (Figure 5.12)—and the catadromous American eel, which spawns in the Sargasso Sea and spends a prolonged juvenile life stage in freshwater.

Anadromous Fisheries

Anadromous fishes, such as members of the trout family, herring family *Clupeidae,* smelt family *Osmeridae,* and the striped bass, migrate into coastal rivers to spawn. At some time after hatching, the offspring migrate back to sea, not returning to freshwater until they themselves are ready to spawn as newly matured adults. Life histories vary greatly among anadromous species, particularly in terms of the following:

1. *The distance that adults migrate into freshwater to spawn.* Rainbow smelt, eulochon, striped bass, and others may stay within or close to the area influenced

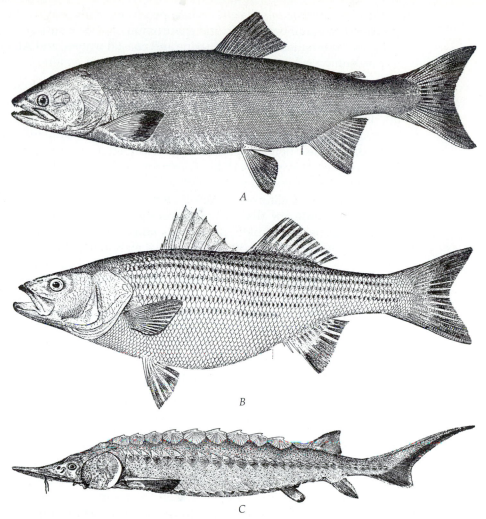

Figure 5.12   Several anadromous species. (*A*) Pink salmon. (*B*) Striped bass. (*C*) Atlantic sturgeon. (From: Henry B. Bigelow and William C. Schroeder, Fishery Bulletin 74. Fishery Bulletin of The Fish and Wildlife Service, Volume 53.)

by the tidal wedge, whereas some species such as chinook and sockeye salmon may migrate hundreds of miles inland to reach spawning grounds.

2. *The amount of time spent in freshwater before migrating into estuaries and the sea.* Eulochon drift into coastal waters shortly after hatching, whereas the American shad migrates to the sea several months and the Atlantic salmon as many as several years after hatching.

3. *The length of time spent at sea before maturing.* This can vary greatly even for individuals within a species. For example, some Atlantic salmon may mature after one year at sea, whereas others may not mature until after 2 or 3 years of ocean life.

4. *Habitat use while in saltwater.* White perch normally stay within estuaries; striped bass; rainbow smelt; and anadromous stocks of cutthroat, brown, and brook trout remain within estuarine and coastal waters, and Atlantic and Pacific salmon exhibit wide migrations across oceanic waters.

Pacific Salmonids.   Six species of Pacific salmonids, the chinook, sockeye, pink, coho, and chum salmon; and the steelhead (an anadromous form of the rainbow trout), historically supported important commercial and recreational fisheries from California to Alaska. There are several hundred separate populations or stocks of salmon because these homing species establish numerous separate gene pools within river basins where they spawn. Habitat destruction, dams, overfishing, and stocking have collectively caused substantial declines in most anadromous salmonid stocks entering rivers from California to Washington. Some native populations have been driven to extinction; a few others have been designated as federally endangered taxa.

Although heavily exploited, Alaska's anadromous salmonid resources are generally in much better condition than those mentioned above. Some stocks are overexploited, causing declining catches in recent years, but many spawning migrations are supporting expansive harvests. The largest recorded harvest in Alaskan waters, 189 million fishes, was taken in 1991 (NOAA 1993). The value of statewide landings in 1990 was $540 million (NOAA 1991). Alaska's salmon resources are in better condition than those of California, Oregon, and Washington in part because many of the river basins that support salmon runs are relatively undisturbed. However, logging and development of coastal wetlands may produce increasing habitat problems for specific runs in the future.

Atlantic Salmon.   The Atlantic salmon is a highly desirable food and sport fish in New England and the Atlantic Maritime Provinces of Canada. Due to more than two centuries of habitat destruction caused by the damming of coastal rivers and development of forested lands for agriculture, few native runs persist in New England. Harvest in U.S. waters is limited to small recreational catches in the state of Maine. Landings have averaged less than 400 fishes per year, which accounts for about 10 percent of the total number of adults migrating into all rivers of that state. In spite of the incredibly small recreational harvests, its mystique as a sport fish makes this species an important recreational fishery resource.

Before juvenile fishes mature and return to their natal rivers to spawn, substantial numbers are harvested in high seas gill-net fisheries. Commercial harvests take about 60 percent of the juveniles that remain 1 year at sea before maturing, and about 80 percent of those that remain 2 years (NOAA 1991). In 1992 the Canadian government placed a fishing moratorium in waters around Newfoundland as part of an international conservation plan developed by NASCO (North Atlantic Salmon Conservation Organization). A major commercial gill-net fishery in Greenland is managed under this plan by use of a quota system (quotas are described in Chapter 7). Restoration efforts, including large-scale stocking and development of fish ladders and other facilities that permit fishes to migrate past dams, have been established in four New England rivers: the Connecticut, Merrimack, Pawcatuck, and Penobscot. To date, these programs have met with limited success in reestablishing salmon runs in those systems.

Striped Bass Fisheries.   Striped bass stocks have supported extremely important recreational and commercial fisheries throughout their native range on the Atlantic coast, and on the West Coast after their introduction in the late 1800s. Spawning on the Atlantic coast is largely restricted to three systems: the Hudson River, Chesapeake Bay, and the Roanoke River. The coastal migratory fishery is supported by the Chesapeake Bay stock and to a lesser extent by the Hudson River stock. The Roanoke River stock remains largely within the estuary and near coastal areas of that basin. Overfishing and deterioration of spawning and nursery habitat led to a major decline in Atlantic striped bass stocks starting in the 1970s. Fisheries declined precipitously for nearly two decades because few new fishes were added to the stocks through natural reproduction. The decline stimulated passage of federal legislation that requires states to cooperate in a multistate management plan for the conservation of these resources (The Striped Bass Act; see Chapter 6). Restrictive fishing regulations and reductions in allowable commercial fishing mortality led to a remarkable recovery of the stocks by the early 1990s (see Chapter 12). West Coast striped bass stocks have suffered similar declines. Habitat deterioration, increases in salinity in the San Francisco Bay estuary caused by water diversions, entrainment and loss of young fishes in diversion canal systems, and toxic pollutants have been detrimental to West Coast striped bass (McGinnis 1984; Stevens et al. 1985; NOAA 1993).

Anadromous Sturgeon.   The white sturgeon on the West Coast and the Atlantic sturgeon on the East Coast have been harvested recreationally and commercially for the marketing of roe—egg-filled ovaries—as caviar and for flesh (Scott and Crossman 1973). These two species reach larger adult sizes than any other species of fishes inhabiting North American freshwaters. The largest recorded white sturgeon weighed nearly 1,400 pounds, and the largest Atlantic sturgeon over 800 pounds. Large adults of both species commonly weigh several hundred pounds or more (Scott and Crossman 1973).

As with sturgeon in general, extremely late age at maturity and nonannual spawning cycles make these two species highly susceptible to overfishing. Declines in abundance also have been caused by habitat deterioration or by dams preventing access to riverine spawning grounds.

Other Anadromous Fisheries.   Other anadromous species are the focus of regionally or locally important recreational and commercial fisheries. The American shad, a member of the herring family, is native to the Atlantic coast from Labrador to Florida and was successfully introduced into West Coast watersheds in the late 1800s. On the Atlantic coast shad spawning migrations historically occurred in virtually every major coastal river system. This species has been harvested for flesh and roe, principally in gill-net fisheries at the mouth of river systems into which shad are migrating to spawn.

Pollution, overfishing, and dams prohibiting migrations into some river systems have caused declines in stock abundances and landings throughout much of this species' range on the Atlantic coast. Restoration efforts involving construction of fish ladders or lifts at dams, habitat restoration, and stocking have led to increased abundances in shad migrations into the Delaware, Connecticut, and Susquehanna Rivers (NOAA 1993).

River herring (two other species of clupeids, the alewife and blueback herring) and the rainbow smelt on the Atlantic coast and eulochon on the Pacific coast are the focus of small-scale local fisheries for human consumption and as live bait.

The Catadromous American Eel

Catadromous fishes spawn in the ocean and spend much of the juvenile growth phase of life in freshwater systems. The catadromous American eel has been the focus of recreational and commercial fisheries in the United States. This species spawns in the Sargasso Sea and is carried by ocean currents as larvae along the coast of North America where individuals enter river systems and metamorphose into the juvenile life stage. Juvenile eels spend many years (for example, as many as 17 years in Lake Ontario) in freshwater before maturing (Scott and Scott 1988).

The American eel has been commercially harvested in traps along much of the Atlantic coast, with the highest commercial landings coming from the Mid-Atlantic states. Elvers, the immature stage that first enters rivers after migrating from the Sargasso Sea; juveniles, called yellow eels; and maturing "silver eels" have been marketed in Europe and Japan, either frozen or densely packed in cool, moist crates. The decline and loss of export markets led to greatly decreased landings by the 1980s.

MARINE FISHERIES RESOURCES

The United States is the sixth largest producer of seafood in the world (NOAA 1993). Coastal and marine commercial fisheries support 300,000 full-time jobs and harvest nearly 5 million metric tons of fish and shellfish resources each year that are worth nearly $4 billion in ex-vessel revenues—the value of the product when brought to the dock before processing and marketing costs are added to the total value. In addition, marine fisheries resources support important recreational fishing activities, subsistence harvest by Native Americans, and nonconsumptive industries such as whale watching. Recreational uses of marine fisheries resources produce economic activity in coastal economies at a level similar to that generated by commercial fishing (see Chapter 4). Numerous stocks are depleted by overfishing, habitat loss, and in many instances incidental mortality caused when they are captured in fisheries not directed toward their harvest. Resources described in the following sections will be categorized by the region of their occurrence and by the type of aquatic communities with which they are associated, generally according to the classification presented in NOAA (1993) and Voorhees et al. (undated).

Harvested Resources of the Atlantic and Gulf of Mexico Coasts

Fisheries resources in coastal regions of eastern United States have supported economically important commercial fisheries since colonial times and important recreational fisheries during the 20th century. In 1991 anglers made more than 54 million fishing trips to waters of these coasts; a trip is defined as fishing during part or all of one day using one type of fishing—that is, party boat, charter boat, private boat, or shore fishing (Voorhees et al. undated). Intensive commercial and recreational fishing pressure has contributed to widespread depletion of the resources supporting these fisheries. Regions of the Atlantic and Gulf of Mexico coasts include the Northeast Atlantic (Maine to Connecticut), the Mid-Atlantic (New York to Virginia), Southeast or South Atlantic (North Carolina to Florida), and the Gulf of Mexico (Florida to Texas).

Principal Groundfish Species of the Northeast and Mid-Atlantic Regions.   Groundfish fisheries harvest demersal species. Principal groundfishes are those finfish taxa that have traditionally been the focus of active commercial and recreational harvesters. This group includes members of the cod family—the Atlantic cod, haddock, pollock, silver hake and red hake; flatfish species—winter, summer, witch, windowpane, and yellowtail flounder, and the American plaice; and the Acadian redfish or ocean perch (Figure 5.13).

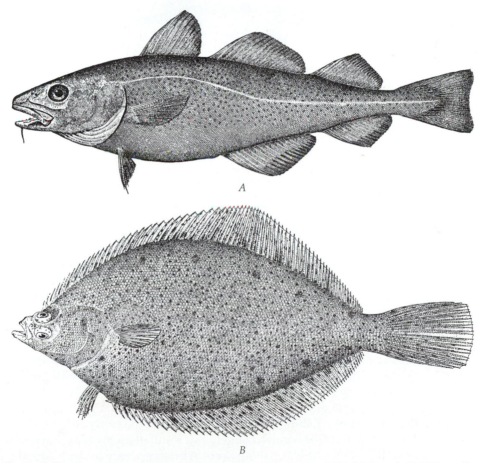

Figure 5.13   Some marine species fished along the Atlantic coast. (*A*) Atlantic cod. (*B*) Yellowtail flounder. (From: Henry B. Bigelow and William C. Schroeder, Fishery Bulletin 74. Fishery Bulletin of The Fish and Wildlife Service, Volume 53.)

Figure 5.13 *continued.*   Some marine species fished along the Atlantic coast. (*C*) Bluefin tuna. (*D*) Atlantic menhaden. (From: Henry B. Bigelow and William C. Schroeder, Fishery Bulletin 74. Fishery Bulletin of The Fish and Wildlife Service, Volume 53.)

Commercial harvests account for most of the landings of this group. About 161,000 metric tons were taken commercially versus 12,000 metric tons taken recreationally in 1990 (NOAA 1991). The otter trawl is the dominant fishing method used by the commercial harvesters (1,056 vessels fished trawls and 258 gill nets in 1992; NOAA 1993). Harvests of principal groundfishes are among the highest-valued commercial fisheries landings in the Northeast. The 1992 commercial landings were valued at $161 million, the same ex-vessel value as American lobster (NOAA 1993). Although representing only a modest proportion of the total harvest, recreational fishing for this group creates significant economic activity in coastal areas. An estimated $196 million of economic activity was associated with expenditures created by angling for summer flounder and winter flounder alone in 1990 (NOAA 1991).

Principal groundfish resources have exhibited significant declines in abundance from overfishing (Figure 5.14). The first major recorded decline in the mid-1960s to early 1970s occurred largely as a result of intensifying harvesting by foreign as well as domestic U.S. fishing fleets. A brief recovery in stock abundance occurred in the mid- to late 1970s as a result of restrictions on harvest in the early 1970s and of the Magnuson Act prohibiting foreign fishing in newly expanded territorial waters in the late 1970s (see Chapter 6). Substantial growth in the domestic fishing

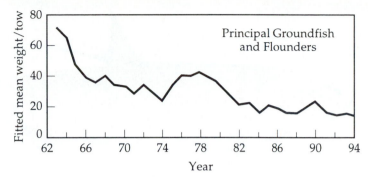

Figure 5.14   The decline in abundance of principal groundfishes of the northeast region. Abundance is an index measured as the average weight of fishes caught per tow in research survey cruises of the Northeast Fisheries Science Center. (Redrawn from NOAA 1993.)

fleet in the 1980s created a second major decline, causing some of the stocks to reach historic low levels of abundance by the 1990s. The 1992 abundance index for all principal groundfishes combined was the lowest in the 30 years for which abundance indices have been calculated (NOAA 1993). Most other Northeast demersal species, such as the goosefish and Atlantic wolffish in the Gulf of Maine and the scup, weakfish, black sea bass, and others in the mid-Atlantic, which are taken largely by vessels fishing for principal groundfish species, show similar declines in abundance. The demersal fish community is now dominated by species of skate and the spiny dogfish, which had been only a minor component of that community before the severe declines of the groundfishes (see Chapter 3).

Southeast and Gulf of Mexico Reef Fishes.   Reef fishes include a variety of species that prefer coral reefs; other hard-bottomed areas with boulders or other natural surfaces providing a complexly structured physical habitat; or habitats created by oil platforms, sunken ships, or other artificial sunken structures. Some agencies have actively pursued the creation of artificial habitats to support reef fish fisheries in the South Atlantic states and Gulf of Mexico (see Chapter 10). Reef fishes have been harvested commercially by a variety of methods, including traps, longlines, trammel nets, spears, and hook and line. More than 100 species of reef fishes support important commercial and recreational fisheries in that region (NOAA 1993).

   Reef fishes comprise numerous species from several families, including snappers (family *Lutjanidae*), sea basses and groupers (*Serranidae*), grunts (*Haemulidae*), porgies (*Sparidae*), and a variety of other taxa (Figure 5.15). The black sea bass supports active recreational and commercial fisheries in the Mid-Atlantic as well as more southern waters, and the scup (*Sparidae*) does so in the Northeast and the Mid-Atlantic. However, most recreational and commercial harvests of reef fishes occur from the Carolinas southward in the Atlantic and in the Gulf of Mexico (NOAA 1993; Voorhees et al. undated).

   In recent years, commercial landings for all reef fish species have averaged about 20,500 metric tons with an ex-vessel value of $48 million. Over 20 million angling trips are directed toward this group of fishes annually (NOAA 1993; Voorhees et al. undated).

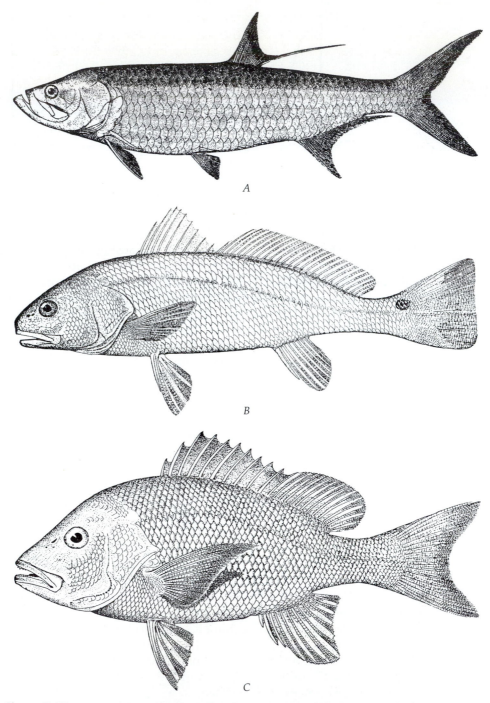

A

B

C

Figure 5.15   Several fishes of the Gulf of Mexico. (*A*) Tarpon. (*B*) Red drum. (*C*) Red snapper. (*A* from Henry B. Bigelow and William C. Schroeder, Fishery Bulletin 74. Fishery Bulletin of The Fish and Wildlife Service, Volume 53; *B* and *C* from D. S. Jordan and B. W. Evermann, *The Fishes of North and Middle America: A descriptive catalogue*. U.S. Government Printing Office: Washington, D.C. 1900.)

Most reef fish stocks are overfished (NOAA 1993). Because of their slow growth and late age at maturity, these species are vulnerable to declines in abundance when fishing mortality is high. Many of the most intensively harvested species have exhibited significant declines in average body length of fish landed in recent years, indicating that the older, larger-size groups of adults in these stocks have generally been fished out. As stock sizes of traditionally important species have declined, harvesters have focused their efforts on other, less utilized and more abundant species. This has resulted in sequential overharvest of an increasing number of stocks (NOAA 1991). Declines in abundance of some stocks may also be the result of the mortality of juvenile fishes incidentally captured in shrimp trawl fisheries.

Drum and Croaker Fisheries.   Members of the drum family *Sciaenidae* have supported important recreational and commercial fisheries in eastern and southern coastal areas since the late 1800s (NOAA 1991). Red drum, spotted and sand seatrout, and spot and Atlantic croaker stocks support active fisheries in the Mid-Atlantic region, the Southeast, and the Gulf of Mexico. Fisheries for other members of this family in the eastern and southern United States are largely centered in one or both of the last two regions. The sport harvest of this group has generally equaled the commercial harvest for human consumption in the Southeast and the Gulf of Mexico. In addition, some species also have supported an "industrial" fishery that focused on the production of pet foods.

The red drum is overexploited, particularly since "blackened redfish" became increasingly popular in the mid-1980s. Red drum harvest is prohibited in federal waters until the Atlantic and Gulf stocks recover. States have generally allocated most harvest in inshore, state-controlled waters to recreational users (NOAA 1993). The condition of most other stocks of this group of species is not known. However, in addition to substantial harvests associated with directed recreational and commercial harvests, high numbers of juvenile spot, sand, and silver seatrout and Atlantic croakers die from their incidental capture in trawls operating in shrimp fisheries. Shrimp trawl mortality averaged about 7.5 billion croakers, 1 billion seatrout, and 500 million spot per year during the 1970s and 1980s (NOAA 1993).

Pelagic Fisheries.   Pelagic fisheries focus on species of smaller-bodied fishes or squids that school and larger-bodied species, some of which school and others that do not. The behavior of these mid- to surface-water species determines their vulnerability to particular types of gears and harvesting methods.

The purse seine fishery for Atlantic and Gulf of Mexico menhaden produces one of the highest catches by weight of any single-species fishery in U.S. waters. Landings in the Atlantic peaked in the 1950s, dropped dramatically in the late 1960s, and then increased somewhat in the 1970s and 1980s. Landings in the Gulf of Mexico peaked in the mid-1980s and declined somewhat by 1990. In spite of recent declines, landings still average about 900,000 metric tons per year (NOAA 1991). Most of the menhaden catch is processed into fish meal, oil, and other products. A small percentage of the harvest is marketed as bait for use in other fisheries. Atlantic herring; Atlantic, king, and Spanish mackerel; butterfish; dolphin fish; longfinned and shortfinned squid; and other small-bodied species also support important coastal pelagic fisheries. These resources have been captured by mid-water trawls, purse seines, gill nets, troll lines, and longlines. Purse seines are prohibited for some of these fisheries under the Coastal Pelagic Fishery Management Plan.

Larger-bodied fishes that support pelagic fisheries are often divided into highly migratory pelagic fisheries and shark fisheries. Highly migratory species are those whose typical migration routes extend well beyond the continental shelf into open-ocean waters. These taxa are fished commercially and for sport. The most important species include tunas (bluefin, yellowfin, bigeye, and skipjack tuna and albacore) and billfishes (swordfish, blue and white marlin, sailfish, and longbill spearfish). Since 1960 the bluefin tuna, swordfish, and yellowfin tuna have provided the largest harvests, each species of this group peaking in landings as the species previously providing the largest catches declined due to overfishing. Commercial harvest of billfishes other than the swordfish is prohibited in U.S. waters, although they are captured as bycatch on tuna and swordfish longlines. Tunas and swordfish are captured in the Atlantic and Gulf of Mexico by longlines, trolling, pole and line chumming, and, to a lesser degree, harpooning. Most of these large-bodied pelagics also are the focus of recreational fisheries.

Overfished bluefin tuna and swordfish stocks have exhibited substantial declines in abundance in recent years. Large bluefin tuna fill a specialty demand in Japan, where they are flown fresh on ice almost immediately after being landed. A commercial harvester may earn as much as $20,000 to $30,000 for each large bluefin tuna landed. Although small vessels fishing for bluefin may land only one fish per week or two of fishing, extremely high dockside prices provide a profit even when severely depleted stocks provide such a low catch rate.

In recent years, sharks have been the target of growing commercial and recreational harvests, as interest has grown in their sporting value and as low-priced markets have developed for human consumption. Sharks are divided into three management groups (NOAA 1993): large coastal sharks such as white, tiger, lemon, great hammerhead, bull, and others; small coastal sharks such as sharpnose, finetooth, bonnethead, and others; and pelagic sharks such as longfin and shortfin mako, thresher, porbeagle, and others. In recent years, harvests have been divided roughly equally between these three groups; about 9,500 metric tons of sharks have been landed annually in Atlantic waters (NOAA 1993). Sharks are harvested in directed longline and gill-net fisheries and are frequently incidentally caught in tuna and swordfish longline fisheries. Only a limited number of species, such as the shortfin mako, are widely marketed for human consumption. Commercial buyers typically market smaller individuals. Larger sharks may be killed to sell only fins, and the carcass often is discarded because of limited market interest. A growing number of species are the focus of expanding recreational fisheries, particularly by tournament anglers. Many sharks landed by anglers are purchased by fish buyers for commercial marketing. In 1986 nearly 10 percent of the "commercial" landings were caught by anglers (NOAA 1993).

Although landings in shark fisheries are very small compared with many other fisheries, the life cycle of these species—very late age at maturity, extremely low fecundity, and slow reproductive schedule—make them very susceptible to overfishing even at relatively low harvest levels. The NMFS implemented a new fishery management plan for sharks in 1993. This plan regulates commercial and recreational harvests, prohibits the practice of finning (cutting off the fins for marketing and wasting the rest of the carcass), and discourages discarding of shark carcasses (NOAA 1993).

Nearshore Fishes.   Many fishes harvested in nearshore coastal waters—here defined as waters from the coastline to 3 miles from shore that are under state rather than federal jurisdiction—such as the winter and summer flounder, reside in stocks that extend well onto the continental shelf. These species have been covered in previous sections of this chapter. However, some species either are restricted in distribution to the coastline or support fisheries that are largely restricted to nearshore areas. Thus management of these species is the responsibility either of individual states or of multiple states acting together through interstate management councils (see Chapter 6).

Some nearshore fish stocks support very active recreational fisheries, such as the bluefish, tarpon, bonefish, snook, and tautog. The bluefish is one of the most avidly pursued recreational species throughout the Atlantic nearshore waters (Voorhees et al. undated). Tautog is avidly pursued by anglers in the Northeast, and tarpon, bonefish, and snook are in southern waters.

Shellfish Fisheries.   A variety of bivalve mollusks and crustaceans support major commercial shellfish fisheries in federally managed waters (3 to 200 miles from the coastline) of the Atlantic coast and Gulf of Mexico, including American lobster, sea scallop, northern shrimp, surf clam, and ocean quahog in the northeast and mid-Atlantic regions, and brown, white, pink, and other shrimps; spiny lobsters; and stone crabs in the southern Atlantic and Gulf of Mexico. In 1992 commercial landings of lobster (which averaged about 27,000 metric tons annually from 1990 to 1992) and sea scallops (which averaged nearly 22,000 metric tons from 1990 to 1992) were the highest ex-vessel–valued fisheries for any single species in the Northeast–Mid-Atlantic region, $161 and $152 million, respectively. About 112,000 metric tons of white, pink, and brown shrimp valued at $405 million were harvested in the Gulf of Mexico in 1990; these numbers declined to 88,000 tons and $367 million in 1992 (NOAA 1991; NOAA 1993). Stocks of some of these resources are depleted due to overfishing, such as American lobster and sea scallops in the Northeast/Mid-Atlantic and all shrimp stocks in the Southeast and Gulf of Mexico, whereas stocks of others such as surf clams and ocean quahogs are being maintained at levels that roughly provide maximum long-term harvests (NOAA 1993).

Most lobsters are captured in baited pots, although American lobsters also are harvested as bycatch in trawl fisheries. About half of the American lobster landings, which exceeded 25,000 metric tons in 1992, come from Maine. Sea scallops, surf clams, and ocean quahogs are harvested with dredges, and shrimp species by trawl.

Shellfishes also support a variety of nearshore fisheries, which are managed by state or local government or by interstate agreement, including Atlantic hard clams, softshell clams, bay scallops, blue mussels, sea urchins and others of the Northeast and Mid-Atlantic, blue crab from the Northeast to Southeast, and oysters and calico scallops of the Mid-Atlantic to the Gulf of Mexico. Chesapeake Bay produces about half of the U.S. blue crab harvest annually (High 1989). Nearshore shellfishes are harvested by a variety of techniques. Oysters, clams, and mussels are collected by hand using rakes or shovels or, in some instances, with dredges. Bay and calico scallops are collected by dredge. Most crabs are collected with pots, trawls, or nets. Sea urchins are trawled in the Gulf of Maine. Recreational harvests of some of these species, such as the blue crab, may constitute a substantial proportion of the total yearly harvest. For example, recreational blue crab harvests in Chesapeake Bay averaged 30 to 40 percent of the total harvests in that system in the 1980s (NOAA 1991; 1993).

Landings of some of the resources that support major commercial fisheries, such as the hard clam, softshell clam, bay scallop, and Atlantic oyster, have declined in recent years. The Chesapeake Bay stock of oysters is considered severely depleted. Because their coastline habitats are often severely modified by human activities, many of these resources are highly susceptible to habitat loss, pollution, siltation, and other types of habitat deterioration. Many shellfish beds are closed to harvest due to sewage contamination. Nutrient pollution can stimulate plankton blooms that cause red tides and paralytic shellfish poisoning. Thus fishing mortality and habitat deterioration both contribute to declines in the viability and quality of these resources.

Harvested Resources of the Pacific

Groundfish Resources.   Over 80 species of demersal fishes in federal waters off California, Oregon, and Washington are managed under fishery management plans of the Pacific Fishery Management Council (Figure 5.16). Of these, 12 are flatfishes

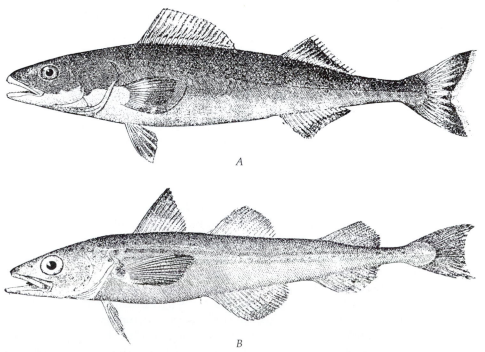

A

B

Figure 5.16   Several Pacific coast species. (*A*) Sablefish. (*B*) Walleye pollock. (From: D. S. Jordan and B. W. Evermann, *The Fishes of North and Middle America: A descriptive catalogue.* U.S. Government Printing Office: Washington, D.C. 1900.)

and 55 are rockfishes (*Sebastes* spp.). In recent years, mid-water trawl fisheries harvested over 286,000 metric tons of Pacific whiting (hake) annually. This biomass, valued at $22 million in 1992, was more than 10 times greater than that produced by any other U.S. Pacific coast groundfish fishery south of Alaska (NOAA 1993). The whiting fishery has been largely a joint venture activity, with U.S. harvesters selling their catch directly to foreign processing vessels by agreement between the U.S. government and foreign countries. Such joint ventures can be established for resources for which there is little or no market in the United States (see Chapter 6). Dover sole, sablefish, and widow rockfish stocks also support substantial harvests (between 11,000 and 18,000 metric tons in recent years). Landings of all groundfishes were valued at $88 million in 1992 (NOAA 1993). Groundfishes in this region are commercially harvested with trawl, trap, or hook and line gear. Most West Coast

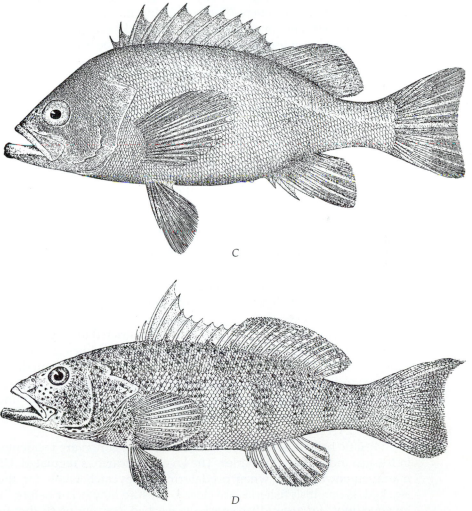

C

D

Figure 5.16 *continued.*   Several Pacific coast species. (*C*) Black rockfish. (*D*) Spotted sand bass. (From: D. S. Jordan and B. W. Evermann, *The Fishes of North and Middle America: A descriptive catalogue.* U.S. Government Printing Office: Washington, D.C. 1900.)

groundfish resources are either harvested at the highest level that the stocks can sustain through time or, in some instances such as the bocaccio and Pacific ocean perch, overharvested.

Groundfishes also support substantial recreational harvests from California to Washington. An average of more than 9 million rockfishes were caught by anglers annually between 1987 and 1989; the bocaccio, black rockfish, and blue rockfish were the most commonly harvested species in this group (Witzig et al. 1992). During the late 1980s dominant species in the recreational catch varied according to subregion within the West Coast states. Sea basses (kelp bass and barred and spotted sand bass), white croaker, and California halibut were dominant recreational species in southern California waters. Blue rockfish, white croaker, and surf smelt dominated in northern California; black rockfish in Oregon; and spiny dogfish and walleye pollock in Washington waters. The greatest recreational landings occurred in southern California waters. Northern California waters produced the next highest landings in this region, whereas Oregon and Washington produced only a fraction of the number of fishes caught recreationally in the more southerly areas (Witzig et al. 1992).

The Alaskan continental shelf supports tremendous groundfish resources that are the focus of valuable groundfish fisheries. The walleye pollock produces the largest single species catch in the United States (NOAA 1993). In recent years an average of about 1.3 million metric tons have been commercially harvested in the Bering Sea–Aleutian Islands region (Figure 5.17). Rather than being directly marketed as fresh seafood, much of this harvest is processed into and marketed as higher-valued imitation crab meat and other processed seafood products. Pacific cod (170,000 metric tons), yellowfin sole (90,000 metric tons), Atka mackerel (30,000 metric tons), and rock sole (30,000 metric tons) stocks also support substantial harvests. Walleye pollock (annual average of 81,000 metric tons), Pacific cod (77,000 metric tons), and sablefish (24,000 metric tons) stocks produce the highest groundfish catches in the Gulf of Alaska. About 34,000 metric tons of Pacific halibut, managed jointly by international treaty of the United States and Canada, were landed commercially by the two countries in 1991; U.S. landings accounted for 31,000 metric tons of this high-valued resource. An additional 4,000 metric tons were taken in a recreational fishery.

Directed fisheries for Pacific halibut are restricted to longline gear. Most other groundfish resources are harvested with trawls. Longlines also are being used for sablefish. Most Alaskan groundfish resources are fully exploited, with some exhibiting declines in abundance in recent years. The population of slope rockfish species—deepwater species that inhabit the outer edge of the continental shelf—has remained depressed since being overharvested by foreign fishing fleets in the 1960s (NOAA 1991).

Pacific halibut, king and tanner crabs, and salmon are incidentally captured in the groundfish trawl fisheries. The level of bycatch is regulated. Under current management regimes, when predetermined bycatch levels for these taxa are reached, groundfish fisheries are closed. Closure may occur before the annual allowable quota for groundfish species is reached, preventing further harvest of the targeted species in the trawl fisheries (NOAA 1991).

Figure 5.17   A 60-metric-ton catch of walleye pollock aboard a U.S. factory trawler in the Bering Sea. (Reprinted by permission of Alaska Fisheries Science Center, Observer Program; National Marine Fisheries Service.)

Hawaiian Bottomfishes.   A variety of bottomfishes, including species of snappers (*Lutjanidae*), jacks (*Carangidae*) and groupers (*Serranidae*), are commercially harvested in the Hawaiian Islands, the territories of Guam and American Samoa, and the Commonwealth of the Northern Marianas Islands. About 90 percent of all landings are taken in Hawaiian waters, a yearly average of over 400,000 metric tons for all species combined. In the early 1990s spawning stocks of several species, including opakapaka, ehu, onaga, and ulua, were depressed to less than 30 percent of their original abundance levels, indicating a level of overfishing that the Western Pacific Fishery Management Council has recommended receive immediate attention (NOAA 1993).

Pelagic Fisheries.   Small-bodied pelagics include several species that have been harvested for food, bait, animal food, and reduction into fishery products such as fish meal, oil, and soluble proteins. Pacific sardine and northern anchovy are harvested largely in coastal waters of California and Pacific herring in waters of Alaska (NOAA 1991).

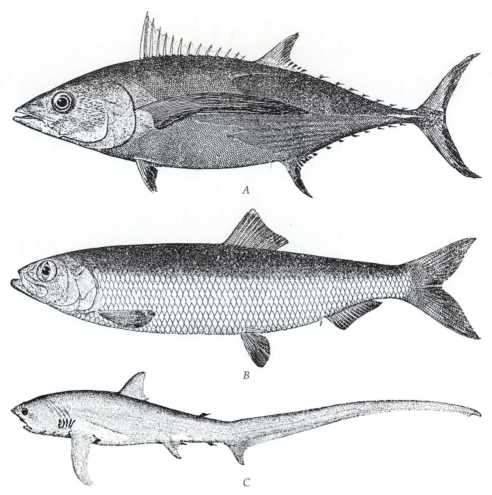

Figure 5.18   Several pelagic species of the Pacific. (*A*) Albacore. (*B*) Pacific herring. (*C*) Thresher shark. (*A* and *B* from D. S. Jordan and B. W. Evermann, *The Fishes of North and Middle America*: *A descriptive catalogue*. U.S. Government Printing Office: Washington, D.C. 1900; *C* from Henry B. Bigelow and William C. Schroeder, Fishery Bulletin 74. Fishery Bulletin of The Fish and Wildlife Service, Volume 53.)

In the 1930s and early 1940s the Pacific sardine (Figure 5.18) supported the largest fishery in the western hemisphere, with California fisheries producing the highest total landings. However, abundance and harvest declined after World War II and the stock collapsed in the early 1960s. A fishing moratorium was placed on this fishery in 1967–1968, and harvest has been highly restricted since that time (NOAA 1993).

In the late 1980s northern anchovy fisheries produced the largest landings of small-bodied pelagics, with annual harvests averaging about 60,000 metric tons. Landings declined to a yearly average of 8,000 metric tons for 1990 to 1992 after the Mexican reduction (industrial) fishery became unprofitable and closed. Pacific herring and chub mackerel landings exceeded those for anchovy in the early 1990s (NOAA 1991; 1993). Northern anchovy and Pacific sardines are harvested by purse seine.

Large-bodied pelagics are the focus of intensive commercial fisheries. Between 1990 and 1992 an average of over 830,000 metric tons of skipjack tuna and 585,000 metric tons of yellowfin tuna were harvested annually in the Eastern Tropical Pacific (ETP) and the Central-Western Pacific (CWP) regions. Most of the catch of skipjack occurs in the CWP, whereas yellowfin catches are generally evenly distributed between the two regions. Mexico is the major fishing nation in the ETP, with a small U.S. fleet harvesting in that region. Japan and the United States dominate purse seine harvests in the CWP (NOAA 1993). Skipjack and yellowfin are captured by purse seine, pole and line gear, longlines, and handlines. Even though more skipjack tuna are harvested than any other tuna species, skipjack are believed to be underexploited—harvested at levels under the maximum that a stock could sustain. Yellowfin are exploited at the maximum long-term potential yield that the stock can support.

There are as many as 2,000 U.S. vessels operating in the North Pacific albacore fishery, which extends form the middle of the North Pacific to North America. Albacore are harvested by drift gill nets, longlines, pole and line gear, and trolling. The recent addition of a drift gill-net fishery has probably led to overexploitation of the North Pacific stock.

A variety of billfishes, sharks, and other highly migratory pelagic species also support Pacific commercial fisheries. Recent total commercial and recreational catches have averaged 90,000 metric tons per year (NOAA 1991). U.S. fleets are dwarfed by foreign longline and drift gill-net fleets in these fisheries. U.S. commercial fishing gear includes handlines, harpoons, longlines, trolling, rod-and-reel, and drift gill nets. Swordfish and thresher sharks are taken by longline in the eastern Pacific, around the Hawaiian Islands, and by harpoon and drift gill net on the west coast of the continental United States (NOAA 1993).

Chub mackerel and Pacific bonito dominated the recreational catch of pelagic species on the west coast of the United States, mostly in waters off southern California.

Shellfish Resources. The continental shelf areas of the West Coast states and Alaska support a variety of invertebrate fisheries in offshore federally managed waters and nearshore waters under state or local jurisdiction.

Three species of king crabs and two of tanner crabs are the focus of pot fisheries in Alaskan shelf waters. The greatest biomass of these shellfishes is taken from the eastern Bering Sea and Aleutian Islands. Because of low abundance levels, king crab fisheries in the Gulf of Alaska generally have been closed to harvest since the early 1980s. King crab resources have exhibited two major declines since 1960. One occurred in the late 1960s to early 1970s due in large part to landings by foreign fishing fleets of Japan and the Soviet Union. A second decline in abundance occurred after the U.S. domestic fleet harvested record landings in the Bering Sea in 1980. U.S. landings in that year surpassed previous high landings of the 1960s that were accomplished by large foreign and domestic fleets.

Alaskan waters also have supported domestic and foreign commercial shrimp fisheries. Shrimp landings in the Bering Sea declined in the mid-1960s. These fisheries ended by 1973. Shrimp landings in the Gulf of Alaska declined precipitously in the late 1970s.

Invertebrate fisheries along the West Coast from California to Washington focus on species that inhabit nearshore areas. Thus most of these fisheries are under state rather than federal jurisdiction for management. Dungeness crabs are harvested in pot fisheries throughout this region. Pacific oysters, Pacific hard clams, Pismo clams (largely from central California), and razor clams (Pacific Northwest states) are harvested by dredging or raking or by hand, abalones are harvested by diving (especially in southern and central California), and sea urchins by trawling and diving (northern Pacific coast; NOAA 1991). Bivalves are harvested recreationally as well as commercially. Recreational harvest can be substantial. For example, there were more than 600,000 people recreationally harvesting razor clams in Washington in the late 1980s. Abalone, razor clams, and Pismo clams have been the subject of overharvesting; because of declining abundance, abalone landings have declined significantly over the last 40 years.

Spiny lobsters and slipper lobsters have been fished in the Northwestern Hawaiian Islands since the late 1970s. About 1 million trap hauls are made in these fisheries annually (NOAA 1991).

Nearshore Finfisheries of the West Coast. A few species of fishes support recreational fisheries largely confined to state territorial waters along the West Coast. Surf smelt was one of the most commonly caught of all recreational species from northern California to Washington in the late 1980s, with average annual harvests of 240 metric tons, nearly 3.5 million fishes in 1989 (NOAA 1991; Witzig et al. 1992). Collectively, surfperch species supported harvests of nearly 400 tons, or an estimated 2 million fishes in 1989, with the greatest harvests occurring in California coastal waters. California corbina and striped bass are also widely fished in California waters.

AQUATIC RESOURCES THREATENED WITH EXTINCTION

Historically, state and federal fisheries agencies focused on production and harvest of commercially or recreationally important species. However, in the last 30 to 40 years a growing awareness of the integral importance of all living resources to the stability and quality of biotic communities led to the passage of legislation that expanded the responsibility of these agencies to include resources that are not harvested—often referred to as *nongame*. A particular focal point of these conservation efforts concerns taxa that are, at some level, faced with the possibility of extinction. A brief review of these taxa is presented here. (Endangered species conservation is presented in detail in Chapter 11.)

Freshwater and Anadromous Resources. As many as 20 percent of freshwater fishes, 55 percent of freshwater bivalve mussels, and 36 percent of crayfish species and subspecies in North America are in some danger of extinction (Warren and Burr 1994). Species and subspecies of trouts and whitefishes, minnows, suckers, killifishes, and darters constitute the greatest number of freshwater fish taxa that are currently imperiled (Williams et al. 1989; Nehlsen et al. 1991). Many of these taxa are warmwater stream or isolated pool species that live in the southeastern and southwestern region of the United States (Warren and Burr 1994). A large number of reproductively isolated stocks of anadromous fishes, particularly Pacific salmon, are also imperiled. A majority of existing species and subspecies of some groups, such as the sturgeon family *Acipenseridae,* the pupfish genus *Cyprinodon,* and the freshwater mussel genera *Epioblasma* and *Pleurobema* face possible extinction.

Marine Mammals and Sea Turtles.   Although once the target of consumptive fisheries, marine mammals and sea turtle populations are protected under several federal acts, the major ones being the Marine Mammal Protection Act of 1972 (MMPA) and the Endangered Species Act of 1973, or ESA (see Chapter 6). The National Marine Fisheries Service (NMFS) has the responsibility of overseeing the protection of the six species of marine turtles that frequent U.S. coastal waters, all of which face some level of threat of extinction. Likewise, the NMFS has the responsibility of protecting all whales, dolphins, seals, and sea lions under the Marine Mammal Protection Act. Eight species of whales, the Stellar sea lion, and several species of seals have been identified under the Endangered Species Act as being in danger of extinction or threatened with possible extinction with relatively minor changes in the condition of their populations. The right whale is perhaps the most seriously endangered species. It is so depleted in both the Atlantic and Pacific Oceans that its continued existence even with protection is in serious question (NOAA 1993).

Protection of marine mammals and sea turtles has led to several major fishery conflicts. Dolphin mortalities in the yellowfin tuna purse seine fishery have led to regulations restricting the manner in which this gear can be fished. Drift gill-net fisheries in waters off New England kill pilot whales and several species of dolphins, and drift gill-net operations in the Pacific also cause marine mammal mortalities. Groundfish gill-net fisheries have killed species of dolphins, seals, and some humpback whales. Large numbers of sea turtles have been killed due to capture in shrimp trawls in the Gulf of Mexico and southern Atlantic states.

6

The Framework for Fisheries Conservation and Management

Conservation of public fisheries resources in the United States is accomplished through the actions of government agencies. The framework for the management process is created by the passage of laws that either establish agencies or other government entities (such as the regional fishery management councils created by the Magnuson Act, discussed later in this chapter) or define the responsibilities of those agencies in the stewardship of fisheries resources. The framework of fisheries conservation in the United States has undergone many changes since the first fisheries agencies were created and their duties defined. These changes were due largely to the ineffectiveness of previous approaches to resolve issues of fisheries conservation and to the changing needs of resources affected by an ever-expanding array of human-created conditions.

The process of carrying out conservation and management responsibilities within the existing framework has also evolved during the 20th century. Early federal laws related to fisheries conservation often listed specific actions that agencies needed to take. For example, federal legislation passed in 1906 to regulate salmon harvest in Alaska included specific fishing regulations, such as prohibiting the use of certain harvest gears. Rather than listing specific actions, more recent laws define broad responsibilities of government agencies. An agency develops programs to meet those responsibilities. Thus efforts to conserve fisheries resources and aquatic systems in which they live may vary from time to time according to the relative emphasis that agencies and the executive branch of government places on these responsibilities.

The development of conservation and management programs is typically a lengthy process. Most government actions start with intensive study to determine the condition of resources and to predict the impact of activities such as harvest or habitat modification on those resources. These studies allow the development of conservation and management strategies. Once developed, such programs are typically posted for public review and discussed at public hearings before they become final. For example, before a species can be listed as endangered under the federal Endangered Species Act, the Department of the Interior must conduct a one-year review of its current status and then must determine whether its listing and designation of habitat critical to its survival is appropriate. Once this is accomplished, a proposed listing is published for public review in the *Federal Register.* Final listing may take another year after a proposal is included in the register. The public has an opportunity to comment on the species review, and on proposals for listing and designating critical habitat before the final listing can occur (C. D. Williams 1994). Most conservation and management programs regulate human behavior or restrict human activities (Chapter 5). Procedures that require public notification and public hearing ensure that those who will be affected by conservation and management programs have an opportunity to participate in their review before final decisions are reached.

In this chapter the origin of fisheries conservation agencies and changes in the structure of the federal agencies responsible for fisheries conservation are briefly reviewed. Laws, treaties, and court rulings that have determined or changed the emphasis of fisheries conservation programs or have facilitated the capacity to carry them out are discussed. This review is not a complete compendium of federal legislation but a synopsis of legislation that reflects the evolution of governmental responsibility. The chapter will focus on federal activity because laws passed at this level reflect regional and national interests, not simply local ones. In addition, a review of legislation that influenced the direction of fisheries conservation in each of the 50 states would be redundant and too cumbersome for presentation here.

ORIGINS OF THE MANAGEMENT SYSTEM

Fisheries Conservation and Management Before the Creation of Fisheries Agencies

Although the management system as currently organized is relatively new, efforts at fisheries conservation in North American waters predate the American Revolution. Historically, fisheries resources and fishing rights on lands in England and Europe were generally under the control of private landowners. However, fishes in colonial North America were considered public resources; ownership of these resources occurred only after their capture (Royce 1989; Nielsen 1993). Charters of many of the colonies granted fishing and hunting rights on public lands to all citizens of the colony. The Massachusetts Bay Colony even passed an ordinance in 1641 granting trespass rights through any private property in order to fish on ponds under common ownership of a town (Redmond 1986).

Concern over dwindling fisheries resources in colonial America led to efforts to control their harvest and use. In 1639 Massachusetts Bay Colony prohibited the use of striped bass and Atlantic cod as fertilizer. Revenues from the sale of striped bass were taxed to support the colony's first public school and to provide aid to widows and children of men who had served the colony (Alperin 1987). In 1652 Massachusetts Bay Colony prohibited fishing for cod, hake, haddock, pollack, and mackerel during the spawning season, and Plymouth Colony restricted early-season mackerel fishing in the 1670s (Redmond 1986; Hennemuth and Rockwell 1987). Fishing methods were also regulated. Middlesex County, Virginia, prohibited the use of jacklights in 1678, Plymouth Colony prohibited the use of seines in 1684, and the Common Council of New York City limited fishing in freshwater ponds to hook and line in 1734 (Redmond 1986; Hennemuth and Rockwell 1987).

The first national concern over fisheries resources dealt with the economic health of the fishing industry, not the biological health of the resources. In 1791 Secretary of State Thomas Jefferson asked Congress to support the Massachusetts salt cod and whale oil industries so that they could better compete with subsidized foreign fishing fleets. Both exports and domestic sales of salt cod and whale oil were suffering from competition with European products. Over 4,000 people were directly involved in both the cod and whaling industries in Massachusetts at that time; this total was more than twice as high as the number of people engaged in commercial fishing in Massachusetts in 1980 (Royce 1989). Federal action well into the 1800s dealt more with international jurisdictional disputes than with domestic disputes over fishing rights or with restrictions of harvest for conservation purposes (Graham 1970).

Creation of Public Fisheries Agencies

Depletion and loss of anadromous stocks in New England in the first half of the 19th century stimulated creation of the first fisheries conservation agencies in the United States. New England's fisheries resources were particularly affected owing to the relatively longer period of European settlement in that region compared with elsewhere in the country. Anadromous fisheries such as American shad, Atlantic salmon, river herring, rainbow smelt, Atlantic sturgeon, and striped bass suffered from overfishing, from habitat degradation caused by deforestation and agricultural and industrial land-use patterns, and from complete elimination of spawning migration pathways caused by the numerous dams in the region's rivers. Overfishing had eliminated striped bass and Atlantic sturgeon from the Exeter River, New Hampshire, before 1762; alewife migrations in the same river were eliminated before 1790 due to dam construction (Bowen 1970). Dams on the mainstem of the Connecticut River, the Merrimack River, and other New England streams and rivers caused elimination or severe depletion of the Atlantic salmon and American shad spawning migrations before 1850 (Stolte 1981; Ross 1991). In response to depletion and loss of anadromous and coastal fish stocks, Massachusetts established a prototype fisheries conservation agency, the Massachusetts Fish Commission, in 1856 (Bowen 1970); New Hampshire and Vermont followed with commissions of their own in 1865 (Redmond 1986). Massachusetts and New Hampshire established the first interstate fisheries conservation effort, jointly attempting to restore anadromous migrations into the Connecticut and Merrimack Rivers in 1866 (Stolte 1981).

All six New England states, New York, and California were among those states that had created fish commissions before the federal government established the first U.S. Office of the Fish Commissioner in 1871 (Ross 1991; Nielsen 1993). This first federal fisheries office was created largely because of the efforts of one person, Spencer F. Baird (Figure 6.1). Baird, at the time the assistant secretary of the Smithsonian Institution, first became aware of the growing problem of depleted fisheries resources while on collecting trips in New England. Prior to the establishment of the U.S. Fish Commission, Baird had already convinced the Smithsonian to support fisheries surveys out of Woods Hole, Massachusetts, in 1870 for the purpose of studying the condition of coastal and anadromous fish stocks (Bowen 1970). After this, Baird lobbied Congress to create a fish commission to determine what "diminution of foodfishes of the coast and inland lakes had occurred" (as reviewed in Ross 1991). Baird was interested in coastal waters, but freshwater systems were included in the legislation that he championed in order to receive support from representatives of midwestern states (Bowen 1970). The legislation required that the commissioner be a federal employee "of proved scientific and practical acquaintance with the fishes of the coast"; such wording guaranteed that Baird would be appointed the first fish commissioner (Bowen 1970). In a matter of months, Baird had successfully lobbied for creation of a federal fisheries agency and had managed to be appointed its first administrator.

The commission's first activities focused on additional surveys conducted out of Woods Hole in 1871 and 1872. From these studies, Baird concluded that the bluefish was the major factor causing the depletion of New England's coastal fish pop-

Figure 6.1 Spencer Baird, the first Commissioner of the U.S. Office of Fisheries, launching a research boat with colleagues circa 1870. (Photograph courtesy of Northeast Fisheries Service Center.)

ulations. He wrote that bluefish have "a voracity and bloodthirstiness which, perhaps, has no parallel in the animal kingdom. . . . The fish seems to live only to destroy, and is constantly employed in pursuing and chopping up whatever it can master. I am quite inclined to assign to the bluefish the very first position among the injurious influences that have affected the supply of fishes on the coast. Yet. . .it is probable that there would not have been so great a decrease of fish as at present but for the concurrent action of man" (Baird 1873). Baird seemed to be crediting the bluefish with a near-human capacity for malicious intent while deflecting most of the blame for historic fish stock depletions from humans to a native fish predator that had existed as part of the coastal fish communities far longer than those communities had exhibited signs of depletion (Ross 1991). Baird did recommend that coastal states, or Congress if the states did not act, control the use of pound and trap nets in coastal waters. Neither the states nor Congress enacted these recommended regulations.

Changes in the Structure of Federal Fisheries Agencies

The structure of the federal management system changed periodically for a century after its creation. The changes were partly due to a redefinition of federal responsibilities in fisheries resource conservation, to a consolidation of federal agencies that had parallel responsibilities, and to the conflicting concerns that arose from groups interested in using fish for recreation or for profit (Table 6.1 outlines these changes; from Ross 1991). In 1903 the Office of the U.S. Fish Commissioner, renamed the Bureau of Fisheries, was moved to the Department of Commerce and

TABLE **6.1** **Placement of the responsibility for conserving public fisheries resources within the federal government system.**

1871	Office of the Commissioner of Fish and Fisheries established by Congress.
1903	Office of the Commissioner is renamed the Bureau of Fisheries, which becomes part of the newly created Department of Commerce and Labor.
1913	Department of Commerce and Labor is divided into separate agencies, with the Bureau of Fisheries remaining in the Department of Commerce.
1939–1940	Bureau of Fisheries and Bureau of the Biological Survey are transferred into the Department of the Interior and then merged into the new Fish and Wildlife Service within that agency.
1956	Responsibility within the Fish and Wildlife Service is divided between the newly created Bureau of Commercial Fisheries and the Bureau of Sport Fisheries and Wildlife.
1970[a]	Bureau of Commercial Fisheries is abolished; federal responsibility for overseeing freshwater fisheries remains within Fish and Wildlife Service, and responsibility for overseeing marine fisheries is returned to the Department of Commerce in the newly created National Marine Fisheries Service within the National Oceanic and Atmospheric Administration.

[a]Conservation of fisheries resources housed within public lands such as national forests, national parks and monuments, and lands of the Bureau of Land Management is the responsibility of the government agencies overseeing the use of these public lands.

Labor. At that time the economic value of fisheries and the health of the commercial fishing industry were clearly a driving force in national efforts of fisheries conservation and management. The Bureau of Fisheries conducted surveys, collected statistical catch information, and propagated fishes for stocking (Chandler 1985).

Under a government reorganization of the Roosevelt administration in 1939, the Bureau of Fisheries was placed in the Department of the Interior. A year later it was combined with the former Biological Survey. This resulting unit within the Department of the Interior was the Fish and Wildlife Service.

The commercial fishing industry, displeased that the fisheries bureau was placed in an agency whose mission was conservation, lobbied for a bureau more attuned to its concerns; conservationists strongly opposed removing fisheries management responsibilities from the Department of the Interior (Chandler 1985). The Fish and Wildlife Act of 1956, a compromise between the two positions, initiated a reorganization of the Fish and Wildlife Service including the creation of the Bureau of Commercial Fisheries and the Bureau of Sport Fisheries and Wildlife; thus the fishing industry's demands were addressed, but the Fish and Wildlife Service still oversaw all fisheries matters. This legislation declared that "the fish, shellfish, and wildlife resources of the Nation make a material contribution to our national economy and food supply, as well as. . .to the health, recreation, and well-being of our citizens. . . . The Congress further declares that the fishing industry. . .can prosper. . . only if certain fundamental needs are satisfied." Along with establishing the two bureau-level agencies within the Department of the Interior, the legislation also defined the responsibilities held by the Secretary of the Interior toward fish and wildlife resources. These included the following:

1. Establishing a revolving loan fund to support owners of commercial fishing vessels
2. Developing measures to ensure maximum sustainable production (yield) of fish and fisheries products
3. Developing the means to stabilize the domestic fishing industry
4. Promoting the consumption of fisheries products
5. Carrying out programs of development and conservation of fisheries resources
6. Carrying out programs of development and conservation of wildlife resources

This reorganization plan clearly emphasized fisheries resources and the health of the commercial fishing industry. The legislation also defined the ultimate goal of fisheries conservation and management: achieving maximum sustainable yield (MSY) or harvest from fish stocks. Sustainability was the key element of this concept. The abundance and age structure of fish stocks were to be maintained at levels that produced the highest yearly harvests that could be sustained through time. Fisheries agencies generally considered MSY the dominant goal of fishery conservation into the mid-20th century (Nielsen 1993). However, as early as the 1950s fisheries scientists and managers began to broaden management goals beyond simply achieving maximum yield. Fisheries economists pointed out that greatest economic benefit comes not only by achieving maximum yield but also by limiting the number of harvesters participating in a fishery: too many harvesters results in too little profit per participant. In addition, fisheries managers began to understand that anglers in recreational fisheries sought numerous personal benefits from fishing that extended well beyond simply catching the maximum number of fishes. Thus by the 1970s the management philosophy that had embraced MSY shifted toward a broader goal, optimum sustainable yield (OSY). Management for OSY entails consideration of fishery-specific sociological and economic characteristics as well as the biological condition and production potential of a fish stock (Nielsen 1993).

In 1970 the Bureau of Commercial Fisheries was disbanded. Federal responsibility for conservation of freshwater fisheries resources was retained within the U.S. Fish and Wildlife Service. The responsibility for overseeing the condition of marine fisheries resources was transferred to the newly created National Marine Fisheries Service within the National Oceanographic and Atmospheric Administration of the Department of Commerce. Thus at that time the government chose to divide its responsibilities according to where its fishes lived. The collapse of commercial fishing in the Great Lakes and the increasing power of the recreational fishing lobby strengthened the emphasis toward the tremendous aesthetic and economic importance of recreational uses of fisheries resources in the freshwaters of the United States. At the same time, the economic viability of commercial fishing industries of the coastlines was becoming increasingly stressed because of the cost of pursuing dwindling resources, competition on many fishing grounds from foreign fleets superior in their capacity to harvest fishes, and competition in the marketplace from countries apparently subsidizing their national fishing fleets. Essentially, the freshwater–marine split in responsibility reflected a recognition of the recreational value of freshwater fisheries resources while acknowledging the urgency of responding to the needs of an ailing coastal fishing industry.

CONSERVATION AND MANAGEMENT OF HARVESTED RESOURCES: EARLY 20TH-CENTURY INITIATIVES

The process of identifying the national responsibility in fisheries conservation was developing by the early 20th century. The first act specifically authorizing the regulation of a fishery was the Sponge Act of 1906, which focused on regulating sponge fisheries in the Gulf of Mexico and the Straits of Florida (Weber 1986). Due to a Supreme Court decision relating to jurisdictional rights of states, Congress amended this legislation in 1914 to apply only to sponge harvest outside state territorial boundaries. Other early legislation either identified federal and state sovereignty, enabled the United States to participate in multinational fisheries agreements, or established specific regulations restricting the harvest and use of particular fisheries resources.

An Alaskan Focus

Alaska's abundant fisheries resources gained particular attention during the early 1900s. An Act for the Protection and Regulation of the Fisheries of Alaska of 1906 placed restrictions on harvest and other activities in rivers of that territory housing significant Pacific salmon spawning runs. This act established daily closures to all but hook and line fishing (from 6 PM to 6 AM); restricted the size of nets that could be placed in salmon streams; and prohibited the erection of dams, traps, and other permanent structures that obstructed the mouths of "red salmon" streams. This act also defined responsibilities of the Secretary of the Department of Commerce and Labor (the parent department of the U.S. Bureau of Fisheries at that time). The secretary could set aside specific watersheds as spawning ground preserves, establish seasonal closures to control harvest levels, and close specific watersheds to fishing for one year at any time population losses caused by harvest exceeded gains created by natural production. The 1906 act was amended in 1924 and 1926 to include specific harvest restrictions. A minimum of 50 percent of fishes in salmon spawning migrations were to escape harvest. If specific stocks declined, the Secretary of Commerce was to increase the level of required escapement for that watershed.

Early 20th-century legislation as above tended to include specific guidelines and regulations restricting harvest, unlike legislation later in the century that focused more on defining responsibilities of fisheries conservation and management and assigning particular government agencies to carry out the responsibilities. The latter approach should be more effective, because specific conservation actions such as harvest restrictions and other regulations can be modified to deal rapidly with the dynamic changes that occur in exploited fish stocks without requiring Congress to pass or amend laws.

Numerous other federal fisheries initiatives in the early 1900s focused on resources in Alaska and adjoining waters. In 1908 a convention (treaty) was signed by the United States and Great Britain, for the Dominion of Canada, that established an International Fish Commission with members from the United States and Canada. This commission was responsible for recommending regulations, closing seasons, restricting gear, and propagating fishes for the "preservation of food fishes in waters contiguous to Canada." Numerous conventions have been established in the 20th century to address conservation of fish stocks housed at least in part outside U.S. territorial waters.

In 1910 Congress passed an Act to Protect Seal Fisheries of the United States. This law prohibited the harvest of fur seals and other fur bearers within Alaskan waters except under the authority of the Secretary of Commerce and Labor. Native inhabitants of the Pribilof Islands were to be employed to conduct any harvest authorized by the secretary. This law not only established protection for declining resources but also recognized the rights of Native Americans to benefit in whatever restricted harvest was allowed. The rights of Native Americans to participate in, or be major beneficiaries of, the exploitation of specific fisheries resources became a recurring issue in the United States later in the 20th century (discussed in Chapter 4 and later in this chapter).

Concern over declining Pacific halibut catches of the Alaskan and Canadian coasts led to the Convention for the Preservation of the Halibut Fishery of the Northern Pacific Ocean of 1923. This convention between the United States and Great Britain (again for the Dominion of Canada) set a closed season for Pacific halibut and established an International Halibut Commission to recommend future regulations to control harvest levels in the fishery. A series of subsequent Pacific

TABLE 6.2   Sequence of conventions that the United States has entered to focus on management of Pacific halibut resources (IPHC 1987).

Convention	Reason It Was Established	Provisions
1923	Initial treaty between United States and Canada.	Established fisheries commission with two members from each country; established 3-month closed season in the winter.
1930	Closure not successful in regulating harvest.	Empowered commission to regulate gear types and limit catch in management areas, prohibit fishing on spawning grounds, and regulate licensing of vessels in the fishery.
1937	Catch limits met very early in year as number of vessels increased annually.	Provided greater control over incidental catch of halibut while fishing during closed season.
1950	Needed to provide reciprocal port privileges to both countries.	Permitted landing of vessels in the other ports without duties; vessels could sell catch as export and obtain supplies and repairs.
1953	Fishing spread unevenly in fishing areas, and quotas still being met too early in season.	Multiple seasons were permitted to spread fishing effort within areas; management goal focused on maximum sustainable yield; commission increased to six members.
1979 Protocol to 1953 Convention	Passage of Magnuson Act required all fishing treaties be renegotiated.	Mandate changed to management for optimum sustainable yield; 2-year phase-out of reciprocal fishing agreements; 60% of catch in area 2 to be taken in Canadian waters.

halibut agreements were signed by the United States and Canada due to the expiration of previous treaties and the need to change the manner in which the member nations participated in the agreement (Table 6.2). Such modifications to international fisheries treaties have been common. For example, continued concern over Pacific salmon stocks led to a series of conventions between the United States and Canada, including the Convention for the Protection, Preservation, and Extension of the Sockeye Salmon Fishery of the Fraser River System of 1930, the Sockeye Salmon Convention of 1937, the International Pacific Salmon Treaty (brought into effect for the United States by the International Pacific Salmon Fish Act of 1950), and the Pacific Salmon Treaty (brought into effect for the United States by the Pacific Salmon Treaty Act of 1985).

Other Early International Agreements

In 1925 the United States signed the International Convention Between the United States and Mexico to Prevent Smuggling and for Other Objects; this treaty established an International Fisheries Commission that would recommend guidelines for the conservation and development of all "marine life resources" of these countries' coastlines in the Gulf of Mexico. Growing fishing activities in these waters led to later treaties such as the Tuna Conventions with Mexico and Puerto Rico in 1950.

Marine mammal resources first became the focus of international management with the signing of the International Whaling Convention of 1931. This treaty was supplanted by a subsequent convention in 1937 and, after a cessation of activities during World War II, by the International Convention for the Regulation of Whaling in 1949. The 1931 agreement was specific and narrowly focused: The killing of all right whales, female whales with calves, calves, and immatures of other species was prohibited. Fifteen countries signed the Convention of 1949, which established the International Whaling Commission (IWC).

The wording of the Whaling Convention, which essentially allows individual nations to ignore the recommendation of the majority of participating countries, indicates why many international fisheries treaties have not facilitated effective conservation. Any amendments specifying regulation of harvest that were adopted by the Whaling Commission could be contested by any participating government. Once a government objected, it was exempted from observing the regulations described by the amendment "until such date as the objection is withdrawn." Further, any government could grant permits to kill, process, and sell whales in order to conduct scientific research; such harvest was "exempt from the operation of the Convention." Generally, international commissions established by treaties have only been given the responsibility to recommend actions or regulations; member countries of the treaty decide whether they will accept commission recommendations. The frequent failure to effectively conserve resources in international waters has often been due partly to the hesitation of member nations to agree to restrictive regulations that might affect their own fishing fleet's success.

Fisheries Initiatives Within U.S. Jurisdiction

National initiatives during this time period regulated specific fisheries, clarified states' rights to manage harvest within their boundaries, and facilitated multiple-state management of fisheries that crossed state borders.

The Black Bass Act of 1926.   This act prohibited the transport, or "the delivery to a common carrier for transport," of largemouth bass and smallmouth bass that had been caught, sold, purchased, or possessed in violation of state, territory, or District of Columbia law. The shipping of live fish or eggs for the purpose of propagation was specifically excluded: The early 20th century marked a time when both state and federal hatchery systems were markedly expanding the range of black bass throughout the United States by stocking programs. This law, later amended to include all species of fishes, noted a state's right to regulate fisheries within its boundaries and supported that right by prohibiting interstate commerce based on shipment and marketing of illegally caught fishes.

Conservation of Fish Stocks Living in Territorial Waters of Multiple States.   Before the middle of the 20th century, legislation was passed that began to define the national responsibility of facilitating the conservation of fisheries resources housed within the territorial waters of more than one state. In 1938 Congress passed An Act for the Conservation of Fishery Resources of the Columbia River (also called the Mitchell Act), which authorized the construction of salmon hatcheries in Oregon, Washington, and Idaho, and the construction of devices to improve salmon migration over dams and other obstructions within the Columbia River watershed. This act was passed during a period when Congress was defining the national responsibility to consider the impacts that dams and other modifications to navigable waters would have on wildlife and fisheries resources (see the section on habitat-related initiatives in this chapter).

In 1940 Congress authorized the creation of compacts or other agreements among Atlantic coast states that would facilitate conservation of fisheries resources that crossed territorial waters of adjoining states. Two years later, Congress established the Atlantic States Marine Fisheries Compact. Concern over substantial fluctuations in the abundance of Atlantic coast striped bass stocks served as the major impetus for passage of this compact (Alperin 1987). All 15 Atlantic seaboard states are represented on the Atlantic States Marine Fisheries Commission (ASMFC) created by this agreement. During early drafting of the legislation, a majority of Atlantic coast states demanded that they retain statutory authority for managing resources within state territorial waters. Thus, as with international commissions, the ASMFC provided advice but did not mandate fishery management programs to member states. Fishery Management Plans (FMPs) supported by a majority of commission members were presented to participating states for consideration. The original compact was amended in 1950 to allow two or more member states to designate the ASMFC as a joint interstate regulatory agency that could manage particular species identified by those states.

This compact provided a formal, organized, and ongoing means for cooperative conservation and placed the responsibility for effective conservation firmly on the shoulders of its member states. Precedent set by this initial multiple-state agreement led to the creation of other compacts to address similar issues, such as the Pacific Marine Fisheries Compact of 1947, the Gulf States Marine Fisheries Compact of 1949, and the Connecticut River Compact for Restoration of Atlantic Salmon of 1983.

REFINING THE SYSTEM OF MANAGING HARVESTED RESOURCES

World War II did not directly influence the process of fisheries conservation other than to diminish offshore fishing activity due to the danger of fishing the high seas. However, because of the country's focus on wartime efforts, legislation initiating changes in the manner in which fisheries resources were utilized and conserved essentially did not occur. Immediately after the war, several factors stimulated what essentially has been half a century of redefining and refining the fisheries conservation and management processes of this country. Growing concern over exploitation and competition with foreign fishing fleets, depletion of traditionally important fish stocks, species in danger of extinction, and habitat degradation, as well as rapidly increasing public interest in angling as a favored leisure-time activity, stimulated numerous national initiatives.

Fisheries Conservation Outside U.S. Territorial Waters

Starting with the International Whaling Convention of 1949, the United States joined a series of international treaties in order to protect the interests of its domestic fishing fleets and ensure the conservation of coastal fisheries resources that were outside its territorial jurisdiction; at that time, territorial waters of any coastal nation extended only 12 miles from its shoreline. Table 6.3 lists many of the major international treaties signed by the United States in the latter half of the 20th century.

Except for a few fleets such as those harvesting Pacific tuna and South Atlantic shrimp, commercial fishing activity of U.S. fleets has never extended far from our coastlines. As in other coastal countries, the U.S. fishing industry generally developed fleets capable of harvesting fisheries resources only along its coasts. Thus most conventions joined by the United States focused on protecting resources easily accessible to its fishing fleets.

Concern over specific resources stimulated the United States' interests in signing international conventions. For example, collapse of the New England haddock fishery in the first half of the 20th century led to the signing of the International Convention for Northwest Atlantic Fisheries (ICNAF). Increasing harvests of Pacific salmon in international waters stimulated signing of the International Convention for the High Seas Fisheries of the North Pacific Ocean. The collapse of lake trout fisheries and the invasion of the sea lamprey in the Great Lakes were major factors leading to the Convention for Great Lakes Fisheries signed with Canada. The numerous conventions established in the latter half of this century followed a standard format

TABLE 6.3   Some post-World War II conventions for fisheries conservation in which the United States has participated as a signatory member.

Convention	Date[a]
International Convention for the Regulation of Whaling	1949
International Convention for Northwest Atlantic Fisheries	1949
International Convention for the Scientific Study of Tropical Tuna, signed with Mexico	1950
Inter-American Tropical Tuna Convention, signed with Costa Rica	1950
International Convention for Pacific Salmon, signed with Canada	1950
International Convention for High Seas Fisheries of the North Pacific Ocean	1952
Convention for the Preservation of the Halibut Fishery of the North Pacific Ocean and Bering Sea, signed with Canada	1953
International Convention on Great Lakes Fisheries, signed with Canada	1954
International Convention for Conservation of Atlantic Tunas	1969
International Convention for Conservation of Salmon in the North Atlantic	1983
United States/Canada Pacific Salmon Interception Treaty	1984

Note: Signing countries are listed only for those conventions that were originally bilateral agreements.

[a]Dates reference when the treaties were signed; in many instances, legislation providing the means to participate in these treaties was passed years after the United States signed the conventions.

typical of earlier fisheries treaties: Permanent international commissions, with members from all countries participating in the convention, were created to recommend strategies for maintaining maximum yield from fish stocks supporting fisheries. Commission recommendations had to be supported and enforced by all member nations.

In 1976 the United States passed the Magnuson Fishery Conservation and Management Act (also referred to as the MFCMA, the Magnuson Act, and the 200-mile-limit law). This act extended the territorial boundaries of the United States from 12 to 200 miles from its coastlines, which incorporated nearly all of its continental shelf areas under its sole jurisdiction. The MFCMA was passed because of the perception that coastal fisheries resources were not being effectively conserved and that countries other than the United States were gaining the greatest economic benefit from the harvest of these resources.

The ICNAF Experience.   A review of the activities of the commission established by ICNAF provides a perspective of the difficulty that multiple governments face when jointly managing a common resource, and why the United States passed the Magnuson Act in 1976.

In the 1930s declines in haddock (Figure 6.2) catches caused some U.S. harvesters to voluntarily increase the mesh size of their trawl nets in order to allow greater numbers of small, immature fish to escape capture. However, these efforts were short-lived because many continued to use smaller-mesh trawls that produced larger catches and higher profits. The haddock fishery progressively worsened, leading nations fishing the offshore banks of the Northwest Atlantic from the Canadian Maritime Provinces to Rhode Island to sign the ICNAF agreement in 1949. Mandatory mesh size regulations were adopted for the Georges Bank haddock fishery within 2 years after ICNAF initiated activities. Like the voluntary measures of the 1930s, these regulations were intended to prevent the capture of small, pre-adult fishes without restricting vessels from fishing where or as often as they

Figure 6.2   The haddock. (From: Henry B. Bigelow and William C. Schroeder, Fishery Bulletin 74. Fishery Bulletin of The Fish and Wildlife Service, Volume 53.)

wished. Because the U.S. fleet was the only one involved in the fishery, passage of this regulation was easy for ICNAF members; no other country had a vested interest in the regulation at that time (Graham 1970).

By the late 1950s distant water fishing fleets from Europe began to fish intensively in the Northwest Atlantic. The arrival and increasing activity of these fleets stimulated an unprecedented growth in levels of exploitation. Distant water vessels harvesting groundfish on Georges Bank were extremely large (Figure 6.3), capable of pulling trawl nets with mouths up to several hundred feet wide (at least several times larger than trawls pulled by much smaller U.S. vessels). These ships processed, froze, and stored large volumes of fishes. In some instances, fishing vessels delivered their catches directly to even larger "mother ships" (huge floating processing and freezing plants). The Spanish fleet employed a paired trawling sys-

Figure 6.3 (*A*) A distant-water harvesting vessel, (*B*) a distant-water "mother ship" receiving the catch of the two harvesting vessels, and (*C*) a typical New England domestic trawling vessel. (Photographs courtesy of Northeast Fisheries Service Center.)

A

B

C

tem, in which two ships cooperatively hauled a trawl nearly twice as large as standard single-vessel trawls in other distant water fleets. Pair-vessel trawls, some more than 500 feet wide, could capture as much as 180 metric tons of fishes in one haul under optimal conditions. In 1971 Spanish pair trawlers made up only 4.5 percent of the total tonnage of foreign vessels fishing in the Northwest Atlantic, yet they harvested 28 percent of the cod catch (Warner 1983).

Such fleets harvested tremendous biomasses of fishes before returning to their distant home ports. Mesh regulations that had been established to protect the haddock stock proved wholly inadequate in the face of uncontrolled increases in fishing effort. The intensifying fishing activity, in concert with unusually successful reproduction of haddock in 1963 (which had temporarily increased haddock abundance), led to a record haddock catch of 155,000 metric tons (341 million pounds) in 1965. This was more than three times the average catch of the previous 30 years (Graham 1970). Such exploitation inevitably led to a collapse of the haddock stock and its associated fishery.

ICNAF established total catch quotas (see Chapter 7) for haddock and yellowtail flounder in 1970 and 1971, respectively. In December of 1971 member nations gave the commission permission to allocate portions of the yearly groundfish catch quotas to fleets from each member nation participating in the groundfish fishery. Catch allocations were intended to divide the catch equitably to all participants while preventing overharvest. By 1973 the program of national allocation of catch levels had expanded to include nearly 20 fish species (Anthony and Murawski 1986). Meanwhile, discontent with ICNAF was rapidly growing within the New England fishing industry.

To control operating costs, fishing vessels from ports of coastal nations are generally small with a short-range harvesting capacity. Thus domestic fleets are dependent on shoreline processing facilities, and they cannot shift easily to other fishing areas (Graham 1970). On the other hand, distant water fleets must be mobile and capable of harvesting over extended periods before returning to home ports. Such fleets can accomplish unusually high levels of harvest over brief time periods. The fishing power of the distant water fleets left the New England domestic fleet feeling outfished along its own coast.

Even with quotas and allocation systems in place, stocks continued to decline in the early 1970s; total catches for heavily exploited species were exceeding calculated allocations due to high levels of bycatch. Thus, if quotas for a particular species are reached in a given year by directed catches, bycatch would cause the total harvest to exceed the desired yearly quota. Allocations were modified in the mid-1970s so that the total directed catch accumulating as the year progressed would be added to predicted levels of total yearly bycatch for particular species. Once the sum of accumulating directed catch and predicted annual bycatch equaled the fishery quota, the directed harvest of that species would be terminated for the remainder of the year. In this way, bycatch would not cause the total harvest of a species to exceed the annual quota.

Once bycatch levels were incorporated into the total allocations of nations, at least some severely depleted New England fish stocks seemed to be recovering. The abundance of principal groundfish species of Georges Bank and the Gulf of

Maine increased annually from 1974 to 1976, the last full year that the United States participated in multinational management of these resources (Anthony and Murawski 1986).

In spite of the improvement in conservation practices of the ICNAF in the mid-1970s, the U.S. fishing industry's discontent with foreign fishing had grown so great that the dissolution of ICNAF had become inevitable. In 1971 Congress passed the Pelly Amendment to the Fishermen's Protection Act, which authorized the president to prohibit the importation of all fisheries products from any country that failed to abide by allocation restrictions. The United States joined the consensus reached at the United Nations' Third Conference on the Law of the Sea in 1974, which proposed that coastal states have sovereign rights over living and nonliving resources of the seas within 200 miles of their coastlines. Worldwide adherence to this resolution eventually nationalized nearly 95 percent of the exploited fisheries resources of the oceans, because nearly all prime fishing areas occur within 200 miles of some coastline (Royce 1987).

During this same period, the United States was attempting to settle arguments over fishing rights of its vessels in waters off the coasts of other nations. The Offshore Shrimp Fisheries Act of 1973 implemented an agreement between the United States and Brazil concerning U.S. shrimp fleet activities off the coast of Brazil. This agreement restricted the number of U.S. vessels that could harvest shrimp in those waters. The act of 1973 created a permit system in which the federal government identified and controlled which vessels participated in that fishery.

Nationalizing Fisheries on the Continental Shelf.   Ultimately, the United States signed the MFCMA into law. The "findings" (justifications for passage of legislation) of the MFCMA identified the economic importance of commercial and recreational fishing to the United States and blamed the depleted condition of numerous coastal resources on the "massive foreign fleets" and the inadequacy of international agreements to prevent overfishing. The findings concluded that a national program "is necessary to prevent overfishing, to rebuild stocks, to ensure conservation, and realize the full potential of the Nation's fishery resources."

The United States assumed exclusive authority for managing all fishes within the fisheries conservation zone, except for highly migratory species such as tunas and billfishes. Management goals are accomplished through fishery management plans developed by regional management councils and approved by the Secretary of Commerce. The focus of management policy is toward optimum yield (OY), or that harvest that provides "the greatest over-all benefit to the nation." OY differs from maximum sustainable yield (MSY) used by ICNAF because relevant social or economic factors as well as the biological condition of the resource are included as criteria for determining harvest regimes. When allowed, foreign fishing is restricted to that portion of the OY that is not harvested by U.S. vessels, which largely limits foreign fishing to species that U.S. industries cannot market.

Regional councils develop the FMPs. The voting membership of each regional council is composed of the principal marine fisheries management official within each coastal state of that region, the regional director of the NMFS, and appointees selected by the Secretary of Commerce from a list of names submitted by the governor of each state. Most appointed members are involved in fishing industries.

Councils may recommend a broad spectrum of conservation measures in their management plans, including limiting types of fishing gear and vessels, limiting the levels of harvest, limiting the times and places where harvest may occur, and limiting the number of vessels that may participate in a fishery.

U.S.–Canadian Boundary Disputes.   Boundary disputes arose as coastal nations extended territorial waters to 200 miles. This was particularly problematic in the Northwest Atlantic, because the extremely productive Georges Bank is within the extended boundaries of both the United States and Canada. A treaty negotiated in the late 1970s proposed joint management of this disputed area. However, in 1980 the United States decided to challenge Canada's jurisdiction over any part of Georges Bank in the World Court rather than sign the draft treaty. The United States claimed that most fisheries resources on Georges Bank were single populations that should be managed throughout the bank under one specific management plan and regime. Canada proposed dividing Georges Bank according to geographical features; each country would have jurisdiction over that area that fell closer to their respective borders. The World Court settled the dispute by splitting the difference between the two proposed lines of demarcation (Figure 6.4). The boundary, often called the Hague Line, after the site of the World Court, placed the northeast section of Georges Bank under Canada's jurisdiction and the rest under jurisdiction of the United States. The most productive scallop fishing grounds of New England were in the northeast section of Georges Bank; the dollar value of scallop landings was higher than any other fishery resource brought to New England ports. Thus the United States found itself losing a portion of the highest-valued commercial fishery in New England.

Effectiveness of the Magnuson Act.   The Magnuson Act changed the direction of conservation of coastal fisheries resources more drastically than any other legislative initiative since the creation of the first U.S. Office of the Fish Commissioner. However, conservation objectives of the legislation frequently have not been met under this management system; many coastal fisheries resources suffered precipitous declines through the early 1990s (see Chapter 12). Although the stated intent of the Magnuson Act was to practice sound conservation to achieve long-term resource stability and human benefit, many who supported its passage viewed it simply as a means of keeping foreign fleets from fishing off the U.S. coast (Royce 1987). Results of studies in the 1970s indicated that removal of foreign fishing would allow growth of domestic fishing industries, creating up to 43,000 new jobs while substantially improving the U.S. trade balance (Royce 1989). In 1980 Congress passed the American Fisheries Promotion Act, which provided increased grants to the fishing industry, directed more funding into fisheries development programs of the NOAA, and provided funds that would allow boat and facilities owners to avoid defaulting on private loans. Thus, although the Magnuson Act promoted conservation action to aid recovery of depleted stocks, the Fisheries Promotion Act supported marked growth of the domestic fleets that continued to harvest the same stocks.

Within a short period of time after passage of the Magnuson Act, the harvesting capacity of domestic fleets within many fisheries had expanded markedly (Figure 6.5). Management councils tended to adopt management plans that minimized regulatory obligations on fishermen, in part due to the viewpoint that broad

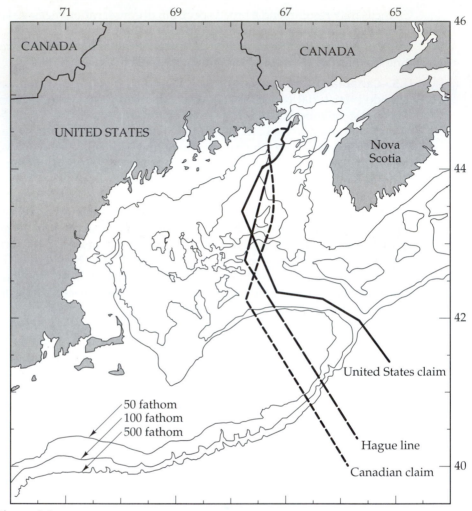

Figure 6.4   The territorial boundary between the United States and Canada in the Northwest Atlantic, as established by the World Court. The original claims by the United States and Canada are also marked. (Illustration courtesy of Amy Lesen, Northeast Fisheries Science Center, NOAA.)

harvest restrictions would drive marginally successful vessels into bankruptcy. Further, enforcement of regulations at sea became progressively more difficult as resources became harder to catch (Campbell 1989). In order to compete for declining resources, vessels owners purchased the newest technology in electrical fish-finding equipment and upgraded the power of their vessels, applying even greater fishing pressure on many fish stocks (as in Chapter 4).

Ultimately, the failure of the management system to conserve numerous coastal resources has occurred when too much focus has been placed on the immediate economic needs and interests of the local community and region, and too little on conservation practices that protect the health of fisheries resources. Man-

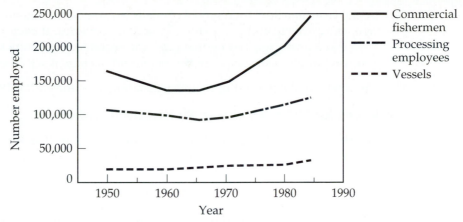

Figure 6.5 The increase in the number of people and vessels involved in the U.S. coastal commercial fishing industry. (Reprinted by permission of The National Audubon Society, Copyright © 1988.)

agement decisions intended to protect the fragile economic health of the fishing industry have led to overfishing and stock collapses that have brought even greater economic hardship to fishing industry participants.

Reauthorization of the Magnuson Act. The MFCMA required congressional reauthorization in 1993 (which was postponed to 1996). By that time, the seriously depleted condition of numerous coastal fisheries resources was causing many people to believe that the management process, particularly the responsibility and functioning of the council system, should be changed. On 18 October 1995, the U.S. House of Representatives passed House Rule 39, proposed amendments to the Magnuson Act. Amendments included:

1. Specific provisions to include the protection of aquatic habitats as an integral means of conserving coastal fisheries resources.

2. A requirement that optimum yield be "prescribed on the basis of maximum sustainable yield from the fishery, as reduced by a relevant social, economic, or ecological factor." Thus, maximum sustainable yield would become the focus of management and conservation efforts.

3. A definition of overfishing as "a level of fishing mortality that jeopardizes the ability of a stock to produce maximum sustainable yield on a continuing basis." Further, the amended legislation requires that resources identified as overfished be rebuilt "to a level consistent with providing maximum sustainable yield."

4. A national focus upon reducing the level of bycatch in fisheries.

5. A rule that prohibits individual members of Regional Fishery Management Councils from voting on any matters "in which the individual. . .has an interest that would be significantly" affected by that vote. Any voting member of a council "shall excuse himself or herself from voting if. . .a personal financial interest [of his/hers] would be augmented by voting on the matter."

6. A one year time limit for regional councils to establish or amend fishery management plans to end overfishing in specific fisheries. If such plans are not developed and approved by that time, the Secretary of Commerce must prepare and implement such plans independent of council activities.

The U.S. Senate had not acted upon this legislation as of the end of 1995.

Initiatives for Fisheries Within Multistate Jurisdiction

Prior to 1984 the federal responsibility was similar in all multistate fisheries compacts: The federal government facilitated creation of fisheries commissions and the development of management plans. Voting membership of the commissions was dominated by or totally composed of representatives from participating states, and states decided whether they would approve recommendations for conservation strategies that were made by the commissions. In 1984 Congress passed the Atlantic Striped Bass Act, which clearly included federal responsibility in the management process of the Atlantic States Marine Fisheries Commission. Congress passed this legislation because four decades of management of this resource under the ASMFC had not prevented severe depletion of these stocks (see Chapter 12).

Several striped bass bills were introduced into Congress in 1984, one of which mandated a multiyear moratorium on all striped bass fishing on the Atlantic Coast; if passed, such action would have identified Congress as the direct "conservers" of this resource. The Atlantic Striped Bass Act was a less extreme divergence from the traditional multiple-state compact approach. This legislation required that all states comply with the provisions of the ASMFC management plan for striped bass. If the ASMFC decides that any state is not in compliance with all management provisions, it must notify the Secretary of Commerce. After investigation, the secretary can impose a total fishing moratorium on striped bass in the territorial waters of that state. Thus adoption of coastwide management plans for this species could be carried out with approval of a majority of states participating in the ASMFC. After passage of this legislation, no member state of the ASMFC held sole statutory authority over management of striped bass resources within its territorial waters.

Congress passed the Atlantic Coastal Fisheries Cooperative Management Act in 1993. This act expanded the management authority of the ASMFC that was established by the Striped Bass Act to all Atlantic coastal fisheries resources that require interstate management. Under this legislation, Atlantic states are required to comply with all management plans developed and voted by the ASMFC. As with the Striped Bass Act, the ASMFC reports noncompliance by any state to the Secretary of Commerce. If after study the secretary makes a determination of noncompliance, a moratorium will be placed on "fishing in the fishery in question within the waters of the noncomplying State." The moratorium will be terminated only upon implementation of regulations by the state.

Funding State Fisheries Conservation Programs

The Dingell-Johnson Act.   In 1950 Congress passed the Fish Restoration and Management Act (also called the Dingell-Johnson or D-J Act), which provided new sources of funding for state fisheries conservation and management programs. This legislation was modeled after the Federal Aid in Wildlife Restoration Act (or Pittman-Robertson Act) of 1937. The Dingell-Johnson Act established a federal excise tax that is applied to the sale of all fishing rods, reels, lures, baits, and flies. Revenues from this excise tax are apportioned to states to manage fisheries having "material value in connection with sport or recreation in the marine and/or freshwaters of the United States."

Forty percent of the revenues are apportioned to states based on the ratio of the land area of each state to the total area of the United States (state territorial waters are included in the calculation of area of coastal and Great Lakes states) and 60 percent based on the ratio of the number of license holders in each state to the total for all states. No state can receive less than 1 percent or more than 5 percent of the total proceeds. The Secretary of the Department of the Interior is responsible for supervision of this program. The Department of the Interior could use no more than 8 percent (later reduced to 6 percent) of the available revenues for administrative costs. Any state that diverts fishing license fees from its fish and game agency to some other use is prohibited from gaining any Dingell-Johnson funds. The Department of the Interior may expend any funds unused by the states. Monies are returned to states on a reimbursement basis. States fund projects from license fees or other state revenues and at the end of each fiscal year are reimbursed 75 percent of the total costs of their approved projects. Approval of projects is the responsibility of the Department of the Interior.

D-J Expansion.   In 1984 the Dingell-Johnson Act underwent extensive expansion under the Wallop-Breaux Amendment. The 1984 legislation did the following:

1. Expanded the excise tax to include all fishing equipment
2. Imposed a tax on electric trolling motors and fish finders
3. Directed a portion of marine fuel tax receipts into the program
4. Diverted duties on imported fishing tackle, boats, and pleasure craft into the program

In addition, states must use a portion of the available funds for recreational boating programs. All coastal states must fund marine as well as freshwater projects, in proportion to the relative importance of saltwater and freshwater angling within the state.

The mandate requiring coastal states to fund marine projects represented a major change in the federal aid program. Although the 1950 legislation mentioned management of coastal resources, coastal states allocated limited funds to marine projects before 1984. Historically, few coastal states required saltwater anglers to purchase a fishing license. The lack of a saltwater license made it difficult for states to measure angling activity in their coastal waters, a mandatory factor in allocation of funds under the Dingell-Johnson formula. Further, because each state is allowed only one designated agency to oversee expenditure of D-J funds, marine fisheries were largely overlooked because in many coastal states they were managed by an agency separate from that administering freshwater fisheries conservation activities.

In the late 1980s the executive branch of the federal government made several attempts to divert some D-J funds to support the Department of the Interior. The Executive Office of Management and Budget argued that some states were not fully using formula funds available to them. Indeed, the Wallop-Breaux Amendment had increased the available funds so rapidly that some states could not fully use their apportionment in the initial years. Passage of the Wallop-Breaux amendment increased available federal aid from only a modest to a substantial proportion of what states had available to expend on fisheries management; some states could not instantaneously increase programs to make use of such large initial increases in funding. This executive branch "federal budget-balancing" approach to use of D-J funds was not allowed by Congress, due in part to strong lobbying efforts and support of national recreational fishing and conservation groups.

Federal–State Partnership and the User Fee Concept.   The Dingell-Johnson Act and Wallop-Breaux Amendment, both strongly supported by organized angling groups, affirmed the concept that individuals directly benefiting from fisheries programs pay for the enjoyment of those benefits. In some states a portion of the funds expended in fisheries conservation programs comes from general revenues. However, in most states the major source of funding comes from fishing license sales and D-J-apportioned funds, both user fees. The federal government's role became one of facilitation, largely allowing the states to determine needs and priorities and to design and carry out programs.

The Anadromous Fish Act.   Although the federal government only facilitates state programs in the Fish Restoration and Management Act, it is a funding partner in state programs intended to enhance anadromous fisheries resources. The Anadromous Fish Conservation Act of 1965 directs the Departments of the Interior and Commerce to enter into cooperative conservation programs with state agencies. Up to 50 percent of the costs of such programs will be provided by the federal government.

Treaty Fishing Rights of Native Americans

Prior to the 20th century, the U.S. government entered into numerous treaties with Native American tribal entities. Many of these treaties identified tribal rights to fish on, and occasionally outside, tribal lands. However, these rights often conflicted with the right of state governments to be solely responsible for determining how fisheries resources within state boundaries would be allocated to all harvesting groups. Throughout the 20th century lawsuits brought against states, in some instances by the federal government, required federal courts to interpret how treaty-guaranteed fishing rights should be addressed by state government. Until pivotal court decisions in the latter half of the century, tribes had little guarantee that treaty-held fishing rights would be observed by state fisheries agencies.

Beginning in the late 1960s, a series of federal court decisions upheld the right of Native American tribes of the Pacific Northwest to an equitable portion of the annual allowable harvest of salmon in specific West Coast river basins. In 1969 federal judge R. Belloni held that the states of Washington and Oregon must allow the Nez Perce, Umatilla, Warm Springs, and Yakima tribes (called the Stevens Treaty tribes, after the U.S. government's representative at the original treaty negotiations) access to a "fair share" of the yearly allowable harvest of anadromous salmonids in the Columbia River basin. This ruling was the first in a series of court cases collectively known as *United States v. Oregon* (Marsh and Johnson 1985). In the 1974 Boldt decision (named after the presiding federal judge), specified tribes were given similar rights to harvest anadromous resources in coastal rivers in the state of Washington. This ruling also defined an equitable share as 50 percent of the total yearly allowable harvest. The 50 percent rule, also adopted in 1974 for chinook salmon in the Columbia River basin decisions, was later upheld by the U.S. Supreme Court.

The Supreme Court also upheld a ruling that required the state of Washington to guarantee tribal harvesters access to their portion of the resource. This decision was based on management policies established by the state of Washington that had allowed removal of most or all of the yearly allowable harvest before the salmon migrations reached "usual and accustomed fishing grounds" of tribal harvesters. Judge Boldt had ruled that the state was responsible for taking "all appropriate steps" within its jurisdiction to control nonnative fishing so that tribes would receive a fair fishing opportunity (Broches 1983). Other Court rulings in the *United States v. Oregon* cases required states to consider the least restrictive harvest alternatives to prevent total closure of tribal harvests, and to provide tribes with a meaningful role in the management decisions associated with these anadromous resources (Marsh and Johnson 1985).

The Salmon and Steelhead Conservation and Enhancement Act of 1980 was passed in part to address the right of Native Americans to harvest these anadromous resources. The 1980 legislation established salmonid conservation areas in the states of Washington and Oregon and directed the Secretary of Commerce to create an advisory committee composed of representatives of the two involved states, the Pacific Fisheries Management Council, the Washington and Columbia River tribal bodies, and the National Marine Fisheries Service. This committee was to recommend management of salmon runs in the conservation areas and to recommend resolution of disputes among treaty and nontreaty fishing groups.

Federal courts were still ruling on legal conflicts between states' authority and tribal treaty rights in the 1990s. For example, in August 1994 a U.S. District Court in Minnesota ruled that state fish and wildlife regulations could not interfere with the harvest privilege afforded the Chippewa tribe by treaty, unless the regulations were clearly related to conservation or public safety. The state retained the right to establish some regulations over the activity of tribal harvesters, but the court reserved the right to judge the scope of that right (Wildlife Law News Quarterly 1994). In December 1994 a federal district court ruling conferred the right to harvest of up 50 percent of allowable harvests of shellfishes in the Puget Sound, including oysters, crabs, and clams, to the Stevens Treaty tribes (*New York Times*, 27 January 1995; page A12).

Federal courts have normally based treaty-right rulings on two basic premises: (1) Treaties must be interpreted as the tribes understood them at the time of signing, and (2) ambiguous terms of treaties are resolved in favor of the tribe (Wildlife Law News Quarterly 1994). Generally, recent legal recognition of Native American fishing rights based on treaty agreements has countered the traditionally held state sovereignty over fisheries resources housed in waterways within state territorial boundaries. By federal court ruling, states must share the responsibility of establishing harvest strategies and regulations for specific resources. In some instances, specific proportions of those fisheries resources must be allocated to tribal entities regardless of other user group interests.

PROTECTING RARE RESOURCES

Endangered Species Legislation

In the 1960s and 1970s the threatened extinction of numerous organisms led to a series of legislative initiatives that define the federal responsibility to protect rare forms of life. On several occasions during that time media attention focused on rare species of fishes influenced the development and modification of federal endangered species legislation.

The Green River Reclamation Project.   After passage of the Colorado River Storage Project Act in 1956, the Bureau of Reclamation initiated a program to impound much of the upper reaches of the Colorado River, including construction of the Flaming Gorge Reservoir on the Green River in southwestern Wyoming and northeastern Utah. In concert with this project, the Wyoming and Utah Fish and Game Departments planned to use toxicants to eliminate native fishes from the Flaming Gorge area in order to create a hatchery-based put-and-take rainbow trout fishery (see Chapter 8) once the reservoir was flooded (Holden 1991). "No apparent thought was given to the fate of native fishes, because they had no value to sportsmen. But that was a prevailing attitude of the fishery profession at that time" (Holden 1991). The American Society of Ichthyologists and Herpetologists (ASIH) opposed this project because of the potential harm to populations of several rare forms of fishes, including the Colorado River squawfish, humpback chub, bonytail chub, and razorback sucker (Figure 6.6). The U.S. Fish and Wildlife Service supported the project, justifying removal of native fishes because "encroachments of civilization's demands on fish environments. . .were soon going to renovate fish populations for us" anyway (as quoted in Holden 1991). However, the U.S. National Park Service became concerned that the use of toxicants such as rotenone would affect rare species found in downstream areas of the Green River that flowed through Dinosaur National Monument (in Utah and Colorado).

Downriver fish kills caused by incomplete detoxification of rotenone received national media attention. Resulting controversy caused the federal government to address its responsibility as steward of all living resources, not just recreationally or commercially important ones. In March 1963 Secretary of Interior Stewart Udall released an agencywide directive that stated in part, "Whenever there is a question

Figure 6.6   The Colorado River Squawfish. (Photograph by Tom McHugh/Steinhart Aquarium; courtesy of Photo Researchers, Inc.)

of danger to a unique species, the potential loss to the pool of genes of living material is of such significance that this must be a dominant consideration in evaluating the advisability of the total project" (quoted in Holden 1991). Thus the Department of the Interior was defining the importance of protecting fragile gene pools several years before the first federal endangered species legislation was passed.

Early Endangered Species Legislation. In 1966 Congress passed the Fish and Wildlife Conservation and Protection Act to "provide for conservation. . .of native species. . .that are threatened with extinction." This legislation consolidated authority for protecting endangered species with the Secretary of the Interior and authorized the secretary to acquire land to carry out the purposes of the act. No other restrictions or guidelines were explicit within this legislation. The Endangered Species Act of 1969 prevented importation of species designated as endangered by the Department of the Interior and prohibited interstate shipment of wildlife taken contrary to state law. Similar prohibition of interstate transport of fish products had been established by the Black Bass Act of 1926, as amended.

The Devil's Hole Pupfish. Water development around a small pool in the Nevada desert in the 1960s also focused attention on the national responsibility to protect endangered life. Withdrawal of groundwater for irrigation in the area around Devil's Hole, a small desert pool that housed the only extant population of the Devil's Hole pupfish (Figure 6.7), threatened to lower the water table of that area below the level necessary to support this rare species. By 1970 the plight of the pupfish had been featured in two television network documentaries as well as numerous major newspaper articles (Pister 1991). In response to growing public interest and pressure, the Department of the Interior established a Pupfish Task Force; Secretary of Interior W. Hickel publicly stated that "the Interior Department will vigorously oppose adverse water use which would endanger the continued existence of these [western desert fishes] surviving species of fish" (quoted in Pister 1991). Starting in 1970 the federal government played an increasingly aggressive role in preventing development that would have destroyed the Devil's Hole pupfish's habitat.

Although this species would not receive effective protection from imminent extinction until the mid-1980s (Pister 1991), strong initial actions taken by the Department of the Interior prior to passage of the 1973 Endangered Species Act signaled a

Figure 6.7 Devil's Hole pupfish. (Photograph by Tom McHugh/Steinhart Aquarium; courtesy of Photo Researchers, Inc.)

growing interest on the part of government to address the problem of species extinction. The publicity given the Devil's Hole pupfish and other desert fishes fueled the climate that led to the passage of the 1973 legislation (Pister 1991).

The Endangered Species Act of 1973.   The 1973 act defined the path that the federal government now takes toward conservation of endangered species. Section 7 of the 1973 act prohibited all federal agencies from either undertaking or writing permits or licenses that allowed private enterprises to undertake any actions that would place an endangered or threatened species in further jeopardy of extinction (Williams and Deacon 1991); the wording allowed no exceptions. Because of the absolute nature of this requirement, this act proved to be one of the most powerful pieces of environmental legislation ever passed by Congress.

Another conflict between watershed development and protection of an endangered fish species was pivotal in the passage of a major modification of the ESA in 1978. Completion of the hydropower Tellico Dam project in Tennessee was jeopardized in the 1970s by the presence of a newly identified endangered species, the snail darter, whose only known population existed in a stretch of the Little Tennessee River including the area to be flooded by the Tellico dam. Hydropower development had historically been viewed as key to economic development in that region of the country; thus Section 7 of the ESA was mandating termination of a politically and economically significant project. The Tennessee Valley Authority, in violation of the ESA, attempted to rush completion of the Tellico project after the snail darter was officially listed as endangered in 1975 (Carlson and Muth 1993). In 1976 the Environmental Defense Fund sued TVA for ignoring the ESA, which ultimately led to the Supreme Court upholding a Circuit Court of Appeals injunction against completing the dam. The snail darter–Tellico dam controversy led to major changes in federal endangered species legislation. Amendments to the ESA in 1978 established a lengthy approval process for listing species, which included requiring an economic analysis of the impact of any new listing of a species and a detailed public hearings process (Carlson and Muth 1993). This legislation also created a cabinet-level Endangered Species Committee that was responsible for considering exemptions from the requirements of Section 7 of the 1973 Act. The Tellico Dam project was one of two that were to receive expedited consideration by this committee. Interestingly, in 1979 the Endangered Species Committee voted unanimously not to exempt the Tellico project because the project was uneconomical. However, later in that year an amendment was attached in Congress to the Energy and Water Development Appropriations Act that exempted the Tellico project from the Endangered Species Act and authorized its completion. Thus the snail darter not only stimulated change of endangered species legislation but also was the first endangered species to be denied protection by the government after passage of the 1978 amendment.

Reauthorization of the ESA in 1982 resulted in a strengthening of the process of conserving endangered resources. A serious backlog of species under review for listing was developing in part due to the complex review process that must occur prior to listing. Amendments of the 1982 reauthorization included provisions streamlining the review process. Further, amendments eliminated the previously held discretion of the administrative branch of the federal government not to list species warranting protection and established that biological considerations would be the sole criteria for determining the status of a species; economic criteria were not to be part of such considerations (C. D. Williams 1994).

Reauthorization Debates of 1994-1995. Congressional debate on the ESA was initiated once more during reauthorization in 1994. After more than 20 years of endangered species conservation under the ESA, many people supported major revision of the law as part of the reauthorization process. Those proposing modification included not only traditional opponents of the law but also elements of the sport fishing community who traditionally have supported numerous state and national conservation efforts (ASA 1994). Dissatisfaction with the application of the ESA has grown within this group because of the conflicts that can arise between the federal responsibility to list and protect species and a state's right to manage recreational and commercial fisheries supported by the same fish stocks (such as Pacific salmon fisheries) or by other species living in the same aquatic systems. For example, the U.S. Fish and Wildlife Service has opposed the stocking of a nonnative strain of largemouth bass by the state of Nevada because of a potential threat to the endangered razorback sucker (ASA 1994).

Congress considered proposals from opponents and proponents of the current act that would substantially change the conservation process of the ESA. Proposals that would restrict or weaken the conservation process include the following (Jones 1994):

1. Requiring submission of an economic analysis for review by the Bureau of Labor Statistics at the time that critical habitat is designated. Because critical habitat is described at the time of listing a species, this essentially is an attempt to reintroduce economic considerations into the listing process.

2. Allowing citizens to bring lawsuits challenging the consideration of species for listing before any final decision for listing has been reached.

3. Removing the mandate of the current ESA that requires federal agencies to ensure that their activities do not harm critical habitat of listed species.

4. Allowing private citizens to sue if suffering "economic injury" as a result of any regulation issued under the ESA.

Congress was still debating reauthorization by the end of 1995.

Marine Mammal Conservation

Frustrated with the inability of past legislation and international treaties to protect marine mammal populations, Congress passed the Marine Mammal Protection Act (MMPA) of 1972 (Weber 1985). This act placed a moratorium on the harvest of marine mammals in U.S. waters and the importation of marine mammal products. The MMPA created a Marine Mammal Commission to review implementation of the responsibilities of this act. The MMPA also preempted state authority for managing marine mammals; several states have since joined with the USFWS and the NMFS to enforce Marine Mammal Protection Act regulations (Weber 1985).

The responsibility for protecting marine mammals is divided between two federal agencies. The U.S. Fish and Wildlife Service is responsible for overseeing conservation of dugongs, manatees, polar bears, sea otters, and walruses, and the National Marine Fisheries Service for cetaceans (whales and dolphins) and pinnipeds (seals and sea lions, excluding walruses).

Amendments to the MMPA allow certain exceptions to the moratorium on harvest. Certain Alaskan natives are allowed to capture marine mammals for subsistence use. In addition, the NMFS sets quotas for the allowable level of incidental capture of dolphins in the domestic eastern tropical Pacific tuna fisheries. In 1986, under threat of a lawsuit from conservation organizations, the NMFS placed restrictions on the domestic tuna fleet once the annual quota was reached (Weber 1987). In the reauthorization of the MMPA in 1984, Congress required that any country capturing tuna in the eastern tropical Pacific can import tuna products to the United States only if its incidental take of dolphins is comparable to that of U.S. fishing operations.

HABITAT-RELATED INITIATIVES

The process for addressing habitat management as an integral part of fisheries conservation was initiated by the early 20th century. The federal government entered the habitat arena with passage of the Rivers and Harbors Act of 1899. This act required a federal permit for all work conducted in navigable waterways of the United States. It also established a mandatory permit system for the discharge of refuse into waterways. The permit program was overseen by the U.S. Army Corps of Engineers. In 1928 Congress passed an Act to Provide for the Conservation of Fish, and for Other Purposes. This act instructed the Secretary of Commerce to determine the best means "to prevent the destruction of fish by ditches, canals, and other works" constructed or maintained by the federal government. The Fish and Wildlife Coordination Act of 1934 authorized the Secretary of Commerce to study the effects of domestic sewage and other forms of pollution on fish and wildlife resources. In 1938 the Rivers and Harbors Act was amended to provide that "due regard" be given to wildlife resources in planning federal projects in navigable waters. Thus, although most major legislation dealing with habitat quality and conservation lay in the future, the process of defining the means for managing and conserving habitat of fisheries resources had begun before the middle of the 20th century.

Habitat-Related Initiatives of the Latter Half of the 20th Century

In the latter half of the 20th century, the United States broadened its responsibility in protecting the quality of natural ecosystems. Some legislative initiatives were broad laws describing the national responsibility to protect the quality of natural systems for the various values they possess, whereas others dealt with mitigating the impact to fisheries resources of specific projects or activities that modify aquatic systems.

Protecting Aquatic Environments. By the end of the 1980s extensive legislation had accumulated that allowed a broad-scale response to the variety of human-initiated disturbances that threatened the health of aquatic habitats. The Water

Quality Act of 1965 established the Water Pollution Control Administration and required all states to develop water-quality standards for interstate waterways. The Clean Water Act (1977) includes provisions that require federal review and permits for the discharge of pollutants into public waterways (Section 402), for dredge and fill of wetlands (Section 404), and for the disposal of sewage sludge (Section 405). Section 401 prohibits the issuance of federal permits for projects that would violate state water-quality standards and laws.

The passage of the National Environmental Policy Act (NEPA) of 1969 was a clear statement by the federal government of its responsibility to address issues related to the quality and health of the nation's natural environments. The act itself was a limited tool for directly controlling environmental degradation. The NEPA requires that a detailed public report outlining the potential environmental impact be written and approved through a rather lengthy process before projects involving federal agencies or federal funding could be started. Thus this law required a clear, public reporting of potential environmental damage but did not provide the means to directly prevent completion of the projects once the reporting requirements were met.

Protection of estuaries and other coastal habitats received particular attention during this time period. The quality and expanse of estuarine habitats have declined markedly throughout this century (see Chapter 9) as extensive commercial development and pollution have taken their tolls (Nixon 1982; Whitlach 1982). The Estuary Protection Act of 1968 provides a balance between the national interest in protecting estuarine natural resources and the local and regional interests in the economic development of these habitats. This legislation instructed the Secretary of the Interior to inventory the nation's estuaries and to enter into agreements with states on the conservation of estuaries and the development of state public areas on a cost-sharing basis. This legislation also specifically required federal agencies to consider the value of estuaries when planning their development and to consider a review and recommendation by the Secretary of the Interior concerning development plans. The 1968 legislation was followed by the Coastal Zone Management Act of 1972, which established a mechanism for funding programs intended to "preserve, protect, develop, and. . .restore or enhance, the resources of the Nation's coastal zone." This legislation authorized the Secretary of Commerce to conduct a grant-in-aid program, which funds state management plans for the coastal zone that are consistent with federal guidelines and objectives.

Congress also passed the Marine Protection, Research, and Sanctuaries Act in 1972. This act created a program within the Environmental Protection Agency to regulate ocean dumping. It also requires analyses of the long-range effects of pollution on ocean ecosystems and authorizes the Secretary of Commerce to designate certain areas as marine sanctuaries to protect them from deterioration caused by human activities.

In 1982 Congress enacted the Coastal Barriers Resources Act. Concern over expansive development, which had already caused notable destruction of coastal barriers, led to this act's passage. The act discourages overdevelopment to the coastal barrier system, in part to reduce damage to wildlife and fish habitats in coastal marshes and estuaries. Section 320 of the 1987 amendments to the Federal Water Pollution

Control Act allow any governor to nominate estuaries within his or her state boundaries for inclusion in a program to protect estuary systems of "national significance."

Water Rights and Protecting Instream Flows. In most instances, states are responsible for regulating the use of water within their boundaries (Lamb and Coughlan 1993). In states where water is an abundant resource, the allocation of its use is governed by what is called the *riparian doctrine:* Persons owning land that is adjacent to a body of water have the right to use the water. When water is less plentiful, its use can be allocated according to the natural flow rule (that is, landowners can use water as long as its quantity or quality is not substantially diminished) or by the reasonable use rule (landowners can use water for any "reasonable purpose"). When the capacity of a basin to provide for all reasonable uses is exceeded, courts may need to determine the priority that is given to potential users (Lamb and Coughlan 1993).

In the West, where water is chronically scarce, water from river basins is allocated by an appropriation doctrine. Western states allocate a certain proportion of the water in a basin to reasonable uses in the order that potential users make application. Once an application is approved, a water right is permanent. A water right is held as private property, which can be transferred or sold by the owner but cannot be removed from that ownership once allocated (Reiser et al. 1989; Lamb and Coughlan 1993). States generally give water resource agencies the responsibility of assigning water rights.

Historically, water withdrawals for irrigation and urban uses, and within-stream uses that support economic development (such as the generation of electricity), took precedence over providing suitable habitat for fishes, particularly in the West, with its "first in time is first in right" doctrine (Reiser et al. 1989). By 1986, however, 15 states had passed legislation recognizing the importance of maintaining instream flows for protection of fish and other aquatic resources. Not surprisingly, the majority of these states were in the western United States, where extreme competition for water and the doctrine of first appropriation seriously threatened aquatic resources in numerous basins. State fisheries agencies typically have advisory responsibility only in instream flow issues, providing recommendations to state water agencies concerning the impact that allocation of water rights might have on the condition of fisheries resources. In states where first appropriation dictates water use laws, instream flow needs can be assessed in several basic ways (Lamb and Coughlan 1993):

1. The state fisheries agency can recommend to the water resources agency that water should be appropriated to maintain instream flows. This creates a "junior" property right controlled by the state that cannot interfere with previously appropriated water rights (as in Colorado).
2. The state fisheries and wildlife agency can review and suggest adjustments to water rights applications (as in California). This system can prevent users from removing or altering flow when it drops below some specified level (as opposed to water users inevitably having the right to use a certain flow volume regardless of flow conditions).

3. In the state of Washington, the Department of Ecology can set aside some portion of flow that is not appropriated and prevent its subsequent appropriation. This decision can be revised by later state administrative action, whereas appropriated water rights are typically held to perpetuity.

Federal legislation has strengthened the power of state agencies to establish instream flow requirements in basins to be altered by federally supported or licensed projects. Section 401 of the Clean Water Act empowers states that have established their own minimum flow standards to require minimum flows as part of the federal licensing process of hydropower dams.

Based on U.S. Supreme Court rulings, the federal government can obtain water rights in states where it has established "reservations" in order to carry out the goals of those lands. Reservations can include lands conferred to Native Americans by treaty or federal holdings such as national parks, national forests and refuges, and military reservations (Lamb and Coughlan 1993). Thus in these instances the federal government can supersede state authority in conferring water to potential users.

Mandating Consideration of Fisheries Resources in Federally Sponsored or Licensed Projects. A variety of laws require the federal government to consider the value of fisheries resources and the impact that habitat modifications carried out or licensed by federal agencies will have on them. The Fish and Wildlife Coordination Act of 1958 requires federal agencies to consult with the U.S. Fish and Wildlife Service and relevant state wildlife and fisheries agencies before federally approved impoundment, diversion, or channelization projects are approved or started.

Section 404 of the Clean Water Act of 1977 requires comment by the EPA, the U.S. Fish and Wildlife Service, and the National Marine Fisheries Service before the U.S. Army Corps of Engineers may issue permits for the disposal of fill in the nation's wetlands and waterways. The EPA can veto the use of any site for this purpose. Other sections of this act also require consultation with natural resource agencies prior to issuance of a permit. A Supreme Court ruling in 1994 (*Jefferson County Public Utility District v. the State of Washington*) upheld the right of states to insist that minimum flow levels that would protect downriver fisheries resources be mandated as part of the federal licensing of hydroelectric dams. Such dams can significantly alter water flow through a river basin (see Chapter 9). This decision was based on Section 401, which requires that federal permits observe state water-quality standards.

The Water Resources Development Act of 1986 includes several provisions that focused on fisheries and other conservation issues. It directs the Army Corps of Engineers to help enhance fisheries and wildlife resources, with 100 percent of the cost of such projects being borne by the federal government when focused on species of national interest. This act also requires that all future projects submitted to the Corps include mitigation plans or determinations that the projects will cause no harm to fish and wildlife. Finally, it has created an Office of Environmental Policy within the Corps to provide guidance on and monitoring of the coordination of the Corps with other federal agencies. Amendments to the Federal Power Act passed in 1986, entitled the Electric Consumers Protection Act, also contain several provisions for fisheries resources. All licenses of power facilities must include conditions to protect and enhance fish and wildlife affected by the project being licensed. Conditions are to be based on recommendations of the U.S. Fish and Wildlife Service, the National Marine Fisheries Service, and relevant state agencies. The Federal Energy Regulatory Commission (FERC) must mandate construction and maintenance of any fish passage facilities deemed necessary by the Department of the Interior or Commerce.

On occasion, conflicts between economic development and fisheries resource conservation have stimulated legislation designed to resolve specific problems. For example, the Pacific Northwest Electric Power Planning and Conservation Act of 1980 was passed "to protect, mitigate, and enhance fish and wildlife [of the Columbia River basin] affected by the development, operation and management of [hydropower] facilities while assuring. . .an adequate. . .and reliable power supply" to the region.

WHERE THE MANAGEMENT SYSTEM STANDS

A century of federal activity has not slowed the process of refining the national management system. In 1993–1995, Congress considered a variety of bills addressing environmental and fisheries/wildlife conservation issues (Table 6.4). Some of these considerations were unresolved by the end of 1995. A major change in the membership of Congress in 1995 strongly suggested that some natural resource and environmental laws, such as the Endangered Species Act and the Clean Water Act, would change significantly through legislative action.

The nation's role as steward of its public fisheries resources has constantly changed due to evolving value systems and to a growing awareness of the impact of many human activities on these resources. Although our capacity to conserve some resources has improved, the condition of many resources has persistently declined. We certainly are not at any end point in the development of the structure and function of the management system. Continued change in the process will be necessary if we are to improve on the impact of our interactions with living resources of our waterways.

TABLE 6.4   Some legislative initiatives considered by Congress in 1993–1995.

Reauthorization of the Magnuson Fishery Conservation and Management Act

Reauthorization of the Endangered Species Act

Reauthorization of the Clean Water Act

Amendments to the Marine Mammal Protection Act

Administrative reorganization amendments to the National Environmental Policy Act

A bill to improve the conservation and management of interjurisdictional fisheries along the
   Atlantic coast

A bill to implement the Convention on Future Multilateral Cooperation in the Northwest Atlantic
   Fisheries

A bill to transfer fisheries responsibilities of NOAA to the Department of the Interior

7

The Use of Regulations

Agencies regulate fishing activity to maintain a fish population's abundance, biomass, and size and age structures at desired levels in the face of harvest. Fisheries operate under sets of natural constraints that determine the productive capacity of the fish population supporting a fishery. These constraints include the productive capacity of the waters; levels of natural predation and competition; and variability of environmental conditions that influence growth, reproduction, and survival. All of these factors can vary greatly among aquatic systems. Management objectives can also vary (for example, achieving maximum sustainable yield, providing a trophy fishery, or achieving fishing satisfaction based on factors other than the level of daily harvest). Natural constraints set limits on what a fish population can provide to a fishery. Agencies establish fishing regulations to maintain the structure of the fish population in a way that meets management objectives for that fishery.

EARLY USE AND DEVELOPMENT OF REGULATIONS

Early Regulations

North American colonies regulated harvest in some fisheries well before the American Revolution (see Chapter 6). Although public agencies initially focused largely on fish culture and stocking (see Chapter 8), by 1877 many states were restricting the types of gear used and the seasons when harvest was allowed in specific fisheries (Redmond 1986). These and other early regulations emphasized the following (Redmond 1986):

1. Preventing fishes from being harvested until they had lived through one spawning season to enhance a population's reproductive effort
2. Allowing small fishes to grow larger before being harvested, thus increasing the weight that a stock could yield to a fishery

Prior to the mid-1900s quotas, size-limit and mesh-size regulations restricted harvest of marine fish populations. Some of these resources recovered from a relatively depleted condition during that time (Russell 1942; Redmond 1986). Many fisheries were regulated simply because it seemed prudent to do so. Analytical methods that allow managers to assess the condition of fish stocks and effects of fishing on them had not been developed. Thus, in the early 20th century, managers often established regulations by guesswork.

In the 1930s fisheries agencies were developing an understanding of (1) limits to the rate at which population additions could occur through growth and recruitment and (2) the impact to populations of losses caused by both natural and fishing mortality (Redmond 1986). Managing for maximum sustainable yield (MSY), or the maximum amount of harvest that can be sustained through time, evolved from these concepts (Nielsen 1976). By the 1940s many state agencies were employing two basic types of regulations: minimum legal size limits set at lengths that allowed most fish to spawn once before being harvested, and restrictive catch limits intended to reduce the yearly level of harvest (Redmond 1986).

The Phasing-Out of Regulations

During the 1940s studies of fisheries in recently flooded reservoirs suggested that harvest regulations had no beneficial effect on fish stocks (Jenkins 1970; Redmond 1986). As knowledge of these studies spread, fisheries agencies entered a period of "liberalization" in which regulations associated with many fisheries were made less restrictive or were eliminated. For example, 34 states had eliminated minimum legal size limits on bass fisheries by 1959; by 1967 only 14 states retained size-limit regulations for any freshwater fisheries (Redmond 1986). Agencies maintained some regulations to meet psychological objectives: Statewide liberal catch limits were applied to some fish species that had previously attracted little recreational fishing effort in order to focus greater attention upon them (Stroud and Martin 1968). Even in the 1970s respected books and review publications still questioned the usefulness of many harvest regulations (Jenkins 1970; Everhart et al. 1975). Statements such as "Fisheries have been regulated on the basis of politics, social pressure, gear competition, prejudice, whim, and sometimes for biological reasons" or "A few laws have been helpful, but of many it can only be said that the fishery survived in spite of the regulation, not because of it" or "Fishing regulations are usually aimed at control of the fishermen with little concern given to the biological health of the fish" (Everhart et al. 1975) probably did reflect the frequent lack of connection between the early use of regulations and effective resource conservation.

Return to the Use of Regulations

By the 1960s rapidly increasing fishing effort and extremely high fishing mortality rates were clearly causing significant declines in abundance of many coastal fisheries resources. Some studies indicated that exploitation rates were extremely high in some freshwater systems, and other studies demonstrated that suitably designed minimum legal size limits improved fisheries and the condition of the fish stocks that supported them (Redmond 1986). In some instances, increased abundance of larger individuals in those stocks provided greater predation pressure on panfish populations, decreasing the abundance of these forage species, increasing their body growth rates, and improving their related fisheries (Redmond 1986). The development of more precise means of analyzing the condition of fish populations and the impacts of harvest on them allowed fisheries scientists to predict the need for harvest regulations more effectively. In recent years agencies have assumed that regulations are necessary to protect the viability of many fisheries resources.

Regulations may be grouped into three basic categories according to the control they are intended to impose upon a fishery:

1. They may directly control the amount or size composition of the harvest
2. They may regulate the level or type of the fishing effort directed toward a fish stock
3. They may influence the species and size composition of the catch by requiring modifications to harvesting gear

The rest of this chapter includes an overview of regulations generally used by fisheries agencies, and the intended impact of these regulations on the resource and fishery being managed.

REGULATING THE COMPOSITION AND AMOUNT OF HARVEST

Size-Limit Regulations

Size limits are applied broadly to a variety of recreational and commercial fisheries in both freshwaters and coastal areas of the United States, and have proven effective in many recreational fishing settings. Anglers often accept size limits more readily than alternative regulations, understanding that small fishes returned to the water will grow, thus making them more desirable for harvest in the future when they are larger (Everhart et al. 1975). Commercial harvesters may consider size-limit regulations an intrusion upon their capacity to catch and sell fishes. Regardless of perception, ultimately acceptance by fishermen is an important factor in the success of any harvest regulation, because the level of compliance influences how effectively harvest is restricted.

Minimum Legal Size Limits. Minimum legal size limits are generally established:

1. To protect the spawning potential of a population
2. To increase biomass production of the population by providing fishes with longer periods of growth before they are harvested
3. To increase catch or harvest rates of larger individuals
4. To create trophy fisheries
5. To increase predation levels on forage species

Many minimum legal size limits still are established specifically to allow fishes to spawn once before harvest. Because most fishes of a species reach sexual maturity at some general average length, it is relatively simple to prevent their harvest until they have completed their first spawning season. However, size limits that are keyed to size at sexual maturity may not be adequate to counteract otherwise unregulated harvest. The youngest, smallest females in a population generally produce fewer eggs than older, larger ones (see Chapter 3). If most of the reproductive effort of a population comes from young, newly matured fishes, the total egg production of the population may be limited by low fecundity of small females. Further, many species of long-lived fishes exhibit marked fluctuations from year to year in reproductive success; high rates of survival of offspring may occur no more frequently than every several years at best. Long life spans and the potential to spawn in multiple years have evolved in many species that face unpredictable rates of offspring survival. If most individuals in populations spawn once and are harvested, several consecutive years of reproductive failure would leave very few individuals alive for future spawning. Thus, if minimum size limits are intended to protect the optimal spawning potential of a population in the face of exploitation, the appropriate size may not always be that which protects individuals through only one reproductive season.

Recent management of striped bass stocks along the Atlantic coast provides an excellent example of the value of using minimum size limit regulations to protect reproductive effort. Protection of the 1982 year class of Chesapeake Bay striped bass until females had reached sexual maturity led to a markedly greater number of spawning females in 1989 than would have been present had harvest of this year class been allowed during its juvenile life stage. The increased reproductive effort created a very abundant 1989 year class, only the second to occur since 1970; subsequent spawning successes have stimulated recovery of the coastal stock (see Chapter 12).

In fisheries experiencing high levels of fishing pressure, most fishes can be harvested shortly after they reach some minimum desirable size. For example, anglers landed up to 74 percent of harvestable-sized largemouth bass annually in Missouri reservoirs and lakes; 69 percent of the available basses were harvested from one new reservoir during just the first four days of fishing (Redmond 1986). Over 40 percent of the harvestable-sized largemouth bass were landed within 6 months after fishing was first allowed in a Mississippi lake, even though catch limits were enforced and refuge areas where fishing was prohibited were established when fishing was opened (Jennings et al. 1986). Thus the size distribution in fish populations unprotected by a minimum legal size limit is often restricted to small length classes. Such populations and the fisheries they support might exhibit less than optimal characteristics. Anglers may not be attracted to specific fisheries if suitable numbers of large fishes are not available. Also, a fish population's role within a fish community can be markedly influenced by such extreme modification of its size structure. Largemouth bass populations with low densities of larger individuals may exert inadequate levels of predation on forage species, such as the bluegill, that are supporting recreational fisheries (Redmond 1986). Under such circumstances the abundance of bluegill populations often increases significantly. This can lead to poor growth conditions and stunting for the bluegill and to substantial reductions in the quality of both the bluegill and the largemouth bass fisheries (see Chapter 8). Thus minimum legal size limits may be established to promote the maintenance of a specific, desirable size structure in the population being regulated. Redmond (1986) believes that a minimum legal size limit large enough to maintain predation on sunfish but not so large that excessive numbers of bass die from natural causes is the best available technique for managing largemouth bass reservoir fisheries. A 12-inch minimum size limit improved the size structure and harvest of bass and bluegills in numerous small lakes in Missouri (Redmond 1986). In Texas a largemouth bass population dominated by small, young fishes shifted to one dominated by large, older ones after the minimum size limit was increased from 10 to 16 inches (Mitchell and Sellers 1989).

Barnhart and Engstrom-Heg (1984) felt that minimum legal size limits nearly always increase catch rates in recreational trout fisheries. Although the use of minimum size limits in trout fisheries leads to greater catches of large fishes, it may reduce growth rates of fishes by increasing the abundance of a population substantially (Barnhart and Engstrom-Heg 1984). Imposition of a minimum size limit has caused similar reductions in growth rates of other gamefishes such as largemouth bass (Novinger 1987).

The opportunity to catch trophy fishes can be provided to anglers by preventing harvest until individuals have grown to a larger size than they might have reached under typical size-limit regulations (Dawson and Wilkins 1981). Trophy size limits address anglers' interests but are not necessarily intended to protect the reproductive capacity of a stock or improve the stock's status in any measurable way over that achieved without the regulation. Thus the fisheries manager is addressing the goals of the user, not necessarily goals of conservation of the resource. However, minimum size limits established to produce trophy fishes typically do not harm the quality of a fishery or the viability of a fish population. Fisheries agencies attempt to manage the interaction between a resource and its human users, so addressing the interests of anglers without negatively affecting the viability of the fishery resource can be a logical objective of a fisheries conservation program.

Catch levels frequently increase significantly after minimum size limits are established in heavily exploited fisheries because smaller fishes can be landed more than once. However, because smaller fishes cannot be harvested, harvest rates may go down, particularly in the first years after the regulation is established (as in Kauffman 1983; Austen and Orth 1984; Dent 1986; Novinger 1986; Richards 1986). Within several years, both number and biomass of fishes harvested can increase from those levels achieved immediately after implementation of minimum size limits (Novinger 1984; Richards 1986).

Slot Limits.   Slot limits protect a specific length range of individuals within a fish population, allowing harvest of fishes smaller or larger than the protected lengths. This type of size regulation is normally applied to populations that exhibit consistently high levels of recruitment of young fishes (Novinger 1984). Allowing harvest of smaller fishes reduces the density of this portion of the population, thus increasing the rate of growth achieved by the remaining young fishes (Summers 1988). Harvest levels of this portion of the population typically are not high, because many anglers do not choose to keep younger, smaller fishes. Thus, when recruitment is high, harvest of this group should not seriously deplete the numbers that survive to grow to larger sizes. The protected range of lengths functions similarly to a minimum size limit. The smallest length in the protected range is that size which would otherwise be kept by a high percentage of anglers; high harvest levels would allow too few fishes from this size to grow larger. The greatest length in the protected range is that size which provides enough large individuals to satisfy angler desires or to impose appropriate predation levels on forage species. Slot limits are often system specific. Slot limits have been established for largemouth bass fisheries in numerous large reservoirs, with protected size limits varying from 11 to 14, 12 to 15, 14 to 18, and 15 to 20 inches (Novinger 1984).

As with many types of regulations, slot limits do not always produce desired results. For example, an 11- to 14-inch slot limit on largemouth bass in Lake Oconee, Georgia, successfully reduced the density of a previously overpopulated stock of gizzard shad and improved the growth of the bass. However, the size structure of the bass population did not shift from one dominated by smaller, young fishes, in part because anglers were not interested in harvesting the small bass that they caught. Because of this, harvest rates in that fishery did not increase even though catch rates did (Martin and Hess 1986). Likewise, anglers in Mississippi did not keep large-mouth bass smaller than 275 millimeters (10.8 inches), causing a slot limit of 275 to 381 millimeters to fail to produce better fishing (Navary 1982, as reviewed in Summers 1988). Up to 70 percent of the annual harvest of largemouth bass in Arbuckle Reservoir, Oklahoma, that were protected with a 300- to 381-millimeter (11.8- to 15-inch) slot limit was composed of fishes less than 300 millimeters in length (Summers 1988). Even though electrofishing surveys indicated that the abundance of fish longer than 381 millimeters increased, harvest rates of that size group did not. Slot limits protecting brown trout between 13 and 16 inches replaced a minimum legal size limit of 12 inches in a reach of the Au Sable River, Michigan, in 1979 in order to reduce the abundance of smaller size classes and increase the abundance of trophy-sized fishes. Ultimate harvests of trout under 13 inches in length were high enough to decrease the number of fishes that entered the protected slot length by nearly 50 percent. The number of fishes harvested increased substantially, but the harvest consisted of these smaller fishes. Growth rates, which apparently were regulated by the limited productivity of the stream, did not increase even after abundance of the population was reduced (Clark and Alexander 1984).

Compliance with Size-Limit Regulations.   Compliance by harvesters is essential to the success of any regulation. It is almost inevitable that sublegal fishes will be harvested after minimum size limits are established; this illegal catch may be minor, or it may constitute a substantial proportion of the harvest. For example, after a 7-inch length limit was established by the state of Connecticut for the scup, an Atlantic coast panfish, less than 1 percent of the recreational catch was smaller than the legal size (Howell et al. 1984). Between 5 and 35 percent of all spotted sea trout harvested by anglers in Texas were smaller than the established 305-millimeter size limit (Meador and Green 1986). Even with these levels of sublegal harvest, both the biomass of the former fish stock and the abundance of adults in the latter increased

TABLE 7.1   Compliance with size limit regulations in reservoir largemouth bass fisheries, measured as a percentage of the harvest represented by sublegal-size fishes.

State	Size Limit	Number of Reservoirs	Percentage of Sublegals
Iowa	14-inch minimum	1	28–39
Oklahoma	14-inch minimum	4	20–42
	14-inch minimum	1	8–67[a]
Kansas	15-inch minimum	2	0–63
Missouri	12- to 15-inch slot	2	14–30

[a]Data are from Glass 1984; all other data from Novinger 1984.

after the size-limit restrictions were established. Sublegal harvest of a particular species can vary markedly among states, or even among waters within the same state. Table 7.1 Lists the level of sublegal harvest of largemouth bass in reservoirs within four central U.S. states.

Noncompliance can reduce the quality of a fishery by causing overharvest of a fish stock. Between 8 and 67 percent of all largemouth bass harvested over several years in Sooner Reservoir, Oklahoma, were sublegal fishes, with the percentage varying by season and type of angling (Glass 1984). Approximately 35 percent of all anglers failed to comply with the minimum legal size limit imposed on that largemouth bass fishery. Most of the sublegals harvested were noticeably smaller than the size limit, indicating that noncompliance was not due to measurement error. Harvest rates in the fishery gradually declined, probably at least in part due to high levels of illegal harvest (Glass 1984).

Not all harvest of sublegals is intentional. Most of the sublegal largemouth bass harvested from Big Creek Lake, Iowa, were close to the minimum legal size, indicating that many sublegals were taken inadvertently due to measurement error (Paragamian 1982). Agencies typically try to minimize sublegal harvest caused by a lack of understanding of regulations by anglers. The last thing a manager should do is stimulate inadvertent violation by establishing a system of regulations that causes confusion. As Everhart et al. (1975) humorously suggest, going fishing should not require "a lawyer in attendance" to interpret regulations.

However simple length-limit regulations are, absolute compliance is rarely achieved. Interestingly, some studies analyzing harvest of sublegal fishes are conducted in fisheries in which the anglers are aware that they will be having their daily catch tabulated at check stations. The well-storied angler's optimistic view of the size of the fish he or she has caught undoubtedly accounts for a portion of the harvest of sublegal fishes. Inevitably, creation of size-limit regulations requires developing a means of enforcing them in order to minimize the proportion of sublegal harvest that is knowing and purposeful.

Mortality of Released Fishes.   When establishing size-limit regulations, agencies must consider death rates of fishes that are landed, handled, and returned to the water. Minimum size limits prevent small fishes from being legally kept but obviously do not prevent them from being caught. Sublegal individuals grab bait, strike lures, and become entangled or entrapped in nets as readily as larger individuals. Even if they are released after capture, such fishes must endure the physiological strain of the capture and the stress and physical damage that can be caused by handling and release. Numerous studies have investigated the severity of angling mortality and the impact that it might have on the effectiveness of size limits. These studies have shown that such mortality is specific to each fishery, depending on factors such as the size and species of fish, the angling equipment used to hook and land the fish and to remove the hook, the handling techniques of the angler, and water temperature.

Mortality of released fishes can be quite low in recreational fisheries. Cutthroat trout mortality in an intensively fished catch-and-release fishery in the Yellowstone River was about 0.3 percent per capture and about 3 percent annually for the population, even though fishes in that fishery are caught an average of nearly 10 times a year (Jones 1984). Less than 5 percent of the spring standing stock in Colorado trout fisheries are estimated to die during the fishing season due to stress and damage caused by capture (Nehring and Anderson 1984).

Different types of artificial lures can cause significantly different rates of hooking mortality. Mortality in a brook trout fishery was over 8 percent for lures with treble hooks and less than 3 percent for the same lures when equipped with single-pointed hooks (Nuhfer and Alexander 1992); differences in mortality rates among different types of lures probably were due to the difficulty in removing hooks once the fishes were landed. Klein (1965) indicated that single-barb hooks can cause greater wounding of rainbow trout than treble hooks if the former are taken further into the mouth or esophagus by the fishes.

Hooking mortality can be higher when anglers use live bait than when they use artificial lures. Fishes tend to engulf live baits more deeply than artificial lures, thus becoming hooked in the gills or in the esophagus more frequently. Resulting tissue damage from removal of the hooks causes the higher mortality. Clapp and Clark (1989) reported an 11 percent mortality rate for smallmouth bass hooked on live minnows, and a 0 percent mortality for those hooked on spinners with treble hooks. Similarly, mortality of walleyes caught on leeches was 10 percent, whereas that of walleye caught on lures was 0 percent (Payer et al. 1989). Schaefer (1989) reported 0 percent mortality for walleye caught on live bait. He attributed the absence of mortality in that fishery to the high frequency of fishes that were hooked superficially in the mouth. Some species suffer very high mortalities when captured on natural baits. Siewert and Cave (1990) reported an 88 percent mortality rate for bluegill sunfishes caught on worm-baited hooks and a 28 percent rate for those caught on artificial lures; Childress (1988) reported a 29 percent hooking mortality for black crappies and white crappies captured on live bait.

Mortality rates in some fisheries apparently are independent of the types of lures or baits used. No significant differences were found between types of hooks or bait types in coastal Texas fisheries for red drum and spotted seatrout (Matlock et al. 1993), in Atlantic coastal river and inland reservoir striped bass fisheries (Harrell 1987), or in some walleye fisheries (Schaefer 1989).

Hooking mortality may vary with season in some fisheries. Bluegill mortality was significantly higher in the summer than winter in Texas, apparently due to increased physiological stress caused by warmer water temperatures (Muoneke 1992).

Mortality of lake trout caught in the summer in recreational fisheries of the upper Great Lakes was somewhat higher than that of lake trout captured through the ice in the winter (Trippel 1993). The mortality of released striped bass increased from an overall yearly average of 15 percent to nearly 40 percent during August (Harrell 1987).

Size of fish can influence rates of mortality. Lake trout greater than 465 millimeters (18.3 inches) in length landed in summer recreational fisheries in the upper Great Lakes suffered significantly lower postrelease mortalities than smaller individuals captured in the same fisheries (Loftus et al. 1988).

Mortality of sublegal fishes is perhaps the greatest problem faced by the manager when imposing minimum legal size limits on some commercial fisheries. For example, trawls can cause high mortality of landed fishes. Numerous factors may influence how high a percentage of sublegals will die:

1.   The size and species of fishes being considered
2.   How long the fishes are entrapped in the trawl before being landed
3.   How long fishes sit on deck during sorting of the catch before they are returned to the water
4.   Air and water temperature conditions on the day of capture (Warmer temperatures, and greater differences between ambient water and air temperatures, generally generate higher rates of mortality.)
5.   Sorting techniques employed by the vessel

Returning undersized fishes to the sea rather than allowing them to be marketed seems wasteful, because their survival cannot be guaranteed. Other types of capture gear may cause similarly high levels of mortality of fishes that cannot be kept legally. Although minimum size limit regulations are used in commercial fisheries, they are usually employed in concert with other restrictive measures that are intended to minimize or restrict capture of small fishes in the first place.

Catch Limit Regulations

Daily Catch, or "Creel," Limits.   Agencies often regulate harvest in recreational fisheries by limiting daily catch, or "possession." Similar regulations, called daily trip limits, are used in conjunction with annual quotas (see next section) in some coastal commercial fisheries.

Agencies establish catch, or "creel," limits in recreational fisheries to reduce yearly harvest by limiting the number of fishes each angler can harvest or possess per day. This type of regulation has also been used occasionally for nonbiological reasons, such as providing anglers with a "satisfaction" goal (Hunt 1975). Quotas are used to control annual harvest in commercial fisheries; agencies apply trip limits in these fisheries to spread the catch more evenly among harvesters and through time.

Daily catch limits cannot directly control the level of yearly harvest in recreational fisheries, because there are no restrictions on how many anglers might participate or on how often each might fish. To control annual harvest via the use of daily catch limits, a fisheries agency must predict what the fishing pressure in that fishery is likely to be (that is, how many anglers will fish and how many total days they will fish). Precision of predictions can be confounded by angler behavior. Surveys have indicated that some anglers become more determined to fish until they catch "their limit" than they might have been before the limit was imposed; many anglers consider the catch limit a goal rather than a restriction (Hunt 1975).

Catch limits are perhaps the most widely used of all regulations by state agencies. All 43 states that manage trout fisheries have regulated harvest by creel limits, even though this has proven to be much less effective in those fisheries that have minimum size limits (Hunt 1975). Of 47 states included in a review by Hebda et al. (1990), only Kansas did not apply daily catch limits to largemouth and smallmouth bass fisheries (Alaska, Florida, and Louisiana were not included in the summary). Daily catch limits have been broadly employed in other freshwater, anadromous, and coastal recreational fisheries.

Daily catch limits have proven ineffective in many fisheries. Evaluations of recreational trout fisheries have indicated that catch limits are usually too liberal to have much value in controlling annual harvest rates (Hunt 1975). Fishing pressure can overwhelm the presumed effectiveness of catch limits. The quality of the largemouth bass population and associated fishery in Clearwater Reservoir, Missouri, declined persistently during years when harvest was limited to six bass per angler per day (Patriarche and Campbell 1957). Reducing the daily catch limit from 10 to 4 bass per day in the heavily fished Little Dixie Lake, Missouri, would have dropped the annual exploitation rate from 72 to 62 percent; this fishing mortality still would have been significantly higher than that population could sustain while continuing to support a productive fishery (Redmond 1986). To be effective the daily catch limit at Little Dixie Lake would have had to have been one or two; anything higher might have actually increased the harvest because of the psychological effect of providing the angler with a goal (Redmond 1986). Fishing pressure was simply too great in the above fisheries for creel limits to realistically change the condition of the fish population supporting the fisheries.

Catch-and-Release Regulations.   Catch-and-release regulations may be the most successful creel limit in terms of protecting the quality of a fishery. Catch-and-release regulations mandate release of all fishes landed by anglers. Fisheries managed by this type of regulation have also been called "fish for fun." Anglers may catch all the fish they wish but must release all that they catch. Catch-and-release regulations require anglers to be satisfied with the angling experience exclusive of harvesting and eating their catch.

Angler compliance is critical to the success of this unusually restrictive regulation. Only a modest proportion of anglers are interested in fishing in catch-and-release waters regularly, for example, perhaps 20 percent of trout anglers (Behnke 1990). The angling public is most supportive when areas managed in this manner represent unusual or unique opportunities, such as catching trophy-sized fishes with relative frequency, experiencing high catch rates in a pristine setting, or fishing for desirable species that are vulnerable to overfishing. Anglers urged Idaho's expansion of the use of catch-and-release regulations in cutthroat trout fisheries previously depleted in part due to overfishing (Hulett and Leider 1990).

Many cutthroat trout fisheries in western river basins are managed with catch-and-release regulations, including the intensively fished stock in Yellowstone National Park and a number of populations in Idaho (Trotter 1990). These stocks are maintained at desirable abundances, with relatively numerous large, older individuals, while supporting very high catch rates (see the section on mortality of released fishes in this chapter). After a decade of catch-and-release in a rainbow trout fishery in Silver Creek, Idaho, angling effort had doubled, catch rates increased from 1.1 to 1.8 per hour of fishing, the percentage of larger fishes in the catch increased, and total annual catch tripled (Riehle et al. 1990).

Gear restrictions are often applied to catch-and-release fisheries (for example, restricting the use of live bait and treble hooks; see the section on gear restrictions in recreational settings). Because catch-and-release areas are so restrictive, states use these regulations to manage only a limited number of sites.

Annual Quotas.   Unlike daily catch limits, quotas do not require predictions of fishing effort to control harvest that accumulates over a season or year. Quotas are typically established by calculating the fishing mortality that a population can sustain while being maintained at an abundance and structure that meets management objectives. Harvest should be prohibited once yearly harvest limits are reached in a fishery to ensure that quotas are not exceeded.

Quotas generally are applied only to commercial fisheries. Enforcement of annual quotas in recreational fisheries would be difficult due to the need to measure cumulative harvest accurately as the fishing season progresses. The time, effort, and cost of determining daily catch rates in recreational fisheries are prohibitive. Also, closing fishing activity once a quota is reached might not engender cooperation among anglers. Such closure would seem an unfair imposition to those who might not typically choose to fish until late in the season. Agencies generally have chosen to manage recreational fisheries in ways that allow anglers to choose when to fish during the year or during the open fishing season.

The use of annual quotas in commercial fisheries is complicated by the way that various gears collect fishes. Some gears, such as the otter trawl, will efficiently catch a variety of species. Thus, when quotas are reached for a particular species and its fishery is closed, further fishing for other species will continue to harvest the protected species as bycatch. In such instances the fisheries agency might choose to do one of the following:

1. Close all fishing with gear to which that species is susceptible, even though other coexisting stocks may not have been fully exploited
2. Allow fishing to continue with the requirement that all captured individuals of the target species be returned to the water

Although potentially effective in protecting stocks, the first option can be politically explosive, because the fishing industry is prohibited from harvesting numerous species to prevent overharvest of one species whose quota has been reached. Many commercial harvesters consider such action an unnecessary economic hardship that leaves potentially underutilized resources unharvested. Due to high mortality rates caused by gears such as otter trawls, the second option may do no more than prevent fishermen from marketing the catch. Because many fishes returned to the water die, it may not effectively prevent overharvest of the species being managed. During the final years of management under the aegis of ICNAF, predictions of bycatch rates of species supporting directed fisheries were subtracted from the

annual allowable quota for the same species to determine the catch levels that were
acceptable for the directed harvest (see Chapter 6). This approach seemed to counter
the negative impact of bycatch levels on the effectiveness of quotas (Anthony and
Murawski 1986). Allowable bycatch quotas of Pacific halibut in groundfish trawl
fisheries of Alaska are subtracted from the yearly harvest quotas allowed in the directed longline fishery for that species. Interestingly, when halibut bycatch quotas
have been reached, groundfish fisheries have been closed (IPHC 1992); thus longline harvesters may feel that the reduction in their directed fishery quota due to a
bycatch allocation is unfair, and trawlers may feel similarly about closures in their
fishery caused when the bycatch allocation is reached.

Quotas have been used extensively in some coastal fisheries and sparingly in
others due more to acceptability than to the potential effectiveness in conserving resources. Many harvesters believe that quotas intrude on the right to access publicly owned resources. Quotas can stimulate harvesters to accelerate their fishing
effort to get a "fair share" before the quota is reached and the fishery closed. Thus
quotas in some fisheries have been reached early in a fishing season or year, stimulating harvesters to argue that they would face bankruptcy if the fishery were
closed for the remainder of the year. In the first years after passage of the Magnuson Act, the New England Fishery Management Council increased quotas levels in
midyear on several occasions to address such controversy, causing overexploitation and potential stock decline. After this experience, the New England Fisheries
Management Council generally did not regulate harvests with quotas through the
1980s to early 1990s. However, other regional councils have employed quotas broadly. The Mid-Atlantic Fisheries Management Council imposed quotas in the surf
clam fishery, allocating specific portions to individual vessels operating in the fishery (see the section on limiting entry). The fishery management plans (FMPs) developed by the North Pacific Fishery Management Council for a variety of Gulf of
Alaska and Bering Sea fisheries resources have included annual quotas. Although
supporting very active fisheries, few of these resources were overexploited during
the 1980s to early 1990s (Loh-Lee Low 1991).

Closures

Closures are typically established to prevent exploitation at some critical phase of
a fish's life cycle. Protection is provided for adult fishes on spawning grounds, for
adults passing through restricted areas during spawning migrations, and for young
fishes on nursery grounds or during periods when they are particularly vulnerable
to fishing gear (Everhart et al. 1975). Many closures prohibit fishing at specific sites
during particular time periods.

Closures may restrict activity in fisheries other than the one being directly
managed. Two general area closures off the New England coast were established
under the Northeast Multispecies Fishery Management Plan of the New England
Fishery Management Council in 1986 to protect the spawning activity of the haddock and yellowtail flounder. Designated spawning areas for haddock on Georges
Bank were seasonally closed to all mobile or fixed commercial fishing gear other
than scallop dredge gear and hooks of a particular size range. An area of the New

England/mid-Atlantic coast was closed to all mobile gear other than specially permitted midwater trawls and sea scallop or surf clam/ocean quahog dredges during the spawning season of the yellowtail flounder. Because many fishing gears such as the otter trawl indiscriminately capture a variety of species, the New England Council found it necessary to restrict the harvest of multiple species to protect two highly valued, and severely depleted, stocks.

Spawning season closures were widely applied to freshwater recreational fisheries for years until studies indicated that prohibiting harvest of spawning adults did not necessarily increase the production of young (Bennett 1970). Table 7.2 lists the states that impose seasonal closures or harvest restrictions on largemouth and smallmouth bass populations to protect fishes during the spawning season. Biologists in states that allow year-round harvest generally believe that recruitment of young fishes is not improved by spawning season closures. Also, the spring spawning season of largemouth bass is the most enjoyable time of year to fish in some southern states; thus anglers in that region might be very displeased with a spawning season closure unless such regulations clearly provided a benefit (Quinn 1993). Some of the states that prohibit fishing or restrict harvest do so to improve reproductive success of bass populations, whereas others do so to reduce harvest of large fishes, which can be extremely high during the spawning season (Quinn 1993).

Agencies have established seasonal closures in part to stimulate interest in recreational fisheries. A springtime opening day following a winter closure supposedly focuses anglers' attention on fishing and increases participation. Such seasonal closures have been used in put-and-take fisheries (see Chapter 8). Stocking typically occurs prior to the first day of legal fishing. Fishes are widely distributed, and anglers don't have to be aware of stocking schedules to fish for "catchables"

TABLE 7.2 **States that close fishing or restrict harvest of largemouth and smallmouth bass during the spawning season of these species. Oregon, Idaho, North Dakota and Missouri have statewide year-round fishing seasons with exceptions on particular waters.**

State	Regulation During The Spring Spawning Season
Maine	Reduced daily catch limits
Maryland	Catch and release
Michigan	Fishing closure; some catch and release waters
Minnesota	Fishing closure
New Hampshire	Catch and release and artificial lures only
New Jersey	Catch and release
New York	Fishing closure
Pennsylvania	Fishing closure in lakes and ponds; reduced catch limits and increased size limits on one river system
Vermont	Catch and release and artificial lures only
Wisconsin	Catch and release

Note: Data are from Quinn 1993; information on exceptions for Michigan and Pennsylvania from Hebda et al. 1990.

before their numbers are depleted. Opening days in fisheries can create intensive participation, but also may not provide a quiet, natural setting for anglers to experience (Figure 7.1). Thus, closures intended to address psychological needs of anglers rather than biological needs of a resource may produce an almost festive opening day that can be enjoyed by some but is objectionable to others.

REGULATING EFFORT

Agencies can influence harvest in a fishery by controlling fishing pressure. The level of fishing pressure, or effort, imposed on fish populations can be controlled by limiting the number of participants in the fishery or by limiting the number or size of fishing units that are being used. In both instances, effort regulations have greater impact on how harvest is spread among participants than they do in directly controlling harvest to specific, desired levels.

Limiting Effort by Controlling Gear

When effort is regulated in recreational fisheries, agencies usually limit the number of fishing units that anglers use rather than denying some portion of the angling public the opportunity to fish. Limiting the number of fishing rods, tip-ups for ice

Figure 7.1   Anglers fishing for lake-run rainbow trout, opening day 1973 on Salmon Creek, Ludlowville, New York. (Photograph courtesy of John Boreman.)

fishing, or handlines can influence daily harvests of individuals without denying participation in a fishery. Agencies may employ these regulations more as a means of spreading harvest more evenly among participants than as a means of restricting levels of harvest, which can be more effectively accomplished by regulations described in the previous section. Limits on the numbers of fishing rods, tip-ups, or handlines may simply stimulate some anglers to fish longer or more often to accomplish a desired harvest; therefore, total harvest levels may not be controlled by such regulations (Everhart et al. 1975).

Limiting Entry and Establishing Individual Quotas

Participation in unregulated commercial fisheries tends to grow until resources become depleted and individual harvesters are unable to earn a living from fishing (see Chapter 5). Restricting participation in commercial fisheries, typically referred to as limiting entry, may be the most effective means of controlling harvest while protecting the economic health of the participants. By limiting the number of participants, agencies can establish quotas that will protect fish stocks while being generous enough to protect the income of the limited number of harvesters participating in the fishery.

Commercial harvesters often have objected vigorously when agencies or fisheries management councils have proposed limiting entry into fisheries. Support for or opposition to limited entry is often related to the "economic self-interest" of those who would be affected by the regulation (Acheson 1980). Harvesters support limited entry to deny "competitors" access to the fishery. They oppose it when they perceive that their own access to the resource is threatened (Acheson 1980). Because of such issues, when limited entry is established, those who previously participated in the fishery are usually allowed further participation. The fishery is simply not allowed to expand effort by the addition of new vessels.

Individual Quotas.   In 1990 the Mid-Atlantic Fisheries Management Council developed a system of individual transferable quotas (ITQs) as part of its amended fishery management plan for surf clams and ocean quahogs. These were the first domestic fisheries to be managed with individual quotas under the Magnuson Act. Within this management plan, the specific portions of the yearly total allowable catch (TAC) of the fishery is allocated to each vessel that has historically participated in the fishery. Vessels can transfer (sell) their individual allocation to other vessels only when they are willing to leave the fishery. New vessels have not been allowed to enter the fishery. After adoption of the ITQ system the number of vessels participating in the surf clam fishery dropped from 128 to 75 between 1990 and 1992 (NEFSC 1992).

The Pacific halibut directed longline fishery is jointly managed by the United States and Canada under auspices of the International Pacific Halibut Commission (IPHC; see Chapter 6). Annual quotas are established for each area of the Pacific Northwest coastline where halibut fisheries operate. In the early 1990s Canada met its allowable quota by establishing IVQs (individual vessel quotas) for the 435 Canadian vessels operating in that fishery. To stay within its annual quota, the United

States established a very limited number of open fishing days (as few as one to several per year) and limits on allowable daily harvest by U.S. vessels, in part because the greater number of U.S. vessels (more than 5,600 fished in Alaskan coastal waters in 1992) would make management by individual quotas difficult. The individual quota system offered Canadian vessels a predictable harvest that could be executed over a period of time, whereas the U.S. fishing season of only one to several days did not. Spreading the catch over a season brought a better price at the dock to Canadian than to Alaskan fishermen based on a simple supply-demand relationship. In Alaska the entire halibut harvest was brought to port during one to several days. Processors would buy the catch, process it, and freeze it for later sale. Because all of the catch arrived at the processors at the same time, vessel captains received less for their catch than if it were delivered over a prolonged fishing season. In part because of the economic and biological benefits experienced by the Canadian quota system, individual fishing quotas (IFQs) will be issued starting in 1995 to manage Pacific halibut and sablefish harvests in Alaskan waters. Quota "shares" in these fisheries will be calculated for each applicant based on the historic level of participation in each fishery.

The allocation of portions of quotas to specific participants in a fishery offers a means of controlling harvest while providing harvesters with a guaranteed catch. It allows participants who are making marginal profits to "sell" their quota to others, thus making a small immediate profit while ending their unprofitable participation in the fishery (Weber 1994). In theory, as quota holders leave the fishery, the number of participants will decline until a point is reached where all remaining vessels will be allocated large enough harvests to make a suitable profit from their participation. Thus the total quota for the fishery will hopefully protect the condition of the resource, and the individual allocations will protect the economic viability of the harvesters.

However, the commitment to an ITQ management system has one potential drawback. By allocating guaranteed portions of a catch to specific harvesters and preventing entry of new participants into the fishery, an ITQ system essentially establishes private ownership of a public resource. In addition, successful participants have the opportunity to "hoard" access to the resource by buying as many individual quotas as possible. To prevent the centralization of most of the resource into the ownership of too few people, the ITQ system being established for Pacific halibut will include restrictions on the number of allocations a single harvester can control (Weber 1994).

REGULATING GEARS TO CONTROL THE SPECIES AND SIZE COMPOSITION OF THE CATCH

Regulating Gear Efficiency

Agencies may require gear modifications that are designed to reduce the capture rates of small fishes or the incidental capture of nontargeted species; agencies often refer to these requirements as gear efficiency regulations. Such regulations may be more effective in some commercial fisheries than are minimum legal size limits, if fishes that are captured suffer high mortality rates due to capture and handling. Agencies also have chosen to regulate harvest by using efficiency regulations in fisheries in which more restrictive regulations such as catch quotas are not politically or sociologically acceptable.

Reducing the Capture of Small Fishes—Mesh-Size Regulations. Agencies commonly apply mesh-size regulations to commercial gill-net or trawl fisheries to reduce capture of small fishes. Fishes are captured in gill nets when they become entangled or wedged in the mesh of the netting. Any specific mesh size will fail to capture fishes so small that they pass through the netting or so large that they cannot pass far enough into the webbing to become entangled or wedged (Hubert 1983). Similarly, the mesh size in trawls should allow small fishes to slip through openings of the mesh and escape while larger individuals are retained in the trawl. Thus agencies can establish mesh-size restrictions that are designed to allow specific size classes of small fishes to escape. However, such mesh-size regulations in trawl fisheries do not alter the size composition of the catch consistently. As trawls are fished, smaller individuals that might otherwise escape through the mesh are entrapped within the expanding load of fishes; thus as time of trawling increases, so does the proportion of smaller fishes in the catch. Minimum mesh requirements in trawl fisheries may reduce the efficiency of entrapping small fishes but do not eliminate their capture altogether. In addition, fisheries managers are becoming increasingly concerned that some fishes squeezed through the mesh of a moving net may be injured and subsequently die.

Effectiveness of Gear Efficiency Regulations. In spite of potential limitations such as those described above, efficiency regulations are occasionally used because they are perceived as a compromise between protecting the viability of a fish stock and being acceptable to the harvesters of that stock. Although biologically effective, controlling harvest directly by using quotas or limiting entry often is deemed unacceptable because commercial harvesters consider such regulations too intrusive upon the right to fish.

Mesh-size regulations did not protect New England groundfish resources from overharvest and stock collapse under the management regimes of ICNAF and the New England Fishery Management Council. Allowing young fishes to escape capture may not protect a stock whose adults are severely overfished. Everhart et al. (1975) suggested that gear restrictions may do no more than stimulate commercial fishermen to fish harder and longer to harvest the same biomass of fishes that would have been landed in an unregulated fishery. Thus harvesters may incur greater costs while the stock may still collapse.

Prohibiting the Use of Gears

Gear prohibitions have been applied to a variety of commercial and recreational fisheries. States prohibited the use of specific gears in coastal commercial fisheries before the beginning of the 20th century. Some of the first federally imposed fishing regulations restricted the use of particular types of traps and weirs in Alaskan salmon fisheries. Agencies occasionally prohibit certain gears to reduce the capture of nontargeted species or age groups. For example, in the early 1980s California prohibited the use of set net fishing gear in specific coastal areas to reduce the frequency of accidental entanglement and strangulation of sea otters (Bricklemeyer et al. 1989). The South Atlantic Fisheries Management Council has prohibited the use of trawl gears in specific snapper and grouper fisheries to protect the condition of critical bottom habitat and to reduce the mortality of incidentally captured juvenile, undersized vermilion snapper (Bricklemeyer et al. 1989). Most mobile gears are prohibited seasonally in specific areas off the New England coast to protect the spawning effort of haddock and yellowtail flounder.

Gear Prohibitions to Prevent Commercial Harvest.   Some gear restrictions are established to favor recreational over commercial use of a resource. For example, striped bass, American shad, and Atlantic salmon fisheries were each restricted to hook-and-line-only harvest by at least one New England state in 1989. In recent years gill-net harvest of walleye was prohibited in Lake Erie to allocate the total harvest of this species in U.S. waters to hook-and-line anglers.

Agencies often establish "no-sale" regulations in concert with gear prohibitions to ensure that specific resources are reserved for recreational use. For example, the state of Texas prohibits commercial harvest of 23 coastal species, including red drum, spotted seatrout, jewfish, tarpon, king mackerel, and billfishes. No-sale and hook-and-line-only regulations are becoming common because of the growing political strength and lobbying power of angling groups (Royce 1987).

Gear Restrictions in Recreational Settings.   As reviewed earlier in this chapter, many gear regulations applied to recreational fisheries (such as restrictions on the use of live bait and treble or barbed hooks) are intended to increase the survival rates of fishes that are returned to the water. However, some gear restrictions in recreational fisheries may serve a greater sociological than biological purpose.

Fly-fishing-only regulations represent one type of gear restriction that may address social pressures more than conservation needs. Agencies have defended fly-fishing-only regulations as a means of reducing harvest by decreasing the number of anglers in a fishery (a kind of limited entry based on gear type) or as a means of reducing mortality of fishes returned to the water after being landed and handled. However, studies have demonstrated that fishes landed on flies may suffer the same rates of mortality as those landed on other artificial lures (Baughman 1984).

Fly-fishing-only regulations do not always improve the condition of a fish stock or its associated fishery. In one study, the productivity of a trout population was not changed by a flies-only regulation, and the quality of the angling experience was different for fly fishermen accessing the restricted area versus other anglers that were required to fish in any-lure sections of the same stream. Fly fishermen caught more fish per hour of angling than did all anglers in the any-lure zone (Hunt 1975). Shetter and Alexander (1962) found that anglers caught the same numbers and biomass of trout in a flies-only zone as a greater number of anglers had caught in years before the regulation was implemented. Further, harvest rose dramatically in the unrestricted zone of the same stream because of the increased number of anglers fishing in that area after the flies-only regulation was instituted in the adjacent stream section. Fly fishing only did not increase survival of trout within a year, nor did it change the population abundance in subsequent years (Shetter and Alexander 1962).

Although fly-fishing-only zones remain popular with some fisheries agencies, fisheries managers must carefully consider the consequences of such regulations. Recreational users should generally be receptive to fishing regulations if agencies can demonstrate that the present and future aesthetic experience of anglers will be enhanced by the restriction. However, the right to access an angling experience is potentially as important to anglers as is the right to make a living to commercial harvesters. Anglers can feel just as disenfranchised as commercial users if they believe that gear prohibitions are denying them experiences that others are having.

Gear Restrictions to Reduce Bycatch

Mortality of bycatch (fishes or other marine taxa incidentally captured in fisheries targeting other species) threatens the viability and productivity of populations of many marine organisms. Numerous commercial fisheries discard substantial biomasses of organisms after capture. Finfishes may be discarded because they are too small to market or are smaller than minimum legal size limits established to protect them. Other marine organisms are discarded because no markets exist for them or because they are protected under guidelines of the Endangered Species Act or Marine Mammal Protection Act (for example, sea turtles and dolphins). In many fisheries, capture and handling of bycatch causes substantial mortality.

Shrimp fisheries of the Gulf of Mexico and the Atlantic coast generate very high levels of finfish and sea turtle bycatch due to the very small mesh of the trawls used in these fisheries. Agencies have supported the development and mandated the use of gear modifications designed to reduce bycatch levels in these shrimp fisheries. Trawling efficiency devices (TEDs; originally called turtle-excluder devices) are required to reduce the frequency of capture of endangered or threatened species of sea turtles as well as immature fishes in the Gulf of Mexico and southeastern Atlantic states. Bycatch reduction devices (BRDs) are required to reduce the capture of immature fishes in the Gulf of Maine northern shrimp fishery.

Bycatch Mortality in Commercial Shrimp Fisheries.   Discarding of bycatch in shrimp fisheries offers one of the most visible examples of waste of living resources caught incidentally in fisheries (Figure 7.2). The total finfish bycatch of the U.S. shrimp fleet operating in the Gulf of Mexico exceeded 300 million pounds yearly during the late 1980s. Nearly 5 billion croaker were killed as bycatch in 1989 alone, and about 12.5 million juvenile red snapper die annually as bycatch (Murray et al. 1992; Perra 1992). The total weight of bycatch exceeds the total weight of shrimp harvested in the Gulf of Maine northern shrimp fishery (Howell and Langan 1992), and discard biomass can be as much as 18 times greater than the weight of shrimp harvested in the Gulf of Mexico (Weber 1986; Murray et al. 1992). For many species the impact of bycatch mortality is unknown; evidence suggests that red snapper, mackerel, and weakfish populations are suffering population declines in part due to bycatch mortality (Murray et al. 1992).

Turtle mortality, although low in absolute terms, is important due to the severely depleted condition of sea turtle populations. During the 1980s approximately 12,600 sea turtles died annually in the Gulf of Mexico as a result of incidental capture in shrimp trawls; all of the involved turtle species are listed either as endangered

Figure 7.2   Fish bycatch and shrimp catch of a Gulf of Maine shrimp trawler. (Photograph courtesy of Steven Hokenson.)

or threatened under guidelines of the Endangered Species Act (Weber 1986). Due to public pressures and its legislated responsibility to protect endangered sea turtles, the National Marine Fisheries Service initiated a research program in the early 1980s to develop effective means of excluding sea turtles and finfish bycatch from capture by shrimp trawls (Figure 7.3). Similarly, technology developed to reduce fish bycatch in Scandinavian shrimp fisheries was adopted by the Atlantic States Marine Fisheries Commission and the New England Fishery Management Council as a required modification to shrimp trawling gear in the Gulf of Maine fishery.

Mandating the Use of TEDs and BRDs.   Commercial harvesters have vigorously opposed regulations that modify gear to reduce bycatch levels. As a result of research efforts administered by the National Marine Fisheries Service, inexpensive excluding devices had been developed by the mid-1980s (Bricklemeyer et al. 1989). NMFS supported the effectiveness of TEDs in excluding bycatch without reducing shrimp catches. However, harvesters strongly objected to the government mandating any

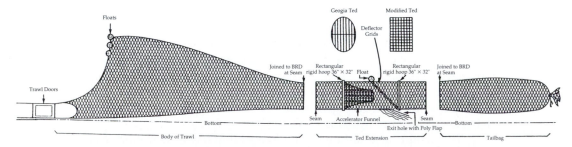

Figure 7.3   Bycatch reduction device installed in a shrimp trawl. Large animals passing through the accelerator funnel strike the deflector grids and pass out the exit hole. Shrimp pass through the grid into the tailbag. (From R. A. Rulifson, J. D. Murray, and J. J. Bahen. 1992. Finfish catch reduction in South Atlantic shrimp trawls using three designs of by-catch reduction devices. *Fisheries* 17(1): 9–20. Reprinted with permission of the American Fisheries Society.)

gear modifications, arguing that shrimp catches and profits would decline. Thus few shrimp harvesters voluntarily used these TEDs. By 1986 conservation groups threatened to take the NOAA to court unless it required the use of TEDs on shrimp vessels in the Gulf of Mexico and southeastern Atlantic states (Chandler 1988; although regulations dealing with harvest of targeted species are developed by regional fisheries management councils, NMFS-NOAA is responsible for protecting endangered sea turtle species under guidelines of the Endangered Species Act). In the fall of 1986 an arbitrated agreement to require TEDs on all shrimp vessels longer than 30 feet by 1989 fell apart when fishing industry representatives that had not been represented on the arbitration panel opposed the agreement. Subsequent negotiations led to an agreement that provided exemptions based on area, time of year, length of boat, and time of tow of a trawl (Bricklemeyer et al. 1989). TEDs have proven to be very effective in preventing capture of sea turtles while minimizing loss of shrimp catches; TEDs reduce the turtle kill by about 97 percent while reducing the shrimp catch by about 5 to 15 percent (NOAA 1991).

Although turtle bycatch was being regulated by 1990, fish bycatch in the Gulf of Mexico and southeastern states shrimp fisheries was not. As part of a revised groundfish FMP, in 1993 the New England Fishery Management Council required the use of a specific bycatch reduction device, the Nordmore grate, in shrimp trawls in federal waters of New England. As in the southern shrimp fisheries, the fishing industry opposed the mandated modification of trawl gears. Some harvesters blamed a reduction in the catch per unit effort of shrimp in 1993 on the Nordmore grate. Interestingly, the reduction in harvest was roughly equal to a decline in abundance that had occurred between 1992 and 1993 for the Gulf of Maine northern shrimp stock; thus reduced harvests were probably due to the lower stock abundance than to the use of a BRD.

WATERSHED SPECIFIC REGULATIONS

Freshwater recreational fisheries typically differ in both the dynamics of the fish population and the characteristics of the fishing activity on practically a watershed-by-watershed basis. The most precise approach toward management of such resources would be to establish specific regulations for each watershed that best suits that fishery. However, this would require precise information on the dynamics of the fish populations and on angling pressure characteristic of each watershed. Such watershed-specific regulations have traditionally been used to manage fisheries that are limited in number and that receive exceptional amounts of fishing pressure or angling interest, such as Great Lakes fisheries, fisheries within large reservoir systems, or anadromous fisheries housed within relatively few river basins. Agencies have the capacity to focus effort on such fisheries because relatively few occur within the boundaries of most states.

Management of fisheries spread throughout a state's boundaries, such as trout fisheries in streams or largemouth bass fisheries in ponds and small reservoirs, are much more difficult to manage precisely on a watershed-specific basis due to their numbers. Historically states have hesitated to establish watershed-specific regulations on a statewide basis, instead opting for single, statewide regulations (Hunt 1975). In the mid-1970s only nine states had adopted regulations tailored to manage trout stream fisheries on a watershed-specific basis (Hunt 1975). Although statewide regulations may not precisely fit the needs of particular fish stocks and the fisheries associated with them, they do provide anglers with a simple system that minimizes misunderstanding and inadvertent violation. Lack of understanding by anglers of watershed-specific regulations can be a daunting problem. Wisconsin manages all trout fisheries on a site-specific basis, which requires a 32-page pamphlet outlining creel limits, size limits, seasons, and gear restrictions applicable to each waterway (State of Wisconsin Trout fishing regulations and guide, 1990; Wisconsin DNR, PUBL-FM-302 90). Such management certainly requires an agency to communicate well with its angling public to minimize confusion and inadvertent lack of compliance. Regardless of such logistical difficulties, in the latter 1900s agencies generally increased the number of watersheds managed as separate fisheries.

Many coastal species do not need to be managed by use of a series of regulations that are specific to local areas, because the fish stocks and fisheries they support are widespread geographically and may cross numerous state boundaries. Effective management of such fisheries typically requires multiple agencies agreeing to establish regionwide regulations rather than developing series of local area-specific harvest restrictions.

THE USE OF MULTIPLE REGULATIONS

No single type of regulation has proven suitable for all management situations, although some have proven to have greater utility than others. Many fisheries are regulated by multiple harvest restrictions. For example, management schemes for many coastal fisheries resources may simultaneously include seasonal and area closures, gear restrictions, minimum legal size limits, and restrictions on the number of participants. The state of Maine manages recreational harvest of its depleted Atlantic salmon stocks by the concurrent use of closures, catch limits, size limits, and gear restrictions on a watershed-by-watershed basis. Atlantic coastal striped bass fisheries are managed with a similar complexity of regulations. Management measures regulating Atlantic surf clam harvest have included total allowable catch quotas, limited entry into the fishery, minimum size limits, and area and time closures. In Columbia River steelhead recreational fisheries, hatchery fish, which were marked by fin clips prior to migrating to the sea, can be harvested during their return migration. Returning wild steelhead, which are not clipped, must be released at capture. Thus this fishery, which is also regulated with other fishing restrictions, allows harvesters to benefit from hatchery fishes while protecting the reproductive potential of the wild stock. In general, the more heavily exploited the fish stock or the more seriously depleted its condition, the greater is the likelihood that the management scheme applied to its fishery will include a set of different types of regulations.

THE EFFECTIVENESS OF REGULATIONS IN FISHERIES CONSERVATION AND MANAGEMENT

Multiple factors affect the condition of many fisheries resources. However, even when habitat deterioration or destruction are not at issue, many fish stocks are depleted and their capacity to support an active fishery is limited because harvest is not effectively regulated.

In fisheries supported by wild stocks, effective regulations obviously must protect the potential of a natural resource to provide the experience expected by human users. Generally, agencies have regulated harvest more successfully in recreational than in commercial fisheries. Many recreational users are receptive to regulations if they understand that their future enjoyment will be enhanced by current constraints on their fishing activities. Commercial harvesters are less patient when their fishing activities are restricted, as their current income often is far more critical to them than is any future prospect of fishing (see Chapter 4). Commercial harvesters often have viewed regulations as a threat to their pursuit of a chosen occupation. Therefore, it is not surprising that commercial harvesters often vigorously oppose restrictive harvest regulations.

Pressure from the fishing public can influence an agency's application of fishing regulations. Occasionally agencies establish regulations in large part because of pressure from users. Some groups may be excluded from participating in a fishery while others are not, because of differing levels of political influence. Agencies may not establish needed regulations due to political pressure from users (as in New England's groundfish fisheries; see Chapter 12). This has occurred relatively frequently in coastal commercial fisheries. Some fisheries become depleted from overfishing because agencies are unable to understand the confounding impacts on resources of fishing, habitat deterioration, and natural variation in environmental conditions. In other instances, fisheries may be ineffectively managed even though the impacts of fishing are well understood; sociological and economic pressures simply do not allow application of effective regulations to prevent overfishing and resource depletion.

8

Fisheries Supported by Hatchery Fishes

Fishes have been reared and stocked into public waterways since the first fisheries conservation agencies were established in the mid-1800s. Agencies have stocked hatchery-reared fishes as a means of enhancing the condition of fisheries resources, as well as creating fishing opportunities that are not supported by wild fish populations. Although stocking programs have been instrumental in establishing and supporting many fisheries, they have failed to enhance the quality of others. Stocking has caused declines in the condition of wild fish populations and in the fisheries that they support. In some instances, objectives of hatchery programs conflict with rather than support effective conservation of wild resources. This chapter provides an overview of the successes, failures, and problems created by the use of hatchery fishes to enhance existing or to create new fisheries. Stocking conducted for other purposes, such as controlling rooted aquatic vegetation through grazing, providing food species for predator populations, and attempting to facilitate recovery of endangered fish populations, will be discussed in Chapters 10 and 11.

EARLY USES OF STOCKING BY FISHERIES AGENCIES

Public agencies generally emphasized stocking over all other activities through the early 20th century. Expenditures of agencies during the late 19th to early 20th centuries clearly display this emphasis. For example, nearly 70 percent of the $295,000 budget appropriated by Congress for the U.S. Office of the Commissioner of Fish and Fisheries in FY 1882 was expended for the rearing and distribution of fishes. The congressional appropriation for FY 1928 included over $260,000 in salaries for hatchery personnel and only $129,000 for all personnel on federal research vessels and at biology field stations. The operating budget for hatcheries, exclusive of salaries, was $487,000, whereas that of all research vessels and field stations was $223,000.

Agencies emphasized stocking in part because fisheries managers knew how to raise fish better than they understood how to regulate harvest and other human interactions with fish populations. Regulating harvest by scientifically analyzing the condition of fish stocks and the impact of harvest on them did not develop until well into the 20th century. Before that time, the dynamic nature of fish reproduction, growth, and mortality was not well understood.

Hatcheries also provided a visible example of agency activity. Thus they were instrumental in gaining public and user group support (Bowen 1970). Indeed, many agencies still make use of the "public relation" value of their hatchery systems, encouraging visits and in some instances maintaining public viewing pools housing exceptionally large fishes as an example of the agency's fish-rearing prowess. Thus hatchery systems represented the major focus of both state and federal agencies well into the 20th century because fisheries managers emphasized the activity they did best—raise fish.

Early hatchery programs focused on introducing species into waters outside of their natural geographic ranges and replenishing fish stocks depleted by harvest or habitat destruction. Within 20 years after it was established, the U.S. Commission of Fish and Fisheries was distributing a variety of native North American fishes and exotic (nonnative to North America) fishes throughout the freshwater and coastal systems of the United States (Table 8.1). Such early stocking programs often were conducted with little consideration given to conditions of the watersheds into which fishes were stocked, to the impact they might have upon native fauna, and to the likelihood of their being successfully established (Redmond 1986). Some of these early introductions failed, such as attempts to establish chinook salmon and pink salmon in East Coast river systems and the Great Lakes (however, see the section on pacific salmon fisheries in the Great Lakes in this chapter). Others produced self-sustaining populations of native fishes well outside their native ranges, such as largemouth bass, striped bass, and American shad in Pacific coast states, and rainbow trout in the Northeast. The European brown trout was established in New York state by 1886, three years after first being brought to the U.S. federal hatcheries. After more than a decade of extremely successful introductions, propagation of the common carp was terminated in federal hatcheries when public enthusiasm for this exotic species turned to criticism because of the carp's tendency to displace native species (Bowen 1970; see the section on declines in native fisheries caused by species introductions in this chapter).

During the late 19th and 20th centuries, federal hatcheries also hatched and reared large numbers of fishes to stock into waters supporting populations of the same species. Such supplemental stocking was intended to replenish depleted fish stocks or simply to provide more fishes for rapidly growing fisheries. At that time, supplemental stocking focused on American shad along the Atlantic coast; coastal species such as Atlantic cod, winter flounder, and lobsters in New England waters; Great Lake species such as walleye and lake whitefish; and Pacific salmon populations in California, Oregon, and Washington.

TABLE **8.1**   **Stocking of fishes reared by federal hatcheries into freshwater and coastal systems of the United States, 1891–1892.**

Species	Number of States Receiving Fishes	Number of Eggs, Larvae, or Older Fishes Stocked
American shad	17, Washington, DC, and the Indian Territories	70,000,000
Common carp[a]	44, Washington, DC, and the Indian Territories	157,000
Goldfish[a]	42, Washington, DC, and the Indian Territories	21,000
Tench[a]	11	36,000
Ide[a]	13	2,000
Channel catfish	14	4,000
Chinook salmon	4	4,581,000
Atlantic salmon	8	1,168,000
Brown trout[a]	27	317,000
Rainbow trout	20	195,000
Brook trout	14	69,000
Lake trout	13	1,424,000
Lake whitefish	7	65,000,000
Lake herring	1	263,000
Largemouth and small-mouth bass	22	20,000
Crappie sp.	16	6,000
Rock bass	24	26,000
Sunfish sp.	11	10,000
White bass	6	2,000
Yellow perch	11	30,000
Walleye	7	94,000,000
Northern pike	7	2,000
Scup	1[b]	35,000
Atlantic cod	1[b]	52,796,000
Pollock	1[b]	2,474,000
Winter flounder	1[b]	6,274,000
Lobster	1[b]	5,799,000

[a]Exotic species from Europe.
[b]Massachusetts.

Supplemental stocking was frequently unsuccessful, often because the cause of the fish population decline was not addressed at the same time. For example, the U.S. Fish Commission stocked more than 8 million shad fry into the Connecticut River between 1872 and 1880 because shad migrations had been declining in that river. New England states markedly exceeded this effort, stocking over 90 million into the Connecticut in 1872 alone (Bowen 1970). Regardless of this and subsequent efforts, migratory runs of shad in the Connecticut did not improve substantially until nearly a century later, when major efforts to provide effective fish passage facilities over dams in this basin opened a greater stretch of the river to migratory and spawning activities of that species.

Expansive stocking programs initiated by the U.S. Fish Commission to increase the abundance of Atlantic coast and Great Lakes fish stocks were similarly unsuccessful. For example, in 1900 over 265 million cod fry, 87 million flounder fry, and 81 million lobsters were released into New England waters (Bowers 1901). In 1938 nearly 0.5 million winter flounder fry and more than 5 billion pollack, cod, flounder, and haddock eggs were stocked out of the federal facility at Gloucester, Massachusetts, alone. Other federal hatcheries in New England contributed similarly large numbers of eggs and fry (Bell 1940, as reviewed in Ross 1991). By 1940 over 98 percent of all eggs and 75 percent of all fry stocked by federal hatcheries were marine species, mostly flounders and members of the cod family (McHugh 1970). Large-scale stocking to replenish depleted stocks in coastal waters ended by the middle of the 20th century because stock abundances had not improved. The reason for the program's failure is obvious. The expansive human effort required to stock several billion eggs and larvae annually was minuscule compared with the reproductive effort generated by even greatly depleted coastal fish stocks. The 5 billion cod, haddock, pollack, and flounder eggs produced by the Gloucester facility in 1938 represent about the same output as that of only several thousand adult females in the ocean, because a general average fecundity for females of these species is in excess of one million eggs. It is not surprising that such minor production of young fish, spread across several species, would have no detectable impact on abundances of fish populations inhabiting an area as vast as the Northwest Atlantic. Early supplemental stocking programs in the Great Lakes ended during this same time period due to a similar lack of noticeable results.

In retrospect, at least one fish husbandry activity of that coastal stocking program seems nearly absurd today. The U.S. Fish Commission's research steamer, the *Fish Hawk,* netted fishes specifically to strip and mix gametes and then to return larvae to the sea immediately after they had hatched on board the vessel (Royce 1989). The *Fish Hawk,* other than protecting eggs from natural mortality while on board the vessel (although excessive handling could have increased both egg and adult mortality), was only doing what the fishes could have accomplished quite well on their own.

Today most agency hatchery production focuses on freshwater and anadromous species. In 1980, 95 percent of the weight of fishes reared at federal hatchery facilities were salmonids and 54 percent of the number of fishes reared were warmwater, freshwater species (Royce 1984). In spite of this shift in emphasis, stocking remains a major activity of fisheries agencies (Keith 1986). The rest of this chapter reviews the role that hatcheries and stocking play in the following activities:

1. Introducing new species into river and lake basins
2. Establishing fisheries in newly created habitats such as small ponds and reservoirs
3. Supplementing populations

4. Restoring fish stocks to basins from which they had been eliminated or severely reduced in abundance

5. Creating and maintaining fisheries supported by nonreproductive stocks of fish

INTRODUCTION OF NONNATIVE SPECIES

Species were being introduced into waters outside their natural geographic range before the first fisheries agencies were created in the United States. Exotic species such as the goldfish and common carp were established in U.S. waters by the late 1600s and 1831, respectively (Courtenay et al. 1984). Introduction stocking and natural dispersal have led to widespread distribution of some exotics; common carp occur in all continental states of the United States other than Alaska, whereas the goldfish and brown trout occur in 42 and 40 states, respectively (Courtenay et al. 1984).

Fauna of many U.S. waterways have been markedly altered by introduction of exotic or North American species that are not indigenous to waterways being stocked. Sixty-seven species of fishes have been introduced into the Colorado River system (Rolston 1991). At least 87 species have been accidentally introduced into the Great Lakes since 1900 (Marsden 1993); additional species have been purposefully stocked by fisheries agencies. Twenty-one and 19 species of exotic fishes are established in Florida and California, respectively (Courtenay et al. 1984). About half of the resident freshwater fish species of Massachusetts are introduced, including most species that support recreational fisheries (Hartel 1992). Although some species have been introduced by commercial aquaculture operations, by aquarists, or as bait fishes, many have been released by state and federal fisheries agencies attempting to create new fisheries. Table 8.2 lists some of the fish species native to North America that have become established well outside their native ranges largely due to the stocking activities of fisheries agencies.

Many have considered the introductions of species such as rainbow trout, brown trout, largemouth bass, smallmouth bass, striped bass, and walleye into new geographic areas to be exceptionally successful. However, introduced species often have created far-reaching problems that may negate benefits gained by their stocking. Introduced species have caused major declines in native fish fauna throughout North America (Minckley and Douglas 1991). Some scientists consider the introduction of nonnative fishes to be one of the great threats to the very fragile native fish faunas of the arid West (Minckley et al. 1991).

Declines in Native Fisheries Caused by Species Introductions

Native populations can be displaced from formerly occupied habitats because of competition with or predation by introduced taxa or because of the modification of habitats caused by the introduced fishes (Krueger and May 1991).

The common carp has proven to be a nuisance introduction in U.S. waters (Radonski et al. 1984). While foraging, the carp can uproot aquatic vegetation, and can cause marked increases in turbidity due to suspension of bottom substrates

TABLE 8.2   A listing of some of the native North American fish species that are supporting fisheries widely outside of their native ranges (Scott and Crossman 1973).

Species	Region of Introduction
American shad	West coast watersheds
Pink, chinook, kokanee salmon	Great Lakes
Coho salmon	Great Lakes, New Hampshire
Rainbow trout	Throughout northeastern and north central United States, southeastern United States in Appalachian Mountains, western watersheds outside of its native range
Brook trout	Western United States
Lake trout	Western United States
Rainbow smelt	Great Lakes, inland waters of northeastern United States
Northern pike	New England, western United States
Channel catfish	East of Appalachian Mountains from New England to Florida, western states
Striped bass	Pacific coast watersheds Oregon to California, in inland reservoirs of more than 30 states
Largemouth and small-mouth bass, rock bass	Eastern coastal plain, western United States
Pumpkinseed	Western United States
Bluegill	Northeast, east of Appalachian Mountains, and western United States
White crappie	West of the Rocky Mountains and New England
Black crappie	Widely in western United States
Yellow perch	Throughout western and southern United States
Walleye	East of Appalachian Mountains and widely in western states

(Taylor et al. 1984). Such increases in turbidity have been implicated in reduced growth rates of fish species in systems colonized by carp. Declines or displacements of native fish assemblages have occurred frequently after carp have been introduced into a fish community (Taylor et al. 1984). The common carp, widely hailed as a new, exciting game and food fish in the late 19th century, is now nearly universally considered a major nuisance in the United States, with very limited value as a fishery resource.

Predation by introduced brown trout has contributed to the decline of native fishes, particularly members of the salmon family (*Salmonidae*) such as the brook trout, golden trout, cutthroat trout, Gila trout, and Dolly Varden (Taylor et al. 1984; Krueger and May 1991). Competition with introduced brown trout has led to displacement of brook trout from prime resting territories in streams in which such habitat conditions are scarce (Fausch and White 1981). Such territories provide critical refuge from fast-flowing currents. The introduction of brown trout has led to the disappearance of brook trout from many stream systems in the eastern United States (Waters 1983; Krueger and May 1991). Brook trout have similarly been replaced by introduced rainbow trout in the southern Appalachian Mountains and by coho salmon in tributaries of the Great Lakes (Krueger and May 1991). In the western United States, cutthroat trout and golden trout populations have been replaced by introduced brook trout, brown trout, rainbow trout, and lake trout. Introduced rainbow trout have replaced or caused significant population declines in native populations of Gila trout and Apache trout in western watersheds (Echelle 1991; Rinne and Turner 1991).

Genetic Problems Caused by Introduced Species

Introduced species can also disrupt the genetic makeup of populations of native species (gene pools). Introgressive hybridization may occur between introduced and native species if the two taxa are closely related. Introgressive hybridization results from the mating of individuals of two species; if fertile, resulting hybrid offspring then may mate with individuals from either of the two species that produced the hybrids. If introgressive hybridization occurs, an influx of genetic material from the introduced taxon into the gene pool of the native population can disrupt the gene sequences that allow the native species to be well adapted to the specific conditions of its environmental niche (Krueger and May 1991).

Secondly, a major decline in abundance of a native population caused by competition with an introduced taxon can reduce the number of alleles, or different forms of the same gene, in the gene pool of the native species. Genetic variability within gene pools that results from multiple genes having two or more alleles present, allows a population to be well adapted to a changing or highly variable environment (Primack 1993). Without genetic variability, populations may not be able to adjust to changing environmental conditions (Krueger et al. 1981). Thus gene pools with reduced allelic diversity can be less adaptive than those with greater diversity (Krueger and May 1991). Very small populations tend to have fewer alleles within their gene pools than do large ones. Because their gene pools are less adaptive, small populations have a greater tendency to become extinct than do large ones (Primack 1993).

Introgressive hybridization has occurred frequently between native and introduced trout species. Allendorf and Leary (1988) believe that introgressive hybridization, particularly with the rainbow trout, is so extensive that cutthroat trout populations are now threatened throughout much of their native range. Gene pools of native golden trout and Apache trout have also been disrupted by introgressive hybridization with the rainbow trout (Krueger and May 1991; Carmichael et al. 1993). Rainbow trout introgression was found in 19 of 31 populations of the Apache trout studied by Carmichael et al. (1993); hybrids were more abundant than Apache trout in 14 of the 19 populations that displayed some level of introgressive hybridization. The introduction of smallmouth bass similarly threatens the Guadelupe bass in the Edwards plateau of Texas (Echelle 1991). Gene pools of stocks or subspecies of largemouth bass, cutthroat trout, and rainbow trout have been modified by reproduction with introduced strains of the same species (Pelzman 1980; Gilliland and Whitaker 1989; Krueger and May 1991). The stocking of hatchery rainbow trout into rivers of the Olympic Peninsula, Washington, since the 1940s has led to the loss of genetic diversity among the gene pools of native rainbow trout stocks (Reisenbichler and Phelps 1989). Introduction of Florida largemouth bass genes into gene pools of the northern largemouth bass subspecies has caused a reduction in growth rates and an increase in mortality rates in populations in Illinois and Tennessee (Smith and Wilson 1982; Isley et al. 1987; Philipp and Whitt 1991), although introgressive hybridization between these two forms produced higher growth rates, larger sizes, and lower mortality of individuals in northern largemouth bass populations in other southern states and California (Pelzman 1980; Wright and Wigtil 1982; Maceina et al. 1988).

Growing Caution Concerning Species Introductions

Introductions have created new fisheries, but they also have created many problems, replacing native resources and causing changes in the makeup of fish communities and even conditions of aquatic systems. Li and Moyle (1993) bluntly state that "introductions [of hatchery fishes] can serve as the perfect alibi for postponing the practice of fisheries management." These authors believe that the use of native fishes, more effective water management, and habitat protection are among the fisheries management alternatives that should be carefully considered when "an introduction is suggested as a solution to a management problem."

The rather long era of expansive introductions without apparent concern over potential consequences is largely over. Aware of the large volume of data that strongly suggests caution, most fisheries agencies have established a protocol for extensively reviewing potential impacts of introductions. Such a process requires that "what might be lost" rather than "what might be gained" be the criterion used to approve or prevent the stocking of nonnative fishes. In recent years, many new introductions into state-managed waters have been made by so-called bucket biologists, individuals in the general public who introduce species into basins due to careless release of bait at the end of an angling trip, or who release fishes purposely without understanding the possible consequences of such introductions. States are actively working to curb such nonagency introductions through legislation and creation of strict regulations.

ESTABLISHING SELF-SUSTAINING FISH POPULATIONS IN NEW RESERVOIRS AND PONDS

The number of reservoirs and ponds increased markedly in the United States during the 20th century in response to needs to produce electricity, control flooding, irrigate farmlands, and create municipal, local, and regional drinking water supplies. When developed on public property, such newly created systems have offered fisheries agencies the opportunity to establish fish species that offer high-quality recreational fishing opportunities. Research information developed for these systems also has influenced the types of fisheries established in ponds constructed by landowners on private property.

Stocking in Large-Volume Reservoirs

In 1900 there were about 100 reservoirs with a surface area greater than 500 acres in the United States (Noble 1980). The number of large reservoirs in the United States grew slowly through the 1950s and then increased rapidly in the 1960s and 1970s (Jantzen 1986). By 1980 there were more than 1,600 large reservoirs in the United States (Summerfelt 1986), many in parts of the country where few large natural lakes occur (Noble 1980). The rate of construction of new systems declined markedly after 1980 (Jantzen 1986). The general distribution of these systems is illustrated in Figure 8.1. Large reservoirs contribute substantially to freshwater recreational fishing opportunities in the United States; about 40 percent of all freshwater recreational fishing occurred on large reservoirs in 1980 (Summerfelt 1986).

The quality of fisheries often declines relatively quickly after they are established in reservoirs. Newly flooded reservoirs often provide excellent growth conditions for stocked fishes, creating very productive fisheries characterized by rapidly increasing yearly harvests. However, within a short period of time, often just a matter of years, catch rates of predator species such as the largemouth bass decline. Subsequent to the decline in abundance of the system's predator species, populations of forage species such as the bluegill reach very high densities, which leads to marked slowing or near cessation of body growth (referred to as stunting). Ultimately these forage populations produce few individuals that are large enough to be of interest to anglers.

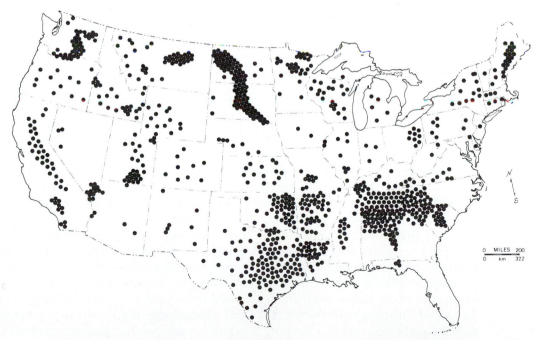

Figure 8.1 The distribution of reservoirs in the United States with greater than 500 acres of surface area, 1970. Each dot equals 10,000 acres. (From R. M. Jenkins. 1970. Reservoir fish management. Pages 173–182 in N. G. Nelson, (editor) A century of fisheries in North America. Reprinted by permission of The American Fisheries Society.)

Several factors may influence the rapid increase and subsequent decline in the quality of reservoir fisheries, referred to as the boom-and-bust cycle, including the following:

1. *Productivity cycles in reservoirs* (see Chapter 2). The general decline in productivity of reservoirs (Kimmel and Groeger 1986) results in a decline in the productivity of fish populations.
2. *Decline in abundance of predators such as largemouth bass due to overharvest.* Overharvest can leave so few large predators that insufficient predation pressure is imposed on forage species such as the bluegill. Inadequate predation pressure allows an increase in the abundance of forage populations, leading to stunting.

Fisheries managers have found that enforcing strict minimum length limits (see Chapter 7), manipulating water levels (see Chapter 10), and following specific stocking strategies can slow the decline in the quality of recreational fisheries in reservoir systems.

Reservoir stocking strategies vary widely among states, depending on specific habitat conditions within the systems, angler preferences for particular species, and other factors. When possible, most states stock predator species as soon as possible after the reservoir fills with water (Keith 1986). Because large reservoirs are created by flooding river basins, such stocking helps to prevent nongame fish species living in the river basin from expanding their populations within the reservoir and dominating the fish community that develops. Often several predator species are stocked into a reservoir to create more complex predator–prey interactions and to increase the total predation pressure over that produced by only one predator species (Keith 1986). The resulting predation pressure helps prevent rapid overpopulation and stunting of forage species. New predator species have been stocked into existing reservoirs after forage species reach high densities and fisheries supported by predator populations have already declined. Striped bass and striped bass–white bass hybrids have been widely stocked into reservoirs that are already overcrowded with pelagic forage species such as threadfin or gizzard shad (as in Lake Texoma, Texas–Oklahoma; described in Harper and Namminga 1986). However, Bailey (1975) felt that most introductions of striped bass or striped bass hybrids, although creating new sport and trophy fisheries, did not substantially reduce the abundance of shad species in southeastern U.S. reservoirs.

Self-sustaining populations of species including the largemouth and smallmouth bass, crappie and sunfish species, yellow perch, catfish species, and other taxa have been established in reservoirs widely after introduction stocking. Fisheries focusing on other species such as the striped bass, walleye, northern pike, muskellunge, and trout species often are maintained through yearly stocking because of the lack of sufficient reproduction. Pike and muskellunge require dense patches of rooted aquatic vegetation or flooding terrestrial vegetation for spawning and nursery areas; walleye and trouts require clean gravel and boulder spawning substrates that are swept by currents; and striped bass lay semibuoyant eggs that typically must be suspended in flowing water before hatching. Such conditions are not characteristic of many reservoir systems.

Stocking in Small Ponds

Although small mill ponds, several acres in surface area or smaller, were common in some parts of the United States by the mid-1800s, the number of ponds in the United States increased rapidly during the mid-20th century as the U.S. Soil Conservation Service established an expansive pond construction program on both publicly and privately owned land (Swingle 1970). Between 1934 and 1965 the number of small ponds in the United States grew from 20,000 to well over 2 million. Figure 8.2 illustrates the distribution of ponds constructed during this period. In addition to construction assistance provided to private landowners by the Soil Conservation Service, the U.S. Fish and Wildlife Service and state agencies stocked many of these new ponds, particularly in the South and areas of the West where small ponds contributed significantly to total recreational fishing activities (Swingle 1970).

Fisheries in small ponds must be managed differently from those in large-volume reservoirs. In order to direct available food resources as fully as possible into fish species supporting recreational fisheries, agencies typically recommend stocking only one to several species of fishes into small ponds. Most state fisheries agencies favor establishing largemouth bass in combination with one to several forage species, particularly sunfishes such as the bluegill, which both serve as food for bass and support panfish fisheries (Modde 1980). A panfish species is desirable because it is easily caught in large numbers and is considered flavorful. However,

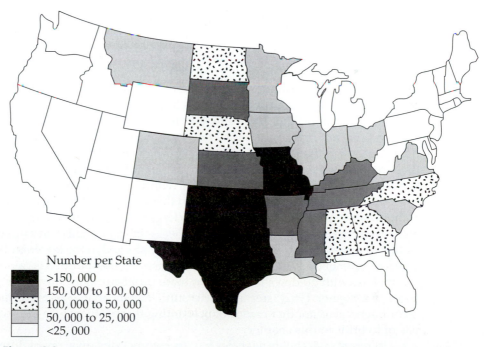

Number per State

- >150,000
- 150,000 to 100,000
- 100,000 to 50,000
- 50,000 to 25,000
- <25,000

Figure 8.2   The distribution of ponds constructed in the United States by the U.S. Soil and Conservation Service. (From T. Modde. 1980. State stocking policies for small warm water impoundments. *Fisheries* 5(5): 13–17. Reprinted by permission of The American Fisheries Society.)

panfish species are typically small bodied. Therefore individuals must reach a relatively large size for that particular species before being big enough to be of interest to anglers.

Due in part to the lack of complex predator–prey interactions, the quality of fisheries can rapidly decline after initial stocking. Within as short a time as 2 to several years, insufficient levels of predation by largemouth bass allows bluegill populations to become extremely abundant. As their abundance increases, growth rates decline until the bluegills exhibit such slow growth that most individuals never reach a size large enough to be kept by anglers. Once bluegills reach high densities, they can disrupt largemouth bass reproduction, causing a decline in the abundance of largemouth bass and a reduction in quality of the bass fishery (Swingle 1970).

The quality of fisheries in ponds is strongly influenced by the level of balance exhibited by the fish community. Predator and forage species are able to reproduce adequately, grow at acceptable rates, and yield satisfactory harvest in balanced fish communities (Swingle 1970). Unbalanced communities characterized by highly abundant, stunted forage and depleted predator populations produce much lower harvest levels. Balance is maintained most effectively when predators such as the largemouth bass combine with harvest to slow the rate of increase in abundance of bluegill populations. Restricting the harvest of predators is widely supported as a means of maintaining high predator density and thus high levels of predation on the forage species. Stocking of predators has also been used as a means of maintaining balance for as long a time period as possible.

Annual growth rates of predator species represent the most important factor influencing the stocking strategies that are used to prolong balance (Modde 1980). Early research suggested that the optimal density and age of fishes to be stocked varied according to geographic region of the country. In southeastern states, bass fingerlings stocked simultaneously with greater numbers of bluegill fingerlings produce prolonged fisheries. In this region, growth of largemouth bass is rapid enough to produce high predation rates and reproduction the year after stocking fingerlings (Swingle and Smith 1947). In more northern areas, bass stocked as fingerlings may not reproduce until 2 or 3 years of age, by which time bluegill abundance is so high that survival of bass offspring is highly limited (Bennett 1970). In these regions, either relatively greater densities of bass fingerlings or larger sizes of bass capable of reaching sexual maturity within one year are stocked.

By the early 1960s research indicated that universal stocking ratios and species combinations were generally ineffective (Regier 1962). By this time, redear sunfish were preferred over bluegill as a forage species in some prairie and upper midwest states, whereas channel catfish were frequently stocked in the lower midwest. Some states recommended stocking largemouth bass or smallmouth bass alone, or largemouth bass with golden shiners to eliminate problems associated with stunting of sunfishes (Regier 1962); sterile hybrid sunfishes incapable of reproduction were recommended as another means of preventing this problem (see the section on the use of hybrids in this chapter).

In recent years, state agencies have recommended an array of different stocking ratios and species combinations (Table 8.3). Most states recommend stocking the same general numbers of bass fingerlings, whereas recommendations for stocking

TABLE 8.3 Stocking recommendations (numbers per hectare) made by state fisheries agencies for small warmwater ponds (Modde 1980).

State	Largemouth Bass	Bluegill	Redear Sunfish	Golden Shiners	Channel Catfish	Bass – Sunfish Stocking Chronology
AL	247[g]				1,236	
	247[f]	2,100–3,151	371–556			Split
AZ[e]	124[g]	1,236				Split
	247[f]	2,471				Split
AR	124[g]	494			124	Simultaneous
	247[f]	988			247	Simultaneous
CA[e]	247–371	2,471–3,707				Split
CO		No recommendation				
CT	[a]					
DE	247	2,471				Split
FL	124[g]	618[c]				Split
	247[f]	1,236[c]				Split
GA	124[g]	988	247		124	Split
	247[f]	1,977	494		247	Split
ID	[a]	[a]				Simultaneous
IL	124[g]	865	371		124	Split
	247[f]	1,730	741		247	Split
IN	247–494	2,471			247	Simultaneous
IA	247	2,471			247	Split
KS	247	1,236			247	Split
KY	297	988			124	Split
LA	124[g]	988	247			Split
	247[f]	1,977	494			Split
ME		No recommendation				
MD	247	2,471			247	Split
MA		No recommendation				
MI	247	1,236				Split[d]
	247			1,236		
					247	
	247[b]					
		(MI also recorded rate of 1,236 fathead minnows)				
MN	594	741				Simultaneous
		741			494	
MS	124[g]	988	247		247	Split
	247[f]	1,977	494		247	Split
MO	247	1,236			247	Split
MT	594–1,198					
NE	247	1,236			247	Simultaneous
NV		No recommendation				
NH		No recommendation				

(Continued)

Tᴀʙʟᴇ **8.3** (continued)

State	Largemouth Bass	Bluegill	Redear Sunfish	Golden Shiners	Channel Catfish	Bass – Sunfish Stocking Chronology
NJ	247	2,471				Simultaneous
NM			No recommendation			
NY	247	2,471				Simultaneous
	247					
				2,471		
NC	247[f]	1,730	741			Split
ND			No recommendation			
OH	247	1,236[c]				Simultaneous
OK	247	247			247	Split
OR	185–371	185–309			124–185	Split[d]
	185–247				124–185	
	185–247				124–185	
		(OR also recorded rate of 185–247 white crappies and 309–618 black crappies)				
PA	247[b]	2,471				Simultaneous
	247			341–988	1,236	
	247				1,236	
		(PA also recorded rate of 2,471 fathead minnows)				
RI	124					
SC	124[g]	927	309			Split
	247[f]	1,853	618			Split
SD	247	741			247	Simultaneous
TN	185[g]	1,236			124	
	247[f]	1,853				Split
TX	247				a	Split
		(TX also recorded rate of 247 hybrid sunfish)				
UT			No recommendation			
VT			No recommendation			
VA	124[g]	1,236			124	Split & simultaneous
WA			No recommendation			
WV	247	247				Simultaneous
WI	247					
	247			988		
					247	
WY	124–247					

[a] Rate not specified.
[b] Adults recommended the following spring.
[c] Combination of bluegill and redear sunfish.
[d] Bass stocked prior to bluegill introduction.
[e] SCS recommendations.
[f] Rates for fertilized ponds.
[g] Rates for unfertilized ponds.

of other species vary widely (Modde 1980). Many states recommend a split stocking schedule for predator and forage species. The most common split involves stocking fingerling sunfishes in the fall and fingerling bass the following spring. This schedule is common in southern and central regions of the country where largemouth bass will normally spawn the year after stocking. In northern states where bass do not mature until the third summer of life, split stocking is not recommended (Modde 1980). Most southeastern states follow the general recommendations originally developed by Swingle and associates in Alabama; however, in no other geographic region do states exhibit similar continuity in stocking recommendations.

Generally, two major factors appear to influence recommendations offered by state agencies: (1) regional characteristics of growth and production and (2) state preferences for optimum harvests of largemouth bass, or optimum total harvests of largemouth bass in combination with forage species. The bass–bluegill stocking ratios recommended by Swingle for southeastern states (100 bass fingerlings to 1,000 bluegill fingerlings per acre) were designed to provide optimum bass biomass in that region. In more northern states, the maximum biomass of harvestable-sized bass is greater in ponds stocked with only largemouth bass or largemouth bass in combination with golden shiners than in ponds stocked with bass and bluegills (Modde 1980). Emphasis in this region is on maximum production of a sport species, the largemouth bass, rather than on providing both sport and panfish fishing opportunities.

Stocking Newly Created Waters as a Management Strategy

Stocking newly constructed reservoirs and ponds probably represents one of the more consistently productive uses of hatchery fishes. Native fish communities are not disrupted by this activity in systems being fed by surface runoff or groundwater percolation because no aquatic system existed previous to construction of the pond. Similarly, many species of riverine fishes are not well adapted to survive in large reservoirs constructed in river basins. Thus any fish community that would develop in the absence of stocking would be very different from the native community present before the reservoir was filled. Although stocking in artificial systems creates few on-site risks to wild resources, the escape of introduced species from reservoirs represents a potential risk to fish communities throughout river basins. As with introduction stocking in natural systems, agencies need to carefully consider the possible impacts of stocked fishes on resident fish communities if they become established in waters outside the reservoirs in which they were stocked.

SUPPLEMENTING POPULATIONS

Although early attempts to replenish depleted stocks of eastern coastal and Great Lakes fishes with hatchery-reared fishes proved fruitless, supplemental stocking continued to be emphasized in many fisheries. Supplemental stocking has been used extensively in freshwater and anadromous fisheries and, to a lesser degree, in coastal fisheries, to address issues of overharvest of wild populations. Supplemental stocking has become an integral part of management of numerous Pacific salmonid populations that are depleted due to the combined effects of habitat deterioration, overharvest, and loss of migratory pathways owing to the construction of dams.

Supplemental stocking programs have stimulated some fisheries associated with depleted stocks. Wild steelhead trout populations migrating from the ocean into Idaho rivers have become so severely depleted that supplementally stocked smolts provide most of the statewide harvest (Thurow 1984). About 50 percent of all steelhead landed in coastal rivers of Washington in the 1970s were hatchery fish; hatchery steelhead accounted for up to 70 to 80 percent of the total catch in specific river systems (Barnhart 1975). Supplemental stocking of the coastal red drum has improved the abundance and harvest of this species in certain Texas estuaries. Up to 20 percent of the juveniles netted by research surveys in estuaries were of hatchery origin. At least 20 percent of the recreational harvest consists of hatchery fishes stocked as fingerlings (SFI 1992). Supplemental stocking of walleye has been successful in reservoirs where natural reproduction is poor (Paragamian and Kingery 1992; LaJeone et al. 1992; Mitzner 1992).

Although many supplemental stocking programs are initiated to improve the abundance of depleted stocks, stocking may continue after recovery has been accomplished. Hatchery-produced pink salmon provided about 30 percent of the total harvest in Alaska from 1987 to 1990, even though on a statewide basis pink salmon stocks were "rebuilt" to or beyond previous high levels (Loh-Lee Low 1991).

Failures to Enhance Fish Populations and Fisheries

Supplemental stocking often has failed to increase harvest levels or to enhance the condition of a wild fish population receiving the stocked fishes. Figure 8.3 illustrates the decline in numbers of migrating adult chinook salmon in the Snake River that occurred during a time period when the numbers of hatchery-reared juveniles released into the river basin increased significantly. An extensive review indicated that only about 5 percent of walleye supplemental stocking programs over the last 100 years have enhanced walleye fisheries and reproduction (Laarman 1978, as reviewed in Koppelman et al. 1992). In many instances, a lack of appropriate planning or understanding of habitat requirements of early life stages of fishes led to failure; Laarman found that density, life stage used (fingerling versus fry), and water temperature and availability of natural foods at the time of stocking all influenced the success or failure of supplemental stocking of the walleye.

Ultimate contributions of stocked fishes to a fishery and to the reproductive activity of a population may not be reflected accurately by densities of hatchery fingerlings shortly after stocking. First-year mortality of stocked walleye fingerlings was 2 to 16 times greater than that of naturally produced fingerlings in an Iowa study (McWilliams and Larscheid 1992). Initial survival of hatchery largemouth bass can be high, but long-term survival and recruitment into the fishery may be negligible. Supplemental stocking contributed substantially to largemouth bass year classes a year after stocking in Colorado (Krieger and Puttman 1986), but densities of stocked largemouth declined throughout the summer after stocking in

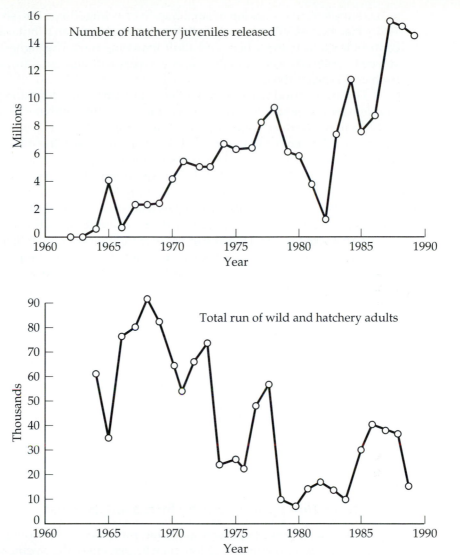

Figure 8.3 The numbers of hatchery-reared chinook salmon released as juveniles into the Snake River, and of adult wild and hatchery chinooks that migrate back into the river system to spawn. (From Ray. J. White, "Why Wild Fish Matter: Balancing Ecological and Aquacultural Fishery Management," TROUT, Autumn 1992. Reprinted by permission.)

Arkansas (Filipek and Gibson 1986). Boxrucker (1986) found that catch and harvest rates in largemouth bass fisheries did not increase with supplemental stocking even though stocked fingerlings accounted for 70 percent of a year class as long as a year after stocking; few of the hatchery fishes ever reached legal size. Supplemental stocking of largemouth bass has generally been unsuccessful (Keith 1986), as has supplemental stocking of smallmouth bass fingerlings into streams (Bennett 1970). Hatchery steelhead trout stocked into Lake Michigan tributaries as yearlings contributed little to the fishery and adult spawning runs. The highest proportion of stocked steelheads were usually found in rivers with marginal trout habitat (Seelbach and Whelan 1988).

During a period when over 3 million unfed fry and 14 million presmolt coho salmon were stocked into Oregon's coastal rivers, the combined densities of wild and hatchery juveniles were 41 percent higher in the stocked rivers than the densities of wild juveniles in the unstocked rivers; however, the density of juveniles that were the result of natural reproduction was 31 percent lower in stocked than unstocked rivers (Nickelson et al. 1986). Apparently, hatchery fishes tended to replace some of the natural reproduction that might have occurred in the absence of supplemental stocking. After these fishes reached adulthood, there were no statistically significant differences between densities of adults migrating back into stocked versus unstocked rivers. Thus the increase in numbers of young fishes noted at the time of migration to the sea did not lead to higher numbers of reproducing adults. Indeed, the release of hatchery presmolts has been associated with a decline in the abundance of native adult fishes returning to rivers to spawn after completing their juvenile life phase at sea (Brown et al. 1994). Survival of hatchery-reared anadromous salmonids may be as low as 1 percent or less in the Columbia River. However, survival has been much higher elsewhere. Between 8 and 20 percent of hatchery smolts return as adults to the Puget Sound; over one-third of all adult spawners are hatchery fishes (Daley 1993).

Decreased survivability of hatchery fishes may be influenced both by the hatchery and stocking experience and by genetics of the fishes themselves. Naturally produced offspring of hatchery steelheads in the Columbia River drainage exhibited lower survival rates than did offspring of wild parents. Reduced offspring survivability occurred even when hatchery fishes interbred with wild individuals, strongly indicating that hatchery fishes were not genetically adapted to the conditions characteristic of the Columbia River (Peven and Hays 1989).

Disrupting Wild Populations by Increasing Harvest

Supplemental stocking can increase fishing pressure and in doing so disrupt the stability of the wild population. More than 97 percent of the steelhead trout caught by anglers in the Alsea River, Oregon, in 1977 were hatchery-produced fish (Schreck 1980). Schreck felt that the additional attention and fishing pressure produced by the presence of stocked fish caused greater mortality, and thus lower reproductive effort, of wild fish in the same river. Annual harvest rates of coho salmon in Canadian waters of Puget Sound may reach 95 percent of available fishes. As hatchery production has increased to support this high level of fishing pressure, wild stocks have declined due to the high harvest rates (Hilborn 1992). Hatchery production can be increased to provide for higher fishing intensities and harvest levels, whereas natural production cannot. A similar decline in wild populations of winter-run and summer-run steelhead trout in the Puget Sound basin also may have been caused by harvest levels that have increased in response to the abundance of hatchery-reared fishes (Daley 1993).

Genetic Problems Caused by Supplemental Stocking

Stocked fishes may contribute substantially to the reproductive effort of a wild population; in western fisheries, up to 80 percent of all spawning adults of specific local populations of coho salmon and rainbow trout have been identified as hatchery fishes (Hindar et al. 1991). If stocked fish participate in the reproductive effort of a wild stock, they may create problems with the fish population's viability, not contribute to a greater reproductive success. The viability of a wild population may be reduced if its gene pool is substantially infiltrated by genes originating from less well-adapted hatchery fishes (Schreck 1980). Long-term stocking of brook trout streams in Wisconsin apparently caused the genetic makeup of wild brook trout populations to become significantly altered (Vincent 1984). Similar changes have occurred in gene pools of chinook and coho salmon and rainbow and brown trout (as reviewed in Hindar et al. 1991). Chinook salmon, steelhead trout, and Atlantic salmon produced by reproduction between wild and hatchery fishes have displayed lower return rates to natal streams as migrating adults than those typical for native strains (Hulett and Leider 1990; Hindar et al. 1991; Hilborn 1992). These authors concluded that interbreeding of hatchery fishes with wild strains of fishes is generally harmful to the adaptiveness of the native gene pools of those basins.

The Northwest Power Planning Council (established by the Pacific Northwest Electric Power Planning and Conservation Act of 1980; see Chapter 6) has established a stocking strategy for the Yakima River of the Columbia River basin that is designed to supplement wild anadromous salmonid stocks without disrupting the adaptiveness of their gene pools (Clune and Dauble 1991). To maintain the genetic integrity of wild populations, all spawners used for hatchery propagation will be naturally produced wild fishes. All offspring released as juveniles will be marked; in subsequent years, only unmarked (and thus naturally produced) adults will be captured for hatchery spawning. In this way, supplementation should enhance harvest and natural reproduction, but all hatchery fishes that are stocked will be the direct offspring of wild fishes (a schematic of this process is illustrated in Figure 8.4). All hatchery fishes will be stocked specifically in areas of the river from which their parents were collected. This should minimize the potential for introducing new genes into a gene pool. In addition, spawners will be collected randomly over the period of migration to limit any loss of genetic variability that occurs due to inbreeding depression, which is the gradual loss of alleles from persistent inbreeding. For example, if all fishes used for hatchery spawning were captured only during the very beginning or end of a migration, and the time of migration were genetically determined, all hatchery-reared fishes that were stocked would be identical for that trait.

Thus this program is designed to minimize the potential loss of genetic diversity within or the addition of genetic traits to the gene pools of the wild populations being supplemented.

Cause for Caution Regarding Supplemental Stocking

Ample evidence indicates that supplemental stocking does not enhance the self-sustaining capacity of wild populations. Factors influencing failure have included the following:

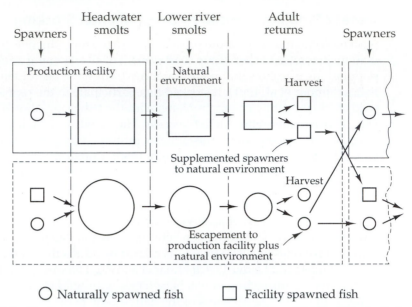

Figure 8.4   Hatchery spawning and stocking strategy of the Yakima/Klickikat Fisheries Project. This strategy requires that all hatchery spawners are wild fishes and all hatchery-reared fishes stocked into wild populations are the offspring of wild fishes from those populations. (From T. Clune and D. Dauble. 1991. The Yakima/Klickikat fisheries project: a strategy for supplementation of anadromous salmonids. *Fisheries* 16(5): 28–34. Reprinted by permission of The American Fisheries Society.)

1. The use of fish ill-adapted to contribute to a wild population's abundance and reproductive success due to genetic factors inbred into the hatchery strains or the hatchery and stocking experience itself

2. A lack of understanding of the capacity of an ecosystem to absorb additional individuals over the numbers of offspring produced naturally by the wild population

3. The inability of fisheries managers to carry out strategies that facilitate stock recovery (as described by Kutkuhn 1980 for the Great Lakes lake trout restoration program through the 1970s)

Even when initially successful, hatchery programs may not effectively sustain fisheries for extended periods of time. Survival rates of hatchery fishes tend to decline through time in long-term stocking programs. Although stocking of chinook salmon in British Columbia increased nearly tenfold from the mid-1970s to the 1980s, the catch of hatchery-reared chinook declined (Hilborn 1992). Figure 8.5 demonstrates the increase in the number of hatchery-reared coho salmon smolts released into the Oregon Production area from 1960 to 1981 and the concurrent decline in the percentage of these smolts that survived to return from the sea as adults. Hatchery smolts greatly outnumbered wild smolts from the mid-1960s onward, so the decline in survival of hatchery smolts correlated closely with a decline in sur-

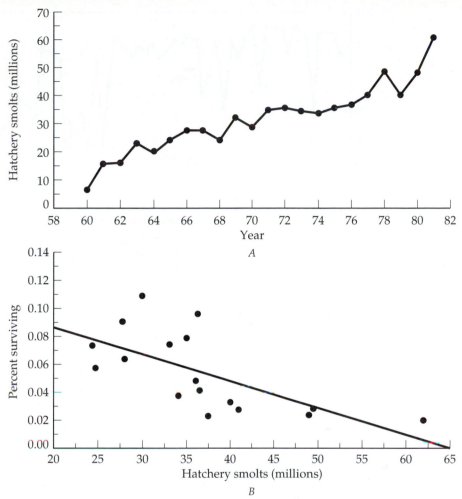

Figure 8.5   *A:* The increase in the number of hatchery smolts stocked into the Oregon Production Area from 1960 to 1981. *B:* The correlation between the number of hatchery smolts stocked and the percentage of hatchery smolts that survive to the adult life stage (regression $r^2 = .43$; $p < 0.01$). Both the numbers stocked and the percentage that survived were relatively low from 1960 to 1964. Figure 8.5B illustrates years after 1964, during which more than 20 million smolts were stocked every year. (From I. E. Nickelson, Influences of upwelling, ocean temperature, and smolt abundance on marine survival of coho salmon (Oncorhynchus kisutch) in the Oregon production area. Can. J. fish. Aquatic Sci. 43:527–535. Reprinted by permission.)

vival of all smolts (Nickelson 1986). Such declines in survival are found "in nearly every [long-term supplemental stocking] hatchery program in North America" (Hilborn 1992).

Supplemental stocking is an intuitively appealing approach to addressing issues of stock depletion. However, such efforts have proven to be more destructive than helpful to many wild fish stocks. Nearly a century of stocking fall-run chinook salmon into the Little White Salmon River contributed to extinction of the native stock because of the introduction of new genetic strains and a management

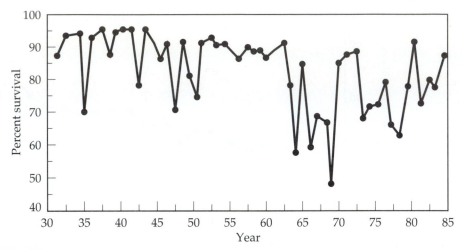

Figure 8.6   The percentage survival of fall chinook salmon stocked out of the Little White Salmon National Fish Hatchery, 1931 to 1984. After 1960 fishes were held longer in the hatchery before stocking in order to take advantage of high river flows. Survival was lower than 85 percent more frequently after 1960 than before. (Data from Nelson and Bodle 1990.)

decision to maintain fishes for a longer time in hatcheries before stocking; the latter action led to higher mortality of hatchery fishes both before and after stocking (Figure 8.6; Nelson and Bodle 1990). Hilborn (1992) believes that hatchery programs pose a substantial threat to the future viability of numerous Pacific salmon stocks. In the long run, supplemental stocking may create a management program in which hatchery-produced fishes become the major resource supporting a fishery rather than supplementing the abundance and viability of wild fish stocks. Daley (1993), while supporting the potential benefits of hatchery programs, also believes that research is needed to provide the information that is critical for making wise decisions about the manner in which hatchery fishes are used to supplement declining fisheries resources.

RESTORATION EFFORTS

Restoration programs involve establishing stable fish populations in waters where wild populations of the same species have been extirpated (driven to local extinction), or where wild populations have collapsed to the point that they are endangered or their fisheries have disappeared. Restoration programs frequently require expansive hatchery production as part of the management effort. When a population still exists, the fisheries manager has a gene pool available for stocking that is presumably adapted to the environment in which the restoration effort is being focused. However, when a population is extirpated, fisheries agencies must stock strains of fishes that are not native to the area. Under these circumstances, restoration programs have focused on one of two stocking strategies (Krueger et al. 1981; Marsden et al. 1989):

1.   Only those strains whose native environment most clearly matches the watershed into which the restoration effort is being focused are stocked.

2.   As many strains as possible are stocked to maximize the genetic diversity that is being introduced into the watershed.

The first method presumes that the strains used are potentially preadapted, and thus that they will have a higher probability of producing a self-sustaining population than would other strains. The latter presumes that a well-adapted strain will develop through differential survival and reproductive success of the various genotypes that are stocked (Kapuscinski and Jacobson 1987).

Genetic Problems Caused by Hatchery Practices

In many restoration programs, particularly those associated with endangered species, the number of adults available for breeding in hatcheries can be quite limited. Under such circumstances, inbreeding depression may occur unless captive breeding programs are carefully planned. Inbreeding depression occurs when so few fishes are used for captive breeding that offspring reared by hatcheries do not carry the diversity of alleles characteristic of the population being restored. Stocking large numbers of fishes that lack allelic diversity can reduce the genetic adaptiveness of the small wild population being restored.

Commonly employed hatchery practices can exacerbate problems of inbreeding depression. Historically, hatcheries often fertilized all of the eggs taken by females with sperm from only a few males. This practice evolved largely for convenience. Because a small number of males can produce enough sperm to fertilize the eggs taken from numerous females, fewer adult fishes have to be maintained and handled for captive breeding if all sperm used is taken from a small number of males. However, combining small numbers of spawners with such unequal sex ratios will increase the potential for inbreeding depression (Kapuscinski and Jacobson 1987). Researchers have recommended that 50 or even 100 fishes evenly divided between the sexes are necessary to minimize the potential reduction in genetic diversity (Kincaid 1983; Kapuscinski and Jacobson 1987).

Other Factors Influencing Restoration

Properly organized hatchery and stocking efforts in a restoration program may require careful consideration beyond developing optimal breeding strategies, such as determining the best size and age of fish to stock, the best time to stock, the best densities to stock, and the best habitats or sites in which to stock. Along with such considerations, restoration programs must also consider poststocking factors such as species interactions in the waters being stocked and appropriate levels of harvest of the target population. For example, control of the sea lamprey was felt to be a critical precursor to successful reestablishment of lake trout in the Great Lakes (Fetterholf 1984).

Efforts to restore self-sustaining lake trout populations were initiated in Lake Superior in the mid-1950s and in the other four Great Lakes between then and the early 1970s. All five lakes had active fisheries by 1980 but almost no natural reproduction; the restoration program had produced hatchery-based fisheries and had not progressed toward establishment of self-sustaining populations. Fetterholf (1984) noted that progress in the lake trout restoration in the Great Lakes was for a long time slowed because "everyone kept hoping that the few strains of lake trout [that were used], reared to a common life stage under similar hatchery regimes, and planted within a narrow range of densities, would somehow suddenly begin to reproduce." Various workers credited the lack of reproduction to (reviewed in Fetterholf 1984):

1. Inadequate numbers of spawners caused by overexploitation, sea lamprey predation, or too few stocked fish finding optimal spawning areas
2. A reduction in egg viability due to pollutants, diseases, or plant toxins
3. Predation on eggs and fry
4. Siltation of spawning habitats
5. Inappropriate stocking techniques, including using fishes of the wrong age or stocking them at the wrong places
6. Maladaptation of hatchery fishes, including an inability to migrate to sites with optimal spawning habitat or to produce viable eggs that show good survival through early life stages.

Kutkuhn (1980) concluded that harvest levels of preadult fish were too high to allow successful reproduction. During the 1970s stocked lake trout supported highly active recreational fisheries. Economically important and actively pursued recreational fisheries based solely on stocking were preventing progress toward the restoration program's goal of establishing self-sustaining fish stocks. By the early 1990s natural recruitment of young lake trout was minimal in all of the Great Lakes other than Superior (Savino et al. 1993).

FISHERIES SUPPORTED BY NONREPRODUCTIVE STOCKS

Certain fisheries are maintained by annual stocking of hatchery-reared fishes rather than by reproduction of wild fish populations. Two basic approaches are taken in maintaining such fisheries: Fishes are stocked at a "catchable" size and thus are immediately available for harvest (a put-and-take fishery), or they are stocked as fingerlings or fry with the expectation that they will survive and grow to a size suitable for harvest (a put-grow-and-take fishery). Put-and-take and put-grow-and-take fisheries are often established in waters unsuited for reproduction of the species being stocked or in waters where natural production cannot support the level of fishing pressure applied to the wild population. Thus these stocking programs focus more on meeting the recreational interests of humans than on the conservation of any natural resources.

Maintenance of fisheries based solely on hatchery production can require substantial funding, particularly in urbanized areas where wild fisheries resources are inadequate to support the level of angling pressure imposed on them. For example, in a densely populated state that has limited levels of natural trout production, the Massachusetts Division of Fisheries and Wildlife may expend up to 70 percent of its total fisheries budget on its hatchery program, most of which supports put–and–take trout fisheries.

Put-and-Take Fisheries

Put-and-take programs are used to create fisheries when a self-sustaining population cannot be established, and when survival of stocked fingerlings and fry is too low or uncertain to maintain a desirable fishery. A fish is considered catchable if it is large enough to satisfy anglers and to exceed minimum size-limit regulations for that species at the time of stocking.

Put-and-take channel catfishes, bullhead catfishes, and common carp are stocked in selected waters of several southern states and in selected urban fisheries (Jenkins 1970; Heidinger 1993). However, trout species are the most widely stocked taxa in put-and-take programs in the United States. Trout are used partly because of angler interest, but more so because they can be reared to a catchable size more quickly and less expensively than many other popular sport fishes. In the northern half of the United States, trout are capable of growing 12 months a year in hatcheries because they are coldwater species; many other species essentially cease growing during the colder months of the year. Fishes stocked when 10 to 12 inches long and one-third to one-half of a pound in weight are much more costly to rear than are individuals stocked as smaller fingerlings or fry. However, when the total cost of a hatchery program is divided by the number of fish harvested after stocking, the cost per fish harvested in put-and-take fisheries often is lower than if fry or fingerlings were used; a much higher percentage of catchables survive to be captured than do fry or fingerlings (Wiley et al. 1993). Generally, the higher the percentage of stocked fish that are harvested, the greater the number of satisfied anglers for any specific level of spending by an agency.

Thus, to be cost effective, a high percentage of the stocked catchables must be harvested by anglers. For example, anglers harvest from 46 to 98 percent of all rainbow trout stocked annually in a put-and-take fishery in the tailwaters of Bull Shoals and Norfolk Dams, Arkansas (Oliver 1984). Return rates vary according to characteristics of particular waters and the angling public. In California higher percentages of catchable trout were landed in lakes than streams; also, 75 percent of the total catch was landed within 11 days after stocking in lakes and within 6 days after stocking in streams (Everhart et al. 1975).

As time elapses after stocking, increasing numbers of fishes will die or will be lost to the fishery by migrating to inaccessible areas (in a fishery sense, the equivalent of dying, because the fish is no longer available to the angler, nor will it reproduce). Because each catchable trout is expensive to grow, those that are not harvested represent a significant cost. Thus put-and-take programs usually stock large numbers of fishes into accessible areas at predictable times of the fishing season to ensure that large numbers of anglers contact stocked fishes soon after stocking.

Management Objectives with Put-and-Take Fisheries

Agencies typically establish put-and-take programs to meet one or more of the following objectives (modified from Everhart et al. 1975):

1. Catchables can be provided in marginal habitats unsuited for long-term survival, growth, and reproduction of young stocked fishes. Many states stock catchable trout in the spring into waters that are too warm in the summer to support trout.

2. Fishing can be provided near large population centers. Many anglers are unable or unwilling to travel far to pristine watersheds that have wild fish populations. Waters near population centers often are not suited to natural production of desirable gamefishes, or natural production is insufficient to satisfy angler demand.

3. Fishing pressure on wild populations can be relieved by diverting anglers to a catchable fishery within the same water system.

4. The catch of stocked fish is spread more evenly among anglers, because stocked fishes are more easily caught than wild ones. Thus a greater number of anglers enjoy the experience of catching fishes.

Some of these objectives typically are not met. Also, wild fish populations residing in the waters supporting put-and-take fisheries can be harmed by such programs.

Problems of Put-and-Take Fisheries

Studies have shown that catch levels are not equally distributed in a catchable fishery; about 10 percent of all anglers catch more than 50 percent of all fishes harvested (Everhart et al. 1975). This distribution of catch is similar to that typical of fisheries supported by wild populations. Unequal distribution of catch in catchable fisheries is probably due to differing levels of angler skill, persistence in determining stocking schedules, and the number of hours and days individual anglers spend fishing.

For decades fisheries managers generally believed that stocking of catchable trout into rivers and lakes would not disrupt the wild populations occupying the same waters (Cooper 1952). Managers theorized that fishing pressure would be diverted toward the more easily caught hatchery fish, thus reducing mortality of wild trout. No disruptive interactions would occur between the two types of fishes because hatchery trout would not adapt well to natural habitats after stocking. In addition, rapid removal of hatchery fishes by anglers would minimize long-term interactions between hatchery and wild trout. Some recent studies have demonstrated substantial impacts when catchable trout are stocked into streams housing wild stocks and others have not.

Experimental work on the Madison River, Montana, indicated that catchables may have a profound effect on wild trout populations (Vincent 1987). The biomass and numbers of wild brown and rainbow trout increased significantly over a 4-year period after stocking of catchable rainbow trout was terminated in certain areas of the Madison River, whereas wild stocks of these species declined marked-ly over the same time period in other areas where catchable trout were stocked for the first time. Movement of wild trout increased significantly during periods of stocking. Such impacts have been noted elsewhere (Snow 1974; Vincent 1984), but other studies have indicated that wild stocks are not significantly affected by put-and-take stocking programs. For example, Petrosky and Bjornn (1988) found that populations of wild cutthroat and rainbow trout were unaffected by the stocking of catchable-size rainbow trout, except when densities of stocked fishes were ex-tremely high. Even in those instances, changes in population levels of wild stocks were modest and inconsistent.

The cost of producing catchable trout, which includes the cost of rearing and the capital cost of constructing, maintaining, or expanding hatchery facilities, can exceed the economic benefits that are received because anglers catch additional fishes over the number that might attract them to the fishery and satisfy them. Johnson et al. (1995) demonstrated that cost exceeded benefit in two river systems in Colorado stocked with relatively high numbers of catchable fishes. Thus many programs may be rearing and stocking greater numbers of fishes than is economically warranted. Johnson et al. (1995) believed that similar discrepancies between costs and benefits of catchable trout programs may exist in other states. These authors suggested that state agencies need to consider whether expenditures for put-and-take programs would produce greater benefit if diverted into other fisheries management options.

Put-and-take fisheries are designed to meet the recreational interests and goals of anglers, not conservation of wild resources. Agencies have occasionally sug-gested that the angling public expects them to provide fishes that nature cannot; thus put-and-take programs meet the angling public's demands. Indeed, agencies need to understand the interests of anglers, because fishing license sales and excise taxes from fishing equipment represent the major source of funding for most state fish-eries management programs (see Chapter 6). However, put-and-take trout pro-grams have frequently been established in waters housing productive wild trout or warmwater fisheries resources. By focusing on such stocking efforts, agencies may ignore conservation strategies best suited to wise use and protection of the wild re-sources, or worse, wild resources may be damaged directly by the put-and-take fishery. Although put-and-take programs are still widely employed, in recent years state agencies have focused increasingly on integrating put-and-take stocking into comprehensive planning that addresses both interests of the angling public and conservation of wild resources.

Put-Grow-and-Take Fisheries

When survival of young fishes is consistent on a yearly basis, and when immedi-ate fishing after stocking is not a management goal, two benefits accrue from stock-ing fry or fingerlings rather than catchable-size fishes. Rearing fingerlings is less expensive than rearing larger fishes. After stocking, fingerlings adjust to natural habitats and behave more like wild fishes than do fish stocked as catchables. Thus anglers seeking a more natural fishing experience may prefer fishing in put-grow-and-take rather than put-and-take fisheries. Survival of stocked fry or fingerlings can be low in diverse fish communities that are relatively undisturbed by human activities. Thus put-grow-and-take fisheries are often established in waters in which the fish community has been altered by humans, particularly if gamefish populations have been severely depleted or eliminated. Put-grow-and-take fisheries are also created in newly flooded reservoirs, where stocked fingerlings do not have to compete with established populations of resident fishes, or in river sections immediately downstream from dams.

The size of fish stocked often is based on the cost of rearing individuals to particular sizes compared with the survival of fishes after stocking. Stocking high densities of walleye fry into Texas reservoirs was more cost effective and produced fisheries more rapidly than stocking lower densities of fingerlings (Kraai et al. 1983). Eight-inch-long rainbow trout were stocked into Laurel River Lake, Kentucky, in order to achieve high return rates typical of the fishery in that system (Jones 1982). Regardless of the size of fish stocked, the average size at capture is significantly larger than the average size at stocking in put-grow-and-take fisheries.

Pacific Salmon Fisheries in the Great Lakes

Pacific salmon fisheries in the Great Lakes represent one of the most spectacularly successful examples of put-grow-and-take programs. Federal and state agencies periodically attempted to establish Pacific salmon fisheries in the Great Lakes beginning in the late 19th century. However, these were unsuccessful other than the establishment of a small self-sustaining population of pink salmon in Lake Superior, which resulted from an accidental introduction in the 1950s (Phillips et al. 1982). Coho and chinook salmon put-grow-and-take fisheries were established in Lake Michigan in the late 1960s. The success of this put-grow-and-take program was rapidly followed by similar efforts in the other Great Lakes. Pacific salmon rapidly became both recreationally and economically important to the Great Lakes region. The success of the stocking program after decades of unsuccessful attempts to establish these species was largely a matter of timing. Native lake trout populations, which had filled a predator niche in the deepwater fish communities of the Great Lakes, were severely depleted or extirpated by the 1960s due to the impacts of overfishing, habitat deterioration, and the invasion of the sea lamprey. At the same time, the forage base that deepwater pelagic predators would feed upon was robust, due in part to invasion of exotic species, particularly the alewife. Pacific salmon essentially filled a predator void in the pelagic fish communities that had developed by that time. Although some reproduction of coho and chinook salmon was occurring in the Great Lakes by the 1980s (Carl 1982), Pacific salmon fisheries were maintained largely by put-grow-and-take management programs. By the early 1990s fisheries scientists were increasingly concerned about the future of the Pacific salmon programs in the Great Lakes. Apparent decreases in survival rates of stocked Pacific salmon (Hilborn 1992) caused in part by a declining forage base, and possible conflict between continued stocking of large numbers of Pacific salmon and restoration of lake trout in the Great Lakes (discussed elsewhere in this chapter), raised questions about future management of the put-grow-and-take fisheries.

Striped Bass and Wiper Fisheries in Reservoirs

One of the most geographically widespread put-grow-and-take efforts has been associated with the introduction of striped bass and striped bass–white bass hybrids (called wipers) into reservoirs. These fishes provided an open-water fishery in systems where many of the fisheries resources consist of shallow-water, inshore species. Although reservoir striped bass often may not reach the sizes that individuals of this species can attain in coastal waters, they frequently exhibit excellent growth, reaching lengths and weights that create trophy fisheries.

By the early 1980s over 30 states had stocked striped bass or striped bass hybrids into reservoirs. These efforts had established fisheries in 100 of the 173 lakes and reservoirs stocked with striped bass and in 179 of 283 of the systems stocked with hybrids (Axon and Whitehurst 1985). Although reproduction occurs in some waters, many of these fisheries are maintained by yearly stocking. State agencies have attempted to establish striped bass fisheries to

1. Create recreational and trophy fisheries
2. Develop a fishery in an otherwise unutilized open-water niche
3. Control undesirable species through predation

Striped bass fisheries contribute an average of nearly 14 percent and a maximum of 66 percent by weight of the total yield in reservoirs where they occur (Axon and Whitehurst 1985).

The success of put-grow-and-take striped bass hybrid fisheries can depend on the presence of a suitable forage base. Optimal forage is typically provided by pelagic species such as gizzard or threadfin shad (Axon and Whitehurst 1985). In more southern reservoirs, striped bass may seek cooler, deeper waters in the summer, where they are separated from warmwater prey species. Under these conditions put-grow-and take rainbow trout may become a major food base (Axon and Whitehurst 1985). Summer mortality of medium- to large-size adult striped bass can be quite high in reservoirs (Matthews et al. 1985). Investigators feel this mortality is due to physiological stress caused when surface water temperatures rise and oxygen levels at great depths in the hypolimnion become too low. Striped bass often become "squeezed" into a very narrow range of water depths that provide the best combination of temperature and dissolved oxygen levels (Coutant 1985; Moss 1985; Matthews et al. 1985). Studies generally indicate that, although these summer die-offs occur commonly in reservoirs, active fisheries can be maintained in systems where they occur.

Tailwater and Two-Story Fisheries

Trout fisheries have been established in many rivers by stocking the tailwater areas below hydroelectric dams. More than half of the 7 million catchable and fingerling trout stocked into tailwaters in 1980 were put into southern and midwestern rivers where natural summer water temperatures are typically too warm for trout survival

(Swink 1983). Deepwater impoundments above dams can possess an expansive cold-water hypolimnion. Release of hypolimnion water through the turbines at the base of a dam creates a coldwater habitat downstream from the dam. This modified habitat is called a tailwater. Native warmwater fish species often fail to reproduce or even to inhabit the tailwater section of rivers (Pfitzer 1975). Because warmwater fisheries are reduced or eliminated, put-grow-and-take trout fisheries provide angling benefits in a habitat that otherwise would offer little fishing opportunity.

The rainbow trout is the most commonly stocked species in tailwater fisheries (Swink 1983). Tailwaters often provide trout with excellent conditions for growth, but the variable flow conditions of these streams generally do not allow successful reproduction (Wiley and Mullan 1975).

A tailwater will support an active put-grow-and take trout fishery if the following conditions are met:

1. The upstream reservoir is deep enough to stratify fully in the summer.
2. The volume of the hypolimnion is great enough to permit coldwater releases by the dam throughout warm periods of the year.
3. Water is released in a manner that prevents long periods of no-flow in the tailwater section.
4. The river has gradients and substrates suitable for trout, for food production, and for increasing dissolved oxygen levels that may have been depleted in the hypolimnion of the reservoir (Pfitzer 1975).

Electricity usage typically varies within and among days of the week and seasonally. When electricity demand is high, peak volumes of water are released through turbines. When demand is low, dam operators tend to store water, reducing releases or eliminating them altogether (see Chapter 9). Periods of no-flow can negatively affect fishes: Pooled, nonflowing water in the tailwater may become too warm for trout, and invertebrate production in the stream may be reduced (Pfitzer 1975).

The nature of water release into the tailwater influences not only the fish population but also the participation of anglers in the fishery. Wade fishing is easiest during low-flow periods, but trout feed most actively in tailwaters during times when water releases are increasing. In many rivers, wading during rapidly increasing flows can be dangerous (Pfitzer 1975). Thus fishing opportunities in tailwater fisheries can be limited temporally.

Put-grow-and-take trout fisheries are also managed within the hypolimnion of reservoirs (Raustron 1977; Braun and Kincaid 1982; Jones 1982; Babey and Berry 1989). Put-grow-and-take trout fisheries are maintained in deep, cold hypolimnetic waters in reservoirs where other gamefishes reside in warmer, surface or shallow inshore waters. Such reservoirs are referred to as two-story systems. A suitable forage base can produce excellent growth and high survival rates of stocked trout in two-story systems. For example, 50 percent of the 8-inch rainbow trout stocked into a two-story rainbow trout fishery in Laurel River Lake, Kentucky, were harvested at an average length of 13 inches by anglers (Jones 1982).

Put-Grow-and-Take in "Reclaimed" Systems

Fisheries agencies have often employed reclamation (or "rehabilitation") programs to stimulate or create recreational fisheries. The typical strategy of reclamation includes using toxicants such as rotenone to remove most or all of the fishes in fish communities that are crowded with nuisance species such as the common carp, or with dense populations of forage species such as various sunfishes whose abundance has led to a decline in the quality of existing fisheries (see Chapter 10). Such waters are typically restocked with species that are intended to develop into self-sustaining populations that support productive recreational fisheries. Chemical reclamation has also been used to reduce the abundance of introduced species that are causing declines of native fish populations (Rinne and Turner 1991; see Chapter 11).

Another management approach that has been identified as reclamation involves removal of all fishes residing in fish communities followed by the maintenance stocking of other species of fishes that will support put-grow-and-take fisheries. Agencies have initiated these programs to create put-grow-and-take trout fisheries in warmwater habitats marginally suited to trout. Removal of the resident fish community eliminates potential competition that the fingerling trout would face, thus improving the opportunity for their survival and growth.

A put-grow-and-take trout fishery in a reclaimed pond that did not need retreatment with toxicants for at least 2 years generally would cost less per fish harvested than a catchable program in ponds where the resident fish community was not removed (Mullan and Tompkins, undated Massachusetts Fisheries and Game Bulletin titled "Trout Pond Management"). The cost of frequent applications of toxicants is lower than the cost of rearing fish to a catchable size. Thus, if a management goal is the maintenance of a hatchery-based trout fishery, reclamation followed by stocking of fingerlings is the less costly of two high-cost hatchery programs.

Ponds and streams managed as reclaimed systems may need to be treated with toxicants every several years. Otherwise, species introduced by anglers or reestablished because they had never been entirely eliminated would return to general prereclamation densities (Rinne and Turner 1991), thus affecting survival and growth of stocked fingerlings.

A project on the Green River in 1962 by the Wyoming and Utah fish and game departments with support of the U.S. Fish and Wildlife Service was the most publicly controversial attempt to create a put-grow-and-take fishery by reclamation. A section of this river was reclaimed to establish a put-grow-and-take rainbow trout fishery in the newly filled Flaming Gorge Reservoir. The resulting die-off of endemic, endangered fish populations stimulated a widespread outcry from part of the scientific community and the general public. This led to a 1963 Department of the Interior directive that required the Fish and Wildlife Service to consider impacts on native fauna before supporting similar programs in the future (Holden 1991; see Chapter 6).

The use of reclamation projects to create put-grow-and-take fisheries has been questioned often. Most negative impacts to wild resources created by hatchery introductions have been due to inadequate planning to achieve projected objectives or to a lack of understanding of the consequences of stocking. However, reclamation followed by maintenance of a put-and-take fishery clearly focuses on creating new angling opportunities while knowingly preventing the production of wild fishes, including species that might also support recreational fisheries. The word *reclamation* is a rather presumptuous term for a program intended to create fisheries dependent on intermittent removal of wild resources and on persistent stocking. In the past, agencies have justified reclamation-based put-grow-and-take fisheries as an economical means of providing anglers with the fish they most desired (usually trout). However, some feel that such programs are simply the result of overemphasizing hatchery production and underemphasizing natural resource conservation.

USE OF HYBRIDS

The production of hybrid fishes (offspring produced from parents of two different species or subspecies) has been an integral part of agency rearing and stocking activities for some time. Numerous species of fishes produce hybrids in nature. Many hybrid forms are rare, apparently resulting from occasional accidental union of gametes from fishes spawning in close proximity to each other (Ross and Cavender 1981). Other factors, such as habitat deterioration, an absence of intraspecific mates, or the introduction of new species, also contribute to the frequent occurrence of some hybrid combinations, such as those between different species of sunfishes (Trautman 1981). Introgressive hybridization has been an important mechanism in the evolution of some species groups, such as the western U.S. genus *Gila* (Dowling and Demarais 1993).

Natural hybridization resulting from the introduction of a species or subspecies into waters outside of its geographic range has threatened the stability of gene pools of numerous native fishes (see the section on genetic problems caused by introduced species in this chapter). However, agencies have stocked hybrids widely because these fishes might offer something to a fishery that is not characteristic of the parental species that are used to produce the hybrid.

Early researchers suggested that hybrid fishes were incapable of reproduction, were extremely vulnerable to angling because of high levels of aggressiveness, and were capable of growing faster and reaching larger sizes than individuals from either parent species. Sterility interested agencies because fishes could be stocked without causing disruptions to resident fish communities due to expansive population growth through reproduction; the maximum number of fishes any particular system would ever have was no greater than the number of fishes actually stocked. Large size, faster growth, and angling vulnerability also were deemed advantageous traits for fishes stocked to create recreational fisheries.

Agency interest in rearing and stocking hybrids has also been based partly on the psychological value of providing the angling public with a new kind of fish that is the creation of the agency. All of these factors have stimulated agency experimentation with the rearing and stocking of a variety of fish hybrids, particularly from the pike (*Esocidae*), sunfish and black bass (*Centrarchidae*), perch (*Percidae*), and trout and salmon (*Salmonidae*) families (Figure 8.7).

Figure 8.7   The splake, a hybrid between the lake trout and brook trout. (Photograph courtesy Tom McHugh/Steinhart Aquarium, Chicago; Photo Researchers, Inc.)

Hybrids have not exhibited superior fishery traits consistently. Although many researchers have shown that hybrids can grow faster and reach larger sizes than either parental species, other workers have not, even when studying the same types of hybrids. Carlander (1969) suggests that many instances of accelerated growth in natural waters may be the result of relatively low densities of hybrids being introduced into new systems rather than the result of any genetic superiority of the hybrids over parental types. The level of sterility has also proven to be highly variable, with many hybrid combinations being as capable of producing viable offspring as are parental fishes.

However, agencies still stock certain hybrids widely because some hybrid combinations do offer advantageous traits. Tiger muskie (northern pike crossed with muskie) are more resistant to thermal stress (high summer temperatures) than either parent species (Carlander 1969); tiger muskie are also reared more easily on dry, processed fish foods prior to stocking than are muskies (Graff and Sorensen 1970). Saugeye (walleye crossed with sauger) are better adapted to littoral areas of ponds than are walleye (Lynch et al. 1982). In reservoirs, wipers (striped bass crossed with white bass) exhibit better survival rates and grow faster in early life stages than striped bass and reach larger sizes than the white bass (Forshage et al. 1986; Ebert et al. 1987). Hybrid sunfishes can reach larger maximum sizes and are more vulnerable to angling than are parentals (Childers 1967; Guest 1984). As such fishery-related traits are identified, in the last 10 to 20 years hatchery systems have moved from experimentation with a wide range of hybrid types to a focus on production and stocking of specific hybrids that do provide advantageous traits for specific fishery situations.

SELECTING STRAINS

Much of the early emphasis concerning the selection of genetic strains of fishes for stocking programs focused on fisheries supported totally by maintenance stocking. Such fisheries may exhibit a variety of characteristics depending on the species and strain of fish used. Calhoun (1966) reviewed the performance of several strains within three species of trout used in California's stocking programs. A threefold difference in percentage of stocked fish harvested and a twofold difference in percentage of fishes creeled in the second year after stocking were noted among different strains of brown trout. Particular strains performed best in specific types of fisheries; the Massachusetts strain of brown trout displayed excellent survival after stocking in productive lakes but did less well than other strains in mountain streams. More recent studies have shown that performance in some maintenance fisheries may differ markedly among strains (as in Braun and Kincaid 1982; Moring 1982; Babey and Berry 1989), whereas in others it does not (as in Hudy and Berry 1983). Domesticated strains tend to provide faster and higher return rates than do strains less influenced by decades of hatchery breeding (as in Fay and Pardue 1986). Domesticated strains typically have passed through multiple generations of selective breeding to establish traits that are highly adaptive to hatchery production, such as exhibiting fast growth rates on diets of processed fish foods or withstanding severely crowded conditions typical of hatcheries without detrimental effect. These traits also tend to make these "domesticated" fishes highly vulnerable to fishing.

The importance of selecting appropriate genetic strains for hatcheries goes far beyond simply using fishes that grow fastest in hatcheries and are harvested most easily after stocking. The success of restoration and supplemental stocking programs strongly depends on selecting and breeding appropriate genotypes. Major problems have arisen when such selection is not planned carefully. Hatchery production frequently is administered by one branch of a fisheries agency whereas programs of conservation and management are by another. The hatchery administrator may select genotypes that are exceptionally productive in hatchery conditions, or those that are most readily available, rather than those that will best support the objectives of the program in which they are being used. The manner in which captive breeding is carried out may be more responsive to efficiency of hatchery operations than to the management of optimal gene pools. "Many hatchery managers still receive salary bonuses based upon [the number of] fish released from the hatchery, regardless of the survival of these releases or the impacts [the hatchery fishes may have] upon wild stocks" (Hilborn 1992). A lack of integrated planning between those rearing fishes and those stocking them can cause disruption or failure of stocking programs.

THE FUTURE OF HATCHERIES IN FISHERIES MANAGEMENT

Fisheries agencies have focused on hatchery production and stocking as integral parts of many management efforts. There appears to be no consensus within review articles concerning the role that hatchery production may play in fisheries management in the future. Keith (1986) enthusiastically projected a marked growth of hatchery production for at least some types of fisheries management programs. Radonski and Martin (1986) considered stocking "an indispensable tool" if it is properly integrated with other "fisheries management protocol." Hindar et al. (1991) and Hilborn (1992) supported establishing strong restrictions on the use of hatchery-reared fishes in fisheries management. Courtenay and Moyle (1992) lamented the loss of native resources that has occurred due at least in part to stocking in a paper entitled "Crimes Against Biodiversity: The Lasting Legacy of Fish Introductions." Martin et al. (1992) believed that experience could guide future planning for uses of hatchery fishes if agencies carefully consider past failures as well as successes.

Hatcheries have provided fishing opportunities in disrupted habitats where they might not otherwise exist (such as in river tailwaters or newly constructed reservoir systems) or have provided significantly greater opportunity for angling success in numerous recreational fisheries. Hatchery production seems critical for recovery of some endangered gene pools that will probably be lost without intervention. However, the introduction of new species into watersheds has been the major cause for the loss of many native fishes and fisheries. A preference for stocking over stricter regulations of harvest, or over mitigating habitat disturbances, has done little to prevent the collapse of many native fish stocks. We may not be able to learn even from the clearest examples of past failures. In 1993 the state of Maine passed legislation to initiate a study of the feasibility of enhancing declining groundfish stocks in the Gulf of Maine with hatchery-reared fishes. Similarly, a century after coastal stocking was first attempted by the U.S. Office of the Fish Commissioner, Congress has given the Northwest Atlantic Ocean Fisheries Reinvestment Program (created in 1992 by amendments to the Fisheries Conservation and Management Act and the Fisheries Promotion Act) the responsibility of evaluating the potential effectiveness of supplementing collapsed New England groundfish stocks with hatchery fishes.

The efficiency with which we can grow fishes still seems to be a motivation toward hatchery production as the answer to a multitude of fisheries resources problems. Although hatcheries offer the ability to address critical fisheries resource issues effectively, they can represent the clearest instances in which agencies have focused on interests of the human "user" and other interest groups while ignoring or excluding effective conservation of wild resources.

Projected growth of the future uses and production levels of hatcheries is frequently based on one of two arguments:

1. Aquatic systems are so perturbed that they cannot provide suitable fisheries resources without hatchery supplementation.

2. Increases in the number of recreational and commercial users require major supplementation in order to meet a harvest "demand."

Although agencies should consider such factors when developing fisheries conservation strategies and programs, they need to be careful that they do not duplicate errors of the past simply because hatchery fishes represent an easy response to issues.

9

Habitat Deterioration and Loss

Viable fish populations depend on the presence of aquatic habitats that provide suitable conditions for successful completion of life cycles. Humans use surface waters in a variety of ways, many of which disrupt the quality of aquatic environments or reduce their accessibility to fishes. Habitat deterioration and loss are serious challenges to the present and future effectiveness of fisheries conservation programs. This chapter addresses some of the more common causes of habitat deterioration and loss and the means that fisheries agencies have to mitigate the effects of environmental disruptions on fish communities. The disruptions reviewed in this chapter are divided into four basic categories:

1. Habitat deterioration or loss caused by land-use activities within watersheds, including wetland development, logging, grazing, and farming
2. Modification of habitats caused by stream channelization, water diversion, and flow regulation at dams
3. Obstruction of migratory pathways in river systems
4. Chemical deterioration due to toxic contaminants, nutrient loading, and acidification

HABITAT DISRUPTION CAUSED BY LAND-USE ACTIVITIES

Elimination of Wetlands

Wetlands include any coastal or inland transition areas between terrestrial and aquatic habitats where the water table is normally at or near the soil's surface or where the land is covered by shallow water (Cowardin et al. 1979; see Chapter 2).

Coastal wetlands (salt marshes and tidal flats) occur in estuaries, and inland wet-lands occur in low-lying areas adjacent to rivers or lakes that are subject to periodic flooding, or in topographic depressions that are shallowly flooded at least period-ically by groundwater seeps (Tiner 1984). Wetlands provide critical feeding, spawn-ing, and nursery habitats for fishes and mediate the effects of flooding, erosion, and chemical contamination of associated waterways (Tiner 1984; see Chapter 2).

Most wetlands in the United States are inland. Alaska, Louisiana, and Flori-da possess the greatest wetlands area, although numerous other states contain sig-nificant wetland habitats (Figure 9.1). Louisiana has nearly one-third of the coastal marsh systems in the United States, and the Chesapeake Bay of Maryland and Vir-ginia is the country's largest estuary (Tiner 1984).

Human activities not only have caused deterioration of the quality of wet-land habitats but have eliminated significant areas of these critically important sys-tems. Of the 220 million acres of wetlands that existed in the 48 conterminous states in the mid-1700s, only 106 million acres remained by the 1970s (Dahl and Johnson 1991). Nearly 300,000 acres of wetlands were lost annually in the United States be-tween the mid-1970s and the mid-1980s. Table 9.1 lists states that have lost more than 50 percent of their wetland area.

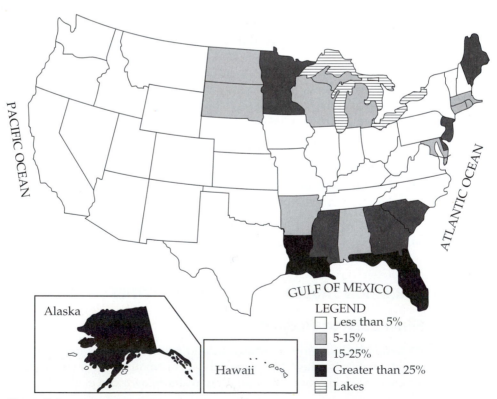

Figure 9.1   The percentage of land consisting of wetlands in the United States, 1984. (From Tiner 1984.)

TABLE **9.1**   States that have lost more than 50 percent of their wetlands between the 1780s and 1980s (Dahl and Johnson 1991).

State	Percentage Lost
Alabama	50
Arkansas	72
California	91
Colorado	50
Connecticut	74
Delaware	54
Idaho	56
Illinois	85
Indiana	87
Iowa	89
Kentucky	81
Maryland	73
Michigan	50
Mississippi	59
Missouri	87
Nevada	52
New York	60
Ohio	90
Oklahoma	67
Pennsylvania	56
Tennessee	59
Texas	52

Historically, in inland areas over 80 percent of all wetlands losses were due to drainage and conversion to farmlands (Dahl and Johnson 1991). Recent losses are more evenly divided between conversion to agricultural uses (54 percent) and conversion for other land development (46 percent). River flood control projects also destroy inland wetlands. These projects typically include extensive channelization and construction of levees that separate a river channel from its backwaters, marshes, oxbows, and floodplain, destroying these wetland habitats and causing declines in many fish populations inhabiting the river system (see the section on stream channelization in this chapter).

Great losses of coastal wetlands also have occurred due to water diversion and flood control programs. Extensive channelization and levee and canal construction of the lower Mississippi River drainage has led to accelerated erosion and saltwater intrusion into Louisiana coastal marshes. Levees also prevent flooding of marsh areas during high-water times of the year, eliminating the rebuilding of marshes through deposition of sediment-rich floodwaters (Gosselink 1984; Reed 1992). Diversion of the natural flow and flooding patterns in the Mississippi River delta causes an annual loss of about 1.5 percent of the delta marsh (Gosselink 1984).

Diversion of natural waterways through canal systems and within levees has destroyed large expanses of wetlands in Florida. Drainage and filling for agricultural and urban development and water withdrawal for domestic uses have caused saltwater intrusion into freshwater marshes of southern Florida (Tiner 1984).

Preventing Wetland Loss.   Federal and state regulations controlling wetlands conversion have reduced the amount of wetland loss in recent years, but large-scale losses still are occurring. Fisheries agencies cannot directly regulate or mitigate the destruction of wetlands. However, two provisions of the Clean Water Act (see Chapter 6) have enabled water resource agencies to prevent some wetland destruction. Section 404 provides opportunity for review and comment by the USFWS, the NMFS, and the EPA before the U.S. Army Corps of Engineers can write permits allowing dredge and fill operations in any wetlands. Further, Section 401 allows states to review requests for federal permits and deny certification of projects that do not meet state water-quality standards or other state laws.

Early analyses indicated that wetlands protection under Section 404 allowed the development of 24 to 50 percent less wetlands area than would occur without that legal constraint on dredge and fill operations (Owen and Jacobs 1992). However, Owen and Jacobs found that wetland loss in Wisconsin was only 15 percent lower than otherwise would have occurred. These authors believed that environmental effects of proposed dredge and fill activities often were not adequately assessed in the permit process. Perceived public benefit was the major criterion used to approve or deny permit applications, with profit from development weighed more heavily than the benefit to the public of preserving wetland systems (Owen and Jacobs 1992). The cumulative effect of multiple dredge operations is not addressed under the traditional site-specific permit evaluation (Griffin 1989). Also, the Corps of Engineers has not always followed recommendations of resource agencies in making final judgments on permits (Griffin 1989). Because the permit system requires a comparison of the relative value of protecting wetlands with the value of development, the effectiveness of the 404 provision is subject to the interpretation of benefits made by the executive branch of the federal government and the emphasis placed by the current administration on environmental quality. Even though the federal government's role in wetlands protection has been inconsistent, relatively few states have established comprehensive wetland protection laws to better protect their wetland areas (Griffin 1989; Figure 9.2). Many of the states that have suffered the greatest wetlands losses (from Table 9.1) have not passed such laws.

Logging

Nearly 500 million acres of land in the United States, most of which are publicly owned, are managed for timber production (Platts and Martin 1980). Logging activities have caused widespread deterioration of riverine habitats due to the following:

1.   Increases in erosion and silt loading into the streambed
2.   Increases in daily and seasonal water temperature fluctuations
3.   Modifications of flow regimes

States with comprehensive wetlands laws

Figure 9.2 States that have established comprehensive laws for wetlands protection. (Reprinted by permission of The Wildlife Management Institute.)

4. Increases in fine woody debris in the streambed
5. Decreases in the number of fallen trees and logs that would provide cover and pool habitat

Habitat deterioration caused by logging is particularly widespread in western states and Alaska, although the long-term effects of logging are also evident in other regions of the United States. Much logging activity in the West, Northeast, and upper Midwest occurs in headwater areas of watersheds that often contain productive salmonid resources. Thus habitat deterioration caused by logging has been detrimental to many populations of these recreationally and commercially important fisheries resources.

Siltation. Forestry harvest practices can accelerate erosion and sediment loading into streambeds (Swanson et al. 1987; Chamberlin et al. 1991). Increased sediment loading often is caused by erosion along logging roads constructed to transport logs out of forests rather than by the actual cutting practices themselves (Platts and Martin 1980; Swanson et al. 1987). Logged, open slopes are a secondary source of sediments (Chamberlin et al. 1991). Most erosion takes place during the first year after road construction (Ketcheson and Megahan 1993; Megahan 1993). In Medicine Bow National Forest, Wyoming, the amount of fine sediments in streams was positively correlated with the number of logging road culverts and the proportion of the drainage that was logged, and negatively correlated with trout standing stocks (Eaglin and Hubert 1993).

Increased loads of sediment can fill shallow pool areas, reducing pool habitat, and can settle into gravel riffles, filling interstitial spaces within gravel substrates (Chamberlin et al. 1991). Increased concentrations of silt settling into the streambed and being transported downstream during periods of high water flows can affect several life stages of fishes. Egg and larval mortality of substrate-spawning fishes can be substantially higher when gravel spawning areas become laden with silt and sand (Shephard et al. 1984; Cederholm and Reid 1987; Chapman 1988). Suspended sediments retard growth of salmonid fry and stress juvenile fishes (Everest et al. 1987). Summer carrying capacity of streams for juvenile trout can decline in streams as silt and sand accumulate in feeding and shelter sites (Platts and Martin 1980). Stream productivity at all levels can decline as sedimentation increases (Everest et al. 1987).

Although sedimentation can affect fisheries resources severely, careful planning of road systems so that stream banks and channels are disrupted as little as possible can reduce much of the effect that siltation might have on stream fisheries resources (Eaglin and Hubert 1993).

Temperature Effects.   Water temperature regimes also can be influenced by logging activities. Removal of cover along stream banks exposes the water's surface to increased solar radiation. Exposure to sunlight increases a stream's mean and maximum daily and seasonal temperatures, as well as the magnitude of daily and seasonal water temperature fluctuations (Platts and Martin 1980; Beschta et al. 1987). Brown and Krygier (1970) reported a 14°F (8°C) increase in water temperature of a stream flowing through a clear-cut area in Oregon. This increase can stimulate stream productivity (Beschta et al. 1987), but temperatures may reach levels that are physiologically stressful to salmonids and other fishes.

Rather simple logging practices can reduce the effects created by increases in water temperatures. Cutting timber in patchwork patterns may have little or no effect on natural temperature regimes (Brown and Krygier 1970) because elevated temperatures in cleared areas become significantly lowered as the water flows back through forested stretches of stream (Platts and Martin 1980). Naturally vegetated buffer strips along stream banks also can minimize temperature effects of logging (Beschta et al. 1987). Buffer strips represent one of the simplest and most practical management tools for protecting the quality of stream habitats in watersheds supporting substantial logging operations (Hall et al. 1987).

Effect of Logging on Flow Regimes.   Extensive logging activities generally increase the total amount of water entering a stream and increase the percentage of the total stream flow that enters the stream as surface runoff rather than underground percolation. Total flow is increased due to a reduction in evaporative loss and in uptake of groundwater by vegetation. Clear-cutting reduces the amount of evaporation that otherwise might occur because the proportion of precipitation that would strike leaves and evaporate in a naturally forested landscape instead reaches the ground (Chamberlin et al. 1991).

Heavy machinery dragging felled trees across slopes can cause soil compaction, reducing the capacity of precipitation to sink into soils and thus increasing surface runoff (Chamberlin et al. 1991). Surface runoff drains into the stream channel more rapidly than underground percolation, so flooding after storms or during snow melt will be greater in logged than in tree-covered watersheds. If soils are not compacted, they will become more saturated with groundwater, potentially increasing flow during drier times of the year. However, if compaction causes an increase in surface runoff, the more rapid movement of water into the river channel can reduce the amount of groundwater that might enter the channel during dry periods.

Therefore, in logged watersheds flow rates will be more variable both over short periods of time and between seasons than in tree-covered watersheds. Increased flooding can increase scouring of the streambed, potentially reducing the invertebrate community that serves as a major food source of stream fishes (Platts and Martin 1980). The impact of increased surface runoff is greatest immediately after extensive cutting and decreases significantly in relatively few years as the vegetation in the watershed regenerates (Platts and Martin 1980).

Woody Debris. Extensive tree removal along stream banks can gradually reduce the amount of fallen trees and brush that lie in or across the streambed (Bilby and Ward 1991). Such large woody debris provides cover for large fishes, aids in the formation of pool habitat, stabilizes gravel bars, and influences downstream movement of sediments and organic matter (Bisson et al. 1987; Bilby and Ward 1991). In Alaskan watersheds, clear-cutting combined with removal of streamside trees has caused the amount of large woody debris in streams to slowly decrease by as much as 70 percent within 75 years. This decline occurs because debris existing before clear-cutting occurs gradually deteriorates and washes downstream, and because additional debris will not fall into the stream until trees along the bank reach a large size and relatively old age. Thus the loss of large woody debris from streambeds is a long-term proposition; levels of large woody debris may not return to precut conditions even after more than 200 years (Murphy and Koski 1989).

At the same time, logging can increase the amount of smaller debris entering the stream. Such debris will accumulate during low flow periods but will break up and be carried downstream during flooding that is typical of logged watersheds. Concentrations of moving debris may greatly increase scouring activity on the streambed and cause increased bank instability and erosion during high-water periods (Platts and Martin 1980; Cederholm and Reid 1987). Accumulation of small debris in areas of quiet, slow flow also can cause declines in concentrations of dissolved oxygen due to aerobic decomposition of these materials by aquatic microorganisms (Platts and Martin 1980). Although oxygen concentrations in these pools are reduced to levels that can affect development of fish eggs and physiologically stress fishes, severe oxygen depletion in streams often may be quite localized. Re-aeration can take place rapidly in water flowing through riffles or rapids.

Woody debris has been removed from streams in numerous West Coast watersheds to enhance salmonid production. Most researchers who support removal consider higher dissolved oxygen levels and the elimination of log dams that impede migration to be the major benefits of this management technique (as reviewed in Bryant 1983). However, removal is not always beneficial to salmonid populations. Removal of log jams releases debris and sediments from behind the jam to wash downstream, altering downstream habitats (Reeves et al. 1991). Elliott (1986) and Dolloff (1986) demonstrated that removal of logging debris was correlated with reductions in the number and size of Dolly Varden and coho salmon in Alaskan watersheds. Elliott (1986) attributed this reduction to changes in river morphology and flow, a temporary reduction in food supply, and a long-term elimination of cover. Bryant (1983) believed that the impact of fallen debris on migration has been overstated; adult and juvenile salmonids will move past all debris but the infrequently occurring massive log jams that prevent movement even during high waters. Thus removal may not substantially improve migration, but it can cause a loss of pool habitat and shelter.

Reducing Habitat Deterioration.   In their review paper, Platts and Martin (1980) indicate that technology is available to minimize the effects of logging on fisheries resources, but the use of sound management practices addressing these impacts "has been slight." In recent years the U.S. Forest Service has placed increasing emphasis on managing public lands under its stewardship for multiple use of resources including fisheries, rather than a more narrow focus of management solely for timber production. As part of this emphasis, the Forest Service nearly tripled the number of fisheries biologists within its staff between 1986 and 1993 (Forsgren and Loftus 1993). In response to growing concern for deteriorating aquatic habitats in West Coast watersheds that support declining anadromous salmonid migrations, by 1994 the Forest Service was implementing a program aimed at protecting stream corridors and controlling the effects resulting from timber harvest activities on adjoining upslope areas (SFI 1994).

Grazing on Public Lands

Livestock grazing is widespread on public lands in the western United States. Nearly 70 percent of the 1.2 billion acres of rangeland in the United States was used for livestock grazing in 1970 (Platts 1991). Although damage to terrestrial habitats caused by grazing has been successfully addressed in recent years, the riparian or streamside habitats in rangelands have continued to be affected by livestock grazing (Platts and Martin 1980). Much of this impact is caused by excessive concentration of livestock within streamside (riparian) zones, because in western rangelands herbaceous vegetation is most productive in these habitats (Armour et al. 1991). Grazing activity has led to major declines in native trout resources in many western watersheds (Platts and Martin 1980).

Excessive streamside grazing and trampling activities by livestock will reduce or eliminate stream bank vegetation, which can cause increased summer water temperatures due to reduction in shading from solar radiation. Stream banks become destabilized, causing slumping of soils into the streambed and increasing silt loading

Figure 9.3 Bank slumping caused by excessive grazing activity can lead to increased silt loading in stream systems. (Photograph reprinted by permission of cover of *Fisheries 1985,* Volume 10, Number 5, American Fisheries Society.)

into the system. Stream channels gradually become wider and shallower (Meehan and Platts 1978; Armour et al. 1991; Platts 1991; see Figure 9.3). Productivity and standing stocks of salmonid species are markedly higher in sections of streams where grazing is not permitted than in other sections in the same systems where it is allowed (Gunderson 1968; Platts and Martin 1980; Stuber 1985; Armour et al. 1991; Platts 1991).

Protecting Streams from Riparian Grazing. Various studies have indicated that stream habitats can recover within a few years once protected from riparian overgrazing. Fish production in Otter Creek, Nebraska; Sheep Creek, Colorado; and Summit Creek, Idaho, increased significantly once overgrazing ended (Keller and Burnham 1982; Armour et al. 1991). However, some researchers suggest that recovery of fish populations in at least some streams subject to overgrazing will be slow (Hubert et al. 1985).

Although the effects of grazing on stream habitats are being identified, rangeland management often has not addressed habitat conditions of streams or the habitat needs of fishes (Meehan and Platts 1978). Land managers have focused on establishing criteria for the intensity of grazing activities on rangelands, whereas fisheries managers have seen a need to address issues of fish harvest and aquatic habitat protection (Platts and Martin 1980). Multidisciplinary planning by both of these groups on the use of rangelands is critical to achieve compatibility between grazing and the maintenance of robust fisheries resources (Platts 1991).

In 1993 the executive branch of the federal government proposed a series of changes in the administration of grazing on public rangelands. Guidelines for range management that would protect riparian and aquatic habitats were among the proposed rangeland reform measures (SFI 1994). However, user-group opposition to proposed increases in grazing fees created passionate discussion in Congress, for the time stalling any significant changes in federal policy that would provide greater protection to aquatic habitats on public rangelands from the negative effects of overgrazing.

Agricultural Development

Converting land to agricultural use has caused widespread removal of natural vegetation within watersheds throughout the United States and has changed the quality of lake and river basins within those watersheds. Farming changes the chemical characteristics of surface water by introducing toxic pesticides, inorganic fertilizers, and organic materials through surface runoff and groundwater leaching. The impact of chemical pollution is reviewed later in this chapter; the following section examines the effect of farming on the rate of sedimentation in waterways, which occurs largely due to removal of natural vegetation from watershed slopes, floodplains, and river banks.

The Effects of Farming on Sediment Loading and Stream Turbidity.   Tilling and row crop farming have increased the rates of erosion and sedimentation in riverine habitats substantially throughout the United States (Karr and Schlosser 1978). Excessive siltation occurs in nearly 50 percent of all streams in the United States (Berkman and Rabeni 1987). Sedimentation has been considered a major factor limiting available fish habitat in stream systems (Rabeni 1993). Excessive sediment loading causes declines in invertebrate biomass production and fish reproduction in gravel and rocky substrates, and concurrent increases in turbidity cause the decline and loss of submergent aquatic vegetation.

Increased erosion and sedimentation in streams has occurred for long periods in some areas of the United States. Trautman (1981) estimates that significant habitat deterioration due to sediment loading occurred on a widespread basis by the mid-1800s in Ohio. Sedimentation and increased turbidity has altered the fish fauna of that state to a greater degree than any other factor. From the mid-1800s to early 1900s, increased sedimentation in Ohio, Indiana, and Illinois led to the decline or loss of numerous species of native fishes that require clear waters or abundant aquatic vegetation, and to the invasion of prairie species that are more tolerant of sediment loading and high turbidity (Trautman 1981).

Slope erosion was once thought to be a major contributor to sediment loading in streams. However, upland depressions, grass fields bordering farmed land, hedgerows, and drainage ditches trap most soils washing from intensively farmed slopes (Wilkin and Hebel 1982). Those soils that are washed completely down to the base of slopes are largely trapped in floodplains, if these lowland habitats are covered with natural vegetation (Roseboom and Russell 1985). The major sources of erosion and sediment loading are farmed floodplains stripped of natural vegetative cover and stream bank erosion (Wilkin and Hebel 1982).

Reducing the Effect of Cropland Erosion.   The extent of sedimentation in streams can be reduced in intensively farmed areas by providing buffers of natural vegetation that trap sediments and reduce the amount of erosion (Wilkin and Hebel 1982; Roseboom and Russell 1985). Protecting trees and shrubs along the banks of streams stabilizes soils and reduces erosion. Likewise, woody and herbaceous vegetation in the floodplain reduces erosion and traps soils coming from adjacent slopes. Wilkin and Hebel (1982) considered the most effective means of reducing sediment loading in streams to include the following:

1. When possible, eliminating farming from the floodplain
2. Maintaining vegetative buffers between slopes and the floodplain
3. Retaining woody and herbaceous buffers from the stream bank into the floodplain

MODIFYING FLOW RATES AND WATER LEVELS

Stream Channelization

Stream channelization has led to the loss of extensive reaches of natural riverine habitats in numerous waterways throughout the United States. Streambeds and riverbeds are channelized to protect agricultural lands from flooding, to facilitate construction of railroads and roads, or to initiate urban and industrial development. Channelization also may be used to create or improve boating channels.

Most channelization projects include altering the streambed and channel in order to allow rapid passage of water through channelized areas with as little effect as possible on the channel banks and surrounding floodplain. To accelerate passage of water through a stream reach and minimize erosion, the stream channel is normally straightened by dredging, because erosional forces in unmodified streams are greatest on the outside banks of meanders and bends (Hynes 1970; see Chapter 2). In addition, the morphology of the stream channel is modified into a smooth-surfaced "tube" or "pipe" to minimize turbulence in water flowing through the channelized section. The cross-section shape of the river channel is modified from one that may exhibit irregularities in depth to one that is a symmetrical trapezoid (Hesse and Sheets 1993). Riffle areas are deepened and pools are partially filled to create a standard water depth. Any within-stream or stream bank structures that may provide cover for fishes, such as boulders, submerged logs, or exposed tree roots on undercut banks, are removed to eliminate turbulence that these structures may create. Stream bank vegetation is often removed in the process of straightening the streambed. Banks on the outside of meanders that cannot be straightened are usually reinforced with boulder "rip-rap." Otherwise, remaining bends would be subject to extreme erosional forces by the faster-flowing water, because the rate of flow in channelized streams is accelerated. Backwaters, side channels, oxbows, and adjoining marshes are often eliminated or are isolated from the main channel, even during high-water periods (Hesse and Sheets 1993; Figures 9.4 and 9.5).

Channelized streams exhibit marked reductions in habitat diversity. Diverse habitat types characteristic of natural systems are changed into one homogeneous substrate and depth without cover. Differences in velocity along a stream reach are

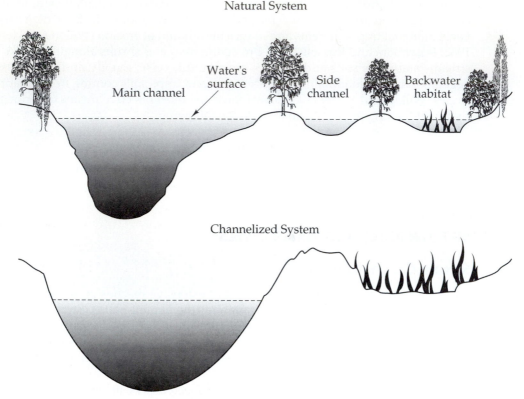

Figure 9.4 Cross section of a stream before and after channelization, demonstrating reduction in habitat diversity. Depth and substrate diversity in the main channel are reduced, and adjoining side channel and backwater areas are lost.

reduced. A series of channelized streams in the state of Washington displayed a 10 percent decrease in sinuosity, a 20 percent decrease in area of wetted streambed, and an 89 percent decrease in overhead tree cover (Chapman and Knudsen 1980). Benthic habitat may be reduced by as much as 65 percent in channelized rivers in which floodplain habitats are cut off from the main channel (Hesse and Sheets 1993). Elimination of stream bank woody vegetation may affect temperature regimes similar to that characteristic of heavily logged watersheds in which no streamside canopy buffer zones are created.

Effects on Fisheries.   Eliminating the diversity of habitat types will lead to changes in the biotic community inhabiting the stream or river. The loss of side channels, backwaters, and other aquatic habitats adjacent to the main channel will cause a reduction in food web productivity and a substantial decline in standing stocks of fish species (Hesse and Sheets 1993). Channelization can cause changes in the species makeup and biomass of invertebrate communities in the main channel, although such communities may begin to recover within several years after channelization (McCarthy 1985). The biomass of all age groups of salmonids in Childs Creek,

Figure 9.5   Channelized portions of the Kissimmee River exhibiting cutoff side channels and back-waters. (Photograph courtesy of Jeff Greenberg; Photo Researchers, Inc.)

Washington, declined immediately after channelization (Chapman and Knudsen 1980). The standing stocks of trout species in 13 streams in Montana were 5.5 times greater in natural channels than in channelized sections of the same streams (Peters and Alvord 1964). Fish densities declined by as much as 80 percent in the Missouri River after channelization and impoundment projects disrupted this system's riverine habitats and flow patterns (Hesse and Sheets 1993). It is not unusual for the abundance of sport fishes to decline as much as 90 percent after stream channelization (Rabeni 1993). Recovery of the standing stocks of specific fish species to levels characteristic of a stream before channelization may take several decades or longer, and in at least some instances only partial recovery of biomass levels may occur even over such extended periods of time (Golden and Twilley 1976; McCarthy 1985). Similarly, the period of time necessary for a fish community to reestablish comparable levels of diversity and productivity may exceed 50 years (Detenbach et al. 1992).

Mitigating the Effects of Channelization.   Fisheries agencies often cannot alter the impact of channelization on fisheries resources because the characteristics that cause a decline in the quality of fish habitat are necessary to meet the objective of flood control. Fisheries agencies might influence the potential impacts of channelization projects by offering advice to the water resource agency that approves such projects. In instances where impacts of past channelization projects can be mitigated in small streams, the installation of instream structures such as deflectors and check dams (see the section on habitat improvement in streams in Chapter 10) can recreate riffle-pool sequences and provide cover that can enhance the condition of depleted fish stocks (McCarthy 1985). Population abundances of smallmouth bass and several other recreationally important species increased substantially in a channelized area of the Olentangy River, Ohio, once a series of artificial riffles and pools were constructed in the affected area (Edwards et al. 1984). Stream fish communities can recover to natural levels of diversity and abundance within as little as 5 years in some channelized streams where siltation levels have been reduced and pool volume has returned to prechannelized levels (Detenbach et al. 1992).

Water Diversion

Surface waters and ground waters are diverted for domestic and industrial use and to irrigate agricultural lands. Diverted water may be returned to the system from which it was withdrawn, or it may be moved outside the boundaries of its basin of origin. The greatest water diversions in the United States have occurred in arid western states (Meador 1992), where local and regional demand for water has greatly exceeded supply. Water is intensively used in that region; for example, the entire volume of water flowing through the Colorado River basin is reused several times before flowing into Mexico (Minckley and Douglas 1991). By the late 1980s over 70 percent of the natural flow of the Platte River was captured by diversion and storage projects for irrigation, domestic use, and generation of electricity (Shoemaker 1988). Western agriculture is particularly dependent on irrigation. One of the most dependent states, California, irrigates over 90 percent of all of its croplands (Saiki and Schmitt 1985).

At its extreme, withdrawal and diversion of surface waters have reduced or even eliminated flow in some river systems. Evaporative loss and diversion allow less than 1 percent of the volume of water entering the Colorado River basin to reach its mouth on the Gulf of California (Moyle 1993). Surface flows in lower reaches of the San Joaquin River, California, can be totally eliminated by withdrawals during the irrigation season (Saiki 1984). Waters flow intermittently in the mid–Rio Grande River above Presidio, Texas (Contreras and Lozano 1994). Water diversion from the Truckee River–Pyramid Lake system in Nevada has dropped water levels in the lake to the point where the endangered cui-ui can migrate into the Truckee to reach traditional spawning grounds only in years of unusually high flows (Scoppettone et al. 1983; Figure 9.6). The Lahontan cutthroat trout was extirpated from Pyramid Lake and most of the Truckee River (Trotter 1990) in part due to this diversion.

When water is diverted for within-basin uses, the quality of irrigation water that is returned to the river channel can exhibit elevated temperatures; high concentrations of suspended sediments, pesticides, and nutrients; and high salinity (Saiki 1984; Saiki and Schmitt 1985). The Colorado and the Rio Grande rivers become highly saline in their lower reaches because impoundments and diversions for irrigation accelerate rates of evaporation, and water returned to the rivers from irrigated fields carries high concentrations of minerals leached from soils (Johnson 1978; Pimentel and Bulkley 1983; Miller et al. 1989).

High salinity and deterioration of other chemical and physical conditions of water used for irrigation can cause considerable changes in the composition and abundance of fish faunas (Saiki 1984). Declines in commercial fisheries in the Gulf of California have been attributed at least in part to severe reductions in the volume, increases in salinity, and decreases in soluble nutrients of the freshwater that enters the gulf from the Colorado River (Johnson and Carothers 1987). Water withdrawals

Figure 9.6 A migration of cui ui moving from Pyramid Lake into the Truckee River, Nevada, that was stranded by a drop in water level when water was diverted from the river for irrigation. (Photograph courtesy of Tom Trelease.)

from the San Joaquin River result in increases in salinity in the Sacramento–San Joaquin estuary due to saltwater intrusion. In low-flow years, high salinities can disrupt spawning activity of the striped bass in that system (Stevens and Chadwick 1979). Diversion-related low flows in rivers entering the Sacramento–San Joaquin estuary increase mortality rates of postlarval striped bass, possibly due to the failure of low flows to flood substantial portions of the estuary that provide nursery habitat for striped bass when inundated (Stevens 1977; Stevens and Chadwick 1979). Production of young chinook salmon, American shad, and longfin smelt is directly correlated with river flow rates in the same river system. Mortality increased in low-flow years in part due to dewatering of egg-bearing nests (of salmon), larval losses in diversion canals, decreased habitat availability, increased predation rates due to crowding, and declines in food productivity (Stevens and Miller 1983). The number of adult chinook salmon returning to spawn in the Sacramento–San Joaquin system is highly correlated with flow rates from March to June 2.5 years prior to spawning, that is, the time when they would have been migrating to the ocean as smolts (Stevens and Miller 1983).

Groundwater withdrawal, particularly when coupled with diversion for out-of-basin uses, is causing precipitous drops in the ground water table throughout the West, as water is being removed from aquifers at rates far faster than it can be replaced (Minckley and Douglas 1991). As major aquifers are drained, characteristics of flow regimes within river basins can be modified due to changes in underground percolation through the watershed.

On a more local basis, groundwater withdrawal for irrigation or domestic use has been a major factor causing or threatening the extinction of a variety of species and subspecies of fishes endemic to spring-fed pools in western desert areas (see

Chapter 11). Seventy-eight percent of the species and subspecies of cyprinodontoid fishes, including members of the families *Cyprinodontidae* (pupfishes, poolfishes, spring fishes, and killifishes) and *Poeciliidae* (live bearers), endemic to arid areas of southwestern states and Mexico face a threat of extinction or have become extinct during this century. Mickley et al. (1991) list groundwater pumping or surface water development as the major threat to or cause of extinction of 25 of these taxa. The plight of one of these, the Devil's Hole pupfish, gained substantial national media attention, which helped fuel support for passage of the first federal endangered species legislation (see Chapter 6). This species ultimately received effective protection from groundwater development when the surrounding Ash Meadows aquifer was designated a national wildlife refuge in 1984.

Riverine Habitat Deterioration Downstream from Dams

Rivers in the United States have been dammed since colonial times to provide water supplies for irrigation, for industrial uses, for domestic consumption, for generation of electricity, and to reduce the effects of flooding. Some river basins have so many dams that they have literally been changed into series of reservoirs connected by limited stretches of flowing water habitats that are highly modified because of the manner in which water is released by the dams. The development of major hydroelectric facilities during the 20th century has affected numerous rivers throughout the United States. Dams cause permanent loss of flowing-water habitats and disrupt riverine fish communities in areas of a basin that are flooded above the dam. Dams also affect river fishes by changing habitat conditions in river reaches below dams due to modification of natural flow patterns. Such effects may be found many miles downstream from dams.

Many reservoir basins fill with silt that settles to the substrate when flow rates abruptly slow as water enters the reservoir. Dams on the Colorado River upstream from Grand Canyon National Park have reduced the sediment load being carried annually down the Colorado within and below the park from 127 million to 18 million metric tons (Johnson and Carothers 1987). This reduction has significantly changed the erosional and depositional processes in that portion of the river system. Further, dredging operations conducted in small reservoirs to deepen a filled basin can cause sediment loading downstream, potentially degrading gravel substrates that may be food-production and spawning habitats of fishes.

Although sediment loading and transport cause major disruptions in habitat conditions in specific river systems, habitat deterioration or change caused by seasonal or daily modifications in temperature and flow regimes is a greater problem in many river systems below dams.

Modification of Temperature Regimes. Natural temperature regimes may be changed somewhat in all reservoir systems, because water held within the reservoir has a large surface area that is exposed to direct solar radiation. However, the modifications of temperature regimes may be most critical to native fisheries resources downstream from reservoirs in which water is released from the base of the dam (that is, water is drawn from the deepest area of the reservoir behind the dam, as in most hydropower facilities). In these cases, if the reservoir is deep enough to stratify and large enough that water below the hypolimnion is released throughout most or all of the spring and summer, water temperatures immediately downriver from the dam are substantially lower than would otherwise occur. For example, average maximum temperature in the Colorado River within the Grand Canyon was 21°C before impoundment of that basin, and midwinter temperatures approached 0°C. Since impoundment, temperature remains close to 10°C throughout the year (Johnson and Carothers 1987). Native warmwater fishes may be eliminated from tailwater areas (the reach of stream modified because of an upriver dam) because waters are too cold for satisfactory growth and reproduction (Funk 1970). Tailwaters may extend for many miles downriver from hydropower dams.

Mitigating disruptions of normal temperature regimes is technically easy; water releases from the hypolimnion need only be supplemented with releases of surface water substantial enough to increase summer water temperatures. However, such releases reduce the amount of water stored to produce electricity. Thus supplementing releases with surface water withdrawals would typically be opposed by electricity producers as counter to efficient energy production. The typical agency response to the loss of native warmwater fisheries has been the establishment of put-and-take or put-grow-and-take hatchery-supported coldwater fisheries (see Chapter 8).

Regulating River Flow.   Although temperature modifications can seriously affect native fishes, in recent years the greatest concern for downstream habitat deterioration caused by dams has been associated with the modification of natural flow patterns. Flow is modified whenever water is diverted for out-of-basin uses (see the section on water diversions in this chapter). Flow within the basin also is modified to control flooding or to generate electricity.

For flood-control purposes, flow is reduced during times of the year when water levels are consistently high and increased later in the year when floodwaters are gradually released. Using water to generate electricity can modify natural flow in river systems on nearly a continual basis. During high-flow periods of the year, hydropower dams retain some water for later use when flow is lower. Thus natural flushing flows typical of particular river basins are reduced or eliminated. Flows may also be modified on a daily basis, as hydropower dam operators release flushes of water called peaking flows downstream through turbines during times of the day or week when electricity demand is highest. Releases are minimized or eliminated during times when electricity demand is low in order to save the water for future electricity generation. Thus flow rates may fluctuate from extremely low or even no flows during nighttime hours and weekends when electricity demand is low to peak flows during daylight hours.

Effects of Regulating Flow.   The modification of natural rates of flow in rivers can increase mortality rates and cause declines in growth rates of fishes. Wide daily fluctuations in flow rates can cause severe icing conditions in the winter, which can increase mortality of incubating eggs or young-of-the-year fishes that seek shelter in gravel substrates during winter months (Peters 1982). If high flows capable of flushing sediments out of gravel substrates are eliminated from streams, fine-particled sediments may accumulate in such substrates, potentially causing mortality of incubating fish eggs (Reiser et al. 1989). Reiser and White (1990) found high mortality rates of chinook salmon eggs in streams in which reduced flows dewatered spawning areas, and in which silt accumulated in spawning gravels due to the elimination of flushing flows. Dewatering of spawning gravels for as short a time as 6 hours caused nearly 100 percent mortality of chinook salmon preemergent alevins (Becker et al. 1983). Paragamian and Wiley (1987) found an inverse correlation between volume of stream flows and growth rates of 1-year-old smallmouth bass.

Highly fluctuating flows can reduce the amount of usable fish habitat in a stream, causing a long-term decrease in population abundances and size and condition of adult fishes (Peters 1982). Increased summer water temperatures brought about by a 90 percent reduction in stream flow in the Trinity River resulted in a reduction in salmonid standing stocks (Harris et al. 1991). The guild of fishes that depended on shallow, slow waters of the stream's edge were reduced in abundance or eliminated from reaches of the Deerfield River, Massachusetts, where flows were most affected by the release schedules of hydropower facilities (Bain et al. 1988). Reduced flows slow the rate of downstream migration of anadromous fishes in the Columbia River basin (Berggren and Filardo 1993), causing higher mortality than might occur under more natural flow regimes (Peters 1982).

River reaches below hydropower facilities that release daily peak and low flows often display reduced diversity, density, and biomass of the benthic invertebrate community. Sediments that accumulate in gravel substrates due to the lack of flushing flows can reduce invertebrate production that serves as a major food base of stream fish communities (Reiser et al. 1989). Reduced summer flows and attenuated spring flood levels in western prairie watersheds have caused substantial declines in riparian poplar forests (Rood and Mahoney 1990). The loss of riparian forests may affect habitat conditions within streams in much the same manner that timber harvest does (see the section on logging in this chapter).

Minimizing Habitat Changes Caused by Regulated Flow.   State fisheries agencies generally recommend instream flow needs to sister water resource agencies who develop flow standards or requirements (see Chapter 6). Recommendations of minimum flow are based on analyses using one of several methodologies. Agencies may employ techniques based largely on historic flow records of watersheds being considered, or they may use computer simulations to predict the effects of various flow reductions on critical habitat characteristics of particular species. Such simulation methods require expansive data sets collected in field surveys (Reiser et al. 1989) and often broad assumptions about habitat characteristics required by different species of fishes.

Few studies have compared long-term standing stocks of fishes in streams before and after mitigation of regulated stream flow. Weisberg and Burton (1993) found that the condition of white perch, yellow perch, and channel catfish was greater, and first-year growth of white perch was higher, in the Susquehanna River below the Conowingo Dam after the establishment of summer minimum flow re-

quirements in 1982 than prior to that year. Harris et al. (1991) found that the standing stock of brown trout increased at one site 7.5 miles below the Rob Roy Dam
(Douglas Creek, Wyoming) after recommended minimum flow levels calculated
to protect brown trout habitat were adopted. The recommended minimum flow
levels were 450 percent higher than the minimum flows allowed in prior years at
this site. Harris et al. (1991) credited an increase in the availability of pools as the
major factor contributing to the increase in standing stock at that site. Another site
of similar distance downstream from the dam showed no increase in standing stock,
apparently because available cover at that site did not increase with the increase in
flow rates. Sites further downstream also showed no increase in standing stock.
Minimum flows at these downstream sites, because of supplemental inflow from
tributary streams, were only 16 to 33 percent above the levels that occurred prior
to adoption of the minimum flow recommendations. Thus minimum flow standards did not substantially improve the quality of fish resources in this system.

Establishing instream flow standards based on within-stream habitat needs of
fishes may not fully protect the quality of riparian zones. Streamside forests can
decline in minimum flows deemed appropriate for maintaining suitable habitat
conditions in the stream channel (Stromberg and Patten 1990).

OBSTRUCTING MIGRATORY PATHWAYS

The construction of dams in river systems inevitably obstructs the movement of
fishes through those systems. Hydroelectric dams have caused declines in populations of freshwater fishes such as the sauger, white bass, and various species of
trout in river systems by preventing their migration to optimal spawning habitats
(Cada and Sale 1993). Winston et al. (1991) demonstrated that dams can fragment
the home range and cause local extinction of populations of minnows that are not
noted for long-range seasonal migrations.
Anadromous fish stocks on the Pacific and Atlantic coasts of the United States have
experienced particularly extreme declines due to the obstruction of migratory pathways by dams. River basins that traditionally served as major spawning grounds
for numerous stocks of anadromous fishes have been extensively dammed. For example, only 8 of the 11 dams on the main channel of the Columbia–Snake River
system have been modified with fish passage facilities to allow some upstream migration (Gessel et al. 1991). A total of 128 hydroelectric and multipurpose dams
store water within the Columbia River basin (Phinney 1986). Because many of these
are impassable, nearly one-third of all anadromous salmonid spawning and nursery habitat in this river system is unavailable (Nehlsen et al. 1991). Impassable dams
on the Sacramento River eliminated chinook salmon spawning migrations, once
the largest migration of that species in California (Williams 1991). Atlantic salmon
stocks in the Connecticut River and Merrimack River of New England were extinct
by the mid-1800s due to the construction of impassable dams (Stolte 1981; Foster
1991). Over 100 populations of Pacific salmon and steelhead trout from West Coast
rivers are extinct, many at least in part due to loss of migratory pathways after construction of dams (Nehlsen et al. 1991).

Effects of Dams on Migration

Dams disrupt populations of all species of anadromous fishes at two critical stages of their life cycles. During upstream migration of prespawning adults and during downstream migration of juvenile fishes. In addition, postspawned adults of iteroparous (see Chapter 3) species such as the Atlantic salmon and American shad must migrate back to the ocean after spawning. Although fish ladders and lifts (elevators) have been engineered and constructed to allow for successful migration of fishes over dams, fish mortality occurs at sites that successfully pass large numbers of fishes. About 5 to 10 percent of the salmon reaching each dam in the Columbia River basin that is equipped with a fish ladder system die either at the site or after successfully negotiating the ladder (Daniel 1993). Apparently some fishes are unable to find or successfully ascend the ladder, whereas others are physiologically stressed sufficiently to cause death.

Even when fishes are successfully passed over dams during upstream migration, downstream migration is a particular problem at hydropower facilities because most of the water released during dry times of the year passes through the dam's turbines. Mortality rates of downriver migrants passing through turbines vary with species and size of fish, type of turbine, and level of operation. Most mortality is caused by extreme changes in water pressure experienced during passage or injuries caused by striking the spinning turbine blades (Stokesbury and Dadswell 1991). Generally, larger fishes suffer higher mortality than smaller ones of the same species, and turbines cause lower mortality when operated at peak efficiency than when operated at less than peak efficiency (EPRI 1992). Mortality can be quite high. For example, mortality rates of salmonid smolts passing through Francis turbines have been reported to be as low as 10 percent and as high as 40 to 50 percent; as many as 46 percent of juvenile clupeids such as American shad die after passing through the Bay of Fundy tidal-flow STRAFLO turbine (Stokesbury and Dadswell 1991; EPRI 1992).

In addition to turbine-related mortality, out-migrating fishes may take significantly longer to reach the ocean because their migration is held up by each dam, and because of the lack of optimal flows both within the reservoir systems and in free-flowing portions of a river that are between dams (Berggren and Filardo 1993). During low-water years, smolts take more than twice as long to descend the Columbia River than they had before its impoundment (Wood 1993). The increased time of downriver migration markedly increases mortality of smolts due to predation, disease, and physiological stress (Wood 1993).

Mitigating the Effect of Dams on Upstream and Downstream Movement

Amendments to the Federal Power Act (the Electric Consumers Protection Act of 1986; see Chapter 6) require the Federal Energy Regulatory Commission (FERC) to consider the impacts that hydroelectric dams have on fish migrations during the licensing of new, or the relicensing of existing, hydroelectric dams. FERC must require fish-passage facilities prescribed by either the Department of Commerce or the Department of the Interior as necessary to mitigate the obstruction of fish migrations (Cada and Sale 1993). Only 9 of 34 states responding to a survey published by Cada and Sale (1993) have developed their own specific policies to mitigate problems of fish passage at hydropower dams.

Mitigation has typically included the construction of fish ladders or fish lifts to allow upstream migration of adult fishes. In recent years construction of turbine bypass systems to facilitate downstream passage of migrating fishes has also been frequently required (Cada and Sale 1993). Several designs of fish ladders and lifts have been developed to provide best upstream passage over dams for particular species of fishes (Figures 9.7, 9.8, and 9.9). Although results have been somewhat inconsistent, design and operation technology has developed to the point where effective passage can be accomplished.

Reducing mortality during downstream passage has proven to be a greater problem. Several basic methods have been used to prevent fishes from passing through turbine systems:

1. Spilling enough water over the dam or through bypass channels to allow fishes to pass by the dam without going through turbines
2. Barging and trucking fishes downriver
3. Using physical or behavioral barriers to divert fishes from the turbine intakes and toward bypass channels

Figure 9.7 Fish ladder at the McNary Dam near the Umatilla River, Oregon. (Photograph courtesy of Calvin Larsen; Photo Researchers, Inc.)

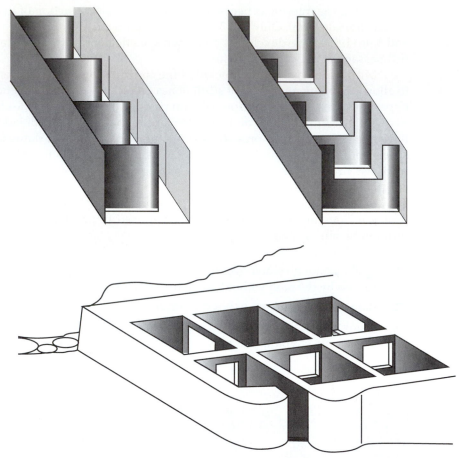

Figure 9.8 Several types of fish ladders. Each type has a specific configuration of baffles that produce characteristic patterns of flow and of quiet, pooled water to enhance the passage of specific fish taxa.

A minimum volume of water can be spilled over the top of the dam or through spillways separate from the turbine intake to move fishes downstream without requiring them to pass through turbines. Fishes passed downriver with spillwater generally suffer lower mortality than those passing through turbines. However, large volumes of water must be spilled to attract substantial numbers of fishes stalled behind a dam. For example, to achieve a 50 percent passage of fishes at Rocky Reach Dam on the Columbia River with spillwater, nearly 40 percent of the spring and summer river flow would need to be spilled (Anderson 1988). Such releases markedly reduce the amount of water that would otherwise be available to generate electricity.

Each year up to 45 million salmon smolts are collected from bypass systems in upper dams of the Columbia–Snake River system and are taken down the Columbia River on barges to be released downstream from the Bonneville Dam (the last downriver barrier to migration in that system; Anderson 1988; Daniel 1993). Essentially,

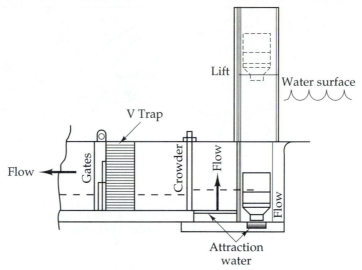

Figure 9.9   Schematic diagram of a fish lift (elevator). Fishes are crowded into the lift compartment, which is raised to the top of the dam where the fishes are released. (From G. F. Cada and M. J. Sale. 1993. Status of fish passage facilities at non-federal hydropower projects. *Fisheries* 18(7): 4–13. Reprinted by permission of The American Fisheries Society.)

barging has developed as an alternative to directly addressing flow and other factors created by hydropower dams that cause high mortality of downstream migrants. Although dam-induced and other within-river sources of mortality are prevented, such action adds another element of artificiality to the conservation of Columbia River salmon resources. Not only are these stocks influenced by the use of hatchery-reared introductions (see Chapter 8), but natural migratory patterns are circumvented by barging them downriver. The ultimate effect of barging on survival and return of these fishes as adults to the upper Columbia basin is not well understood.

A variety of systems have been tested to divert fishes away from turbine intakes and into bypass chutes or channels:

1. Behavioral barriers that drive fish away, such as electrical screens, curtains of air bubbles, curtains of hanging chains, strobe or mercury lighting, and sound
2. Physical barriers that prevent their entrance, such as bar racks, screens, barrier nets, or rows of louver panels (Figure 9.10; EPRI 1986)

In spite of a variety of techniques that have been tested, no single device has proven to be "biologically effective, practical to install and operate, and acceptable to regulatory agencies under a wide variety of site conditions" (Cada and Sale 1993).

Mitigation can also include provision of appropriate flows through impounded river systems to facilitate downriver migration of anadromous species. The Northwest Power Planning Council is required to provide a water budget through the Columbia basin that is of sufficient quantity to enhance out-migration of Pacific salmon smolts. The water budget sets aside water to be released during smolt migration to ensure that fishes are able to reach the Pacific Ocean in suitable periods

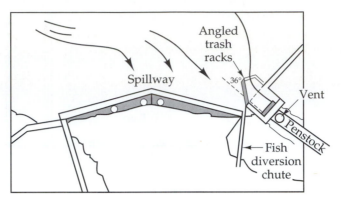

Figure 9.10 An angled trash rack installation for diverting fishes away from the turbine intake and toward a bypass or diversion chute. (From G. F. Cada and M. J. Sale. 1993. Status of fish passage facilities at non-federal hydropower projects. *Fisheries* 18(7): 4–13. Reprinted by permission of The American Fisheries Society.)

of time. However, disagreement among fish and wildlife agencies, the Army Corps of Engineers, and the electric utility industry about the relative importance of providing flow for fisheries resources versus retaining water for generation of electricity has retarded the development of an effective water budget program for salmonid migration in that watershed (Wood 1993).

CHEMICAL POLLUTION

Humans have contaminated public waterways with enormous amounts of chemicals. Industry has traditionally used rivers and lakes as a means of dissipating almost innumerable chemical waste materials, and agricultural practices have led to the introduction of a variety of toxic substances via the runoff of pesticides that have been applied to farmlands. Waterways have become highly enriched with nutrients due to the addition of organic and inorganic sources of nitrogen and phosphorus from agricultural and suburban fertilizers, sewage treatment plant effluents, and septic system outflows. In addition to these chemical pollutants derived from direct input, many waterways have become acidified due to precipitation carrying high concentrations of atmospheric acids, particularly sulfuric and nitric acids, that originate from the burning of fossil fuels (petroleum products and coal). This section reviews the effects of such chemical pollution on aquatic systems and the fisheries resources they house.

Toxic Contaminants

Chemical contaminants can affect the quality of fisheries resources by affecting the health and survival of exposed organisms, and by threatening their safety as human foods (Varanasi et al. 1993). Chemically long-lived toxic substances such as chlorinated hydrocarbon pesticides, industrial chemicals such as PCBs (polychlorinated biphenyls), and heavy metals have been released directly into aquatic systems or have entered surface waters via surface runoff or groundwater percolation in many North American waters. Aquatic organisms ingest and absorb these chemicals. Fishes and macroinvertebrates can develop significantly higher concentrations of contaminants than those present in the water column, because the concentration of contaminants stored in tissues of living organisms increases as the contaminants pass through a food chain. Researchers have concluded that the greatest proportions of the concentrations of various chlorinated hydrocarbons and heavy metals stored in the body tissues of fishes come from dietary sources rather than direct absorption from water (Merna 1986; Woodward et al. 1994). The large-bodied, long-lived predator species that tend to accumulate contaminants at the highest levels are frequently the same species that are the focus of the most active recreational and commercial fisheries. For example, high levels of contaminants such as DDT and PCBs have been identified in salmonids and other important recreational taxa in the Great Lakes (Merna 1986; Reinert et al. 1991; Williams and Geisy 1992).

Effects on Fisheries Resources.   Identifying the impact of specific contaminants on fish health has often been daunting due to the great number of contaminants that fishes are exposed to in many polluted waters. For example, 167 organic contaminants were identified in one lake trout captured in Lake Michigan (Evans et al. 1990). Contaminants in fishes have been associated with (Varanasi et al. 1993; Woodward et al. 1994)

1. A high frequency of liver lesions, including cancers
2. Degeneration of liver tissues
3. Reductions in growth rates
4. Alterations in immune systems
5. Declines in reproductive success

The frequency of liver lesions in the English sole, starry flounder, white croaker, and winter flounder captured in contaminated waters in urban areas of both coasts is higher than frequencies in fishes of the same species captured in nonurban areas (Mix 1986; Varanasi et al. 1993). Schwiewe et al. (1991) exposed English sole from clean waters to chemical extracts of sediments from Puget Sound; these fishes developed liver lesions similar to those described for fishes captured from Puget Sound. Rainbow trout fry carrying high concentrations of several heavy metals exhibited liver cell degeneration, which has been associated with poor growth and death in severe cases (Woodward et al. 1994). Growth rates of rainbow trout fry and juvenile chinook salmon and rainbow trout have been reduced in fishes exposed to chemical contaminants (Mayer et al. 1985; Varanasi et al. 1993; Woodward et al. 1994).

Concentrations of heavy metals such as zinc, copper, and cadmium are correlated to an increase in susceptibility to a variety of infections in fishes (as reviewed in Mayer et al. 1985). The immunocompetence of migrating chinook salmon is reduced in heavily polluted urban estuaries; immune systems of channel catfish injected with PCBs in the laboratory have been similarly affected (Varanasi et al. 1993).

Female English sole taken from contaminated waters had reduced levels of estradiol, a female sex hormone, and exhibited inhibited ovarian development. High percentages of these females either failed to spawn or produced larvae with significantly higher frequencies of abnormalities that may affect survival (Varanasi et al. 1993). Coho salmon eggs exposed to PCB contamination exhibited reduced hatchability (Weis and Weis 1989).

Similarly, survival after hatching is influenced by exposure to contaminants. Westin et al. (1985) demonstrated that survival of striped bass larvae was inversely correlated to concentrations of chlorinated hydrocarbons found in yolk and oils of the eggs from which they hatched. Survival of lake trout fry was inversely correlated to levels of exposure to chlorinated hydrocarbons (Berlin et al. 1981). Between 60 and 70 percent of rainbow trout fry that hatched from eggs carrying high concentrations of PCBs showed a variety of major morphological deformities.

Eating Advisories.   State and federal fisheries agencies generally cannot directly mitigate chemical contamination of fishes because these agencies have no direct authority to prevent pollution of public waters. However, they do advise sister agencies that are empowered to address issues related to chemical contamination of the country's waterways under auspices of the Clean Water Act and other legislation (see Chapter 6).

States generally protect the public from consuming chemically contaminated fishes by issuing eating advisories or bans (Reinert et al. 1991). These advisories typically recommend limits on the amount of a particular fish species from specific waters that can be consumed over a specific period of time. The state fisheries agency may not establish an advisory but may serve to advertise it to the public. For example, in Massachusetts the Department of Public Health and the Department of Environmental Protection have the legal authority to test fishes for contamination and issue advisories; the role of the Division of Fisheries, Wildlife and Environmental Law Enforcement is largely limited to public awareness. Guidance and technical assistance are provided to states by the U.S. Environmental Protection Agency (EPA). The U.S. Food and Drug Administration (FDA) is responsible for regulating the marketing and consumption of contaminated fishes sold in interstate commerce (Reinert et al. 1991).

A 1989 survey showed that advisories restricting the consumption of certain fisheries resources were established in 37 states. PCBs, pesticides, and heavy metals were the focus of the majority of these advisories (Reinert et al. 1991). Because of widespread mercury contamination, in 1994 New Hampshire announced a statewide advisory urging children and women of childbearing age to eat freshwater fish from waters of that state no more than once every two months. Similar widespread contamination led New York to list general advisories recommending a maximum weekly consumption of fishes from all waters while providing stricter guidelines for waterways that are highly contaminated. Interestingly, states may use standards of either the EPA or the FDA, even though the EPA risk assessment methods generally estimate a much higher level of risk for consumption of specific levels of certain contaminants than do those of the FDA.

The Effect of Advisories on Angling.   Advisories or bans on consumption can affect the interaction between angler and resource. Kauffman (1980) found that fishing pressure decreased by 74 percent for a year after a consumption ban due to mercury contamination was established for a branch of the Shenandoah River. Economic losses to local communities due to the decline in fishing activity was well over $400,000. However, fishing pressure returned to preban levels by the following year. Kauffman (1980) believed that fishing pressure had increased because (1) some anglers who ate fish prior to the ban with no visible effects probably ignored the ban after some time; (2) anglers may have been attracted to fishing opportunities on the Shenandoah River regardless of whether they ate the fish they caught; and (3) anglers may have forgotten about the ban after the first year of publicity. "Reliance on advisories rather than regulations to protect human health presumes anglers are aware of the recommendations in an advisory and understand the consequences of their decisions to follow or disregard the advisory" (Reinert et al. 1991). Based on Kauffman's study, agencies should not expect all anglers to make long-term decisions concerning the risk associated with consuming contaminated fishes even when advisories are well advertised.

Nutrient Enrichment: Cultural Eutrophication

Cultural eutrophication, the rapid buildup of organic and inorganic nutrients that come from human sources, can have multiple effects on living aquatic resources. Numerous human activities accelerate nutrient loading into waterways well beyond levels that would occur naturally. Rural land-use activities, particularly agriculture, can substantially increase concentrations of soluble forms of phosphorus and nitrogen, critical nutrients for plant production, via runoff of fertilizers and erosion of soils. Domestic sewage and organic compounds from industrial sources contribute substantially to nutrient loading in urbanized areas of the United States. Untreated or partially treated sewage historically was dumped directly into rivers to be carried away from urban centers. For example, until metropolitan Detroit significantly improved treatment of its domestic waste in the 1970s, an average of 1.4 million pounds of suspended organic solids were discharged into the Detroit River daily (Beeton 1969).

Dumping of sewage sludge at sea has been particularly harmful to coastal waters. Prior to passage of the Ocean Dumping Ban Act of 1988, the New York–New Jersey coastal metropolitan area dumped an average of 5.8 million metric tons of sewage sludge annually into a coastal dump site 12 miles from shore in the New York Bight. Before New York City phased out ocean dumping entirely in 1992, six million metric tons of sludge were dumped into a deepwater site 106 miles off the New Jersey coast annually between 1986 and 1992 (Chang 1993).

Oxygen Depletion and Anoxia.   Decomposition of organic materials in sediments by aerobic organisms—those that use oxygen in their metabolic processes—can result in anoxia, the depletion of dissolved oxygen to levels that are insufficient to support fishes and sediment-dwelling invertebrates. Similar effects may occur in waters receiving inorganic forms of nitrogen and phosphorus from fertilizers and other sources. The increased nutrient concentrations produce substantial increases in production of phytoplankton, rooted aquatic vegetation, and benthic algae. As these plant taxa die, they contribute to organic materials that settle in bottom sediments. Thus, as organic materials produced within the aquatic system accumulate in sediments, rates of microbial decomposition increase, potentially producing the same declines in dissolved oxygen as in waters receiving domestic organic wastes. The entire hypolimnion of bodies of water that stratify (see Chapter 2) can exhibit declines in dissolved oxygen to levels insufficient to support at least some forms of aquatic life throughout the period of stratification each year.

Oxygen depletion and sediment anoxia were thought to be responsible for a major change in the bottom fauna in the western basin of Lake Erie in the 1950s. The *Hexagenia* mayfly, a large-bodied aquatic insect that had been a dominant member of the benthic invertebrate community and an important food of numerous fish species, nearly disappeared and was replaced in importance by small-bodied oligochaetes and midge larvae after the first two periods of recorded anoxia in that basin in the early 1950s (Beeton 1969; Britt et al. 1973). Hayward and Margraf (1987) related the decline in growth rate and stunting of yellow perch to the loss of large-bodied benthic prey in the western basin. Although its extinction may have been caused by multiple factors, the Lake Erie population of blue pike (a subspecies of walleye endemic to that lake) suffered a major decline in abundance during the time when oxygen depletion was first recorded and was spreading throughout that lake's central basin, the major spawning area and center of abundance of this taxon (Smith 1972).

Oxygen depletion and sediment anoxia have caused declines in fisheries resources in other systems. As the area of culturally produced anoxia spread within the upper Chesapeake Bay in recent decades, several major changes occurred in traditionally important fisheries (Officer et al. 1984). Blue crabs became increasingly restricted to shallower waters of the upper bay where dissolved oxygen levels remained suitable. Greater percentages of oysters harvested from deeper waters were dead at capture, and the abundance of bottom-feeding fishes such as the spot and croaker declined while that of the plankton-feeding menhaden increased (Officer et al. 1984). The abundance of numerous species of fishes declined significantly at the deep water sewage sludge disposal site off of the New Jersey coast during years of sludge dumping (Chang 1993).

Oxygen depletion in the hypolimnion is particularly harmful to fish species that dwell there in the summer because surface waters are too warm. To find appropriate dissolved oxygen and temperature conditions, striped bass can be forced into a narrow depth zone just below the thermocline (see Chapter 2) or into very restricted areas of large systems where oxygen depletion may not be as severe and where surface waters are cooler (Cheek et al. 1985; Matthews et al. 1985; Price et al. 1985). Physiological stress associated with suboptimal temperature and oxygen conditions can lead to adult mortality, increased occurrences of diseases due to crowding, and reduced reproductive output in the following year (Coutant and Benson 1990). The lake cisco and other coldwater-adapted species are similarly limited to narrow zones in enriched lakes due to temperature and dissolved oxygen conditions (Larkin and Northcote 1969).

Changes in Primary Production.   Nutrient loading can produce changes in habitat conditions through increases in the levels of primary production and modification of the taxa that predominate the plant community. Rooted aquatic vegetation can become dominant throughout shallow water ponds. Declines in the quality of fisheries caused by disruptions of predator–prey interactions and difficulties in angling within dense aquatic vegetation frequently lead to "weed control" programs intended to reopen areas of the water column (see Chapter 10). Expansive growths of rooted vegetation in small ponds often result from nutrient loading caused by fertilizers from agricultural and suburban uses, or from outflow from septic systems of housing development's on the water's edge.

Some declines or disappearances of rooted aquatic vegetation in deeper bodies of water also have been associated with nutrient loading. Accelerated nutrient input promotes dense growths of phytoplankton, which reduce light penetration through the water column, and of epiphytes (microscopic plants that grow on the leaf surface of other plants) and filamentous algae that shade the leaf surfaces of rooted aquatics. Photosynthesis of the rooted aquatics is reduced due to reduction in the intensity of light reaching the leaf surface. Such conditions have led to the loss of rooted aquatic vegetation in both freshwater and estuarine systems (Phillips et al. 1978; Anderson 1989).

Estuarine eelgrass beds provide critical nursery, feeding, and shelter habitats for numerous fish and shellfish species and contribute substantial organic materials for associated food webs as well as for export to food webs in adjacent coastal ecosystems (Thayer et al. 1984; Phillips 1984; see Chapter 2). A decline in the abundance of eelgrass due to nutrient loading is associated with a decline in the blue crab fishery in Chesapeake Bay (Anderson 1989). Similarly, declines in abundance of rooted aquatics is thought to be responsible for the decline or loss of numerous freshwater fish species from specific waters (for example, redfin pickerel, yellow perch, and black crappie, as reviewed in Trautman 1981).

Mitigating Cultural Eutrophication.   The dumping of sewage sludge in coastal waters has been severely restricted by the Ocean Dumping Ban Act. The Clean Water Act (see Chapter 6) mandates that urban areas develop sewage treatment facilities and provides grants to help accomplish that goal. This act has been effective in reducing the amount of organic pollution entering waterways from urban areas. For example, upgrading the efficiency of sewage treatment facilities in Detroit and Chicago has significantly improved water quality in the western basin of Lake Erie and the Illinois River between Chicago and Peoria (Roseboom 1993). Cities have historically gathered organic wastes in sewer systems and delivered them to waterways from central piping systems. Thus proper sewage treatment facilities could be included easily into the gathering system, markedly reducing organic enrichment of adjoining waterways that historically served to carry the organic wastes away.

Even though the effect of organic pollution from many urban centers has been significantly reduced, nutrients entering aquatic systems from widespread areas rather than single sources has proven much more difficult to control. Such nonpoint sources of nutrients, such as fertilizer runoff from agricultural lands or septic effluents and lawn fertilizers from shoreline areas surrounding small lakes, continue to be a major cause of water-quality degradation (Roseboom 1993). Land-use practices must be modified to reduce the source of nonpoint enrichment; this has proven difficult to achieve.

Acidification of Surface Waters

Acidification (the buildup of high concentrations of hydrogen ions) of surface waters has eliminated fish populations or caused declines in their viability in various waters of the United States, particularly in much of the Northeast and to a lesser extent southward along the Atlantic coast (Haines 1981). Acidic precipitation causes the chronic low-pH characteristic of these waters. Although precipitation can become acidic from the solution of natural atmospheric gases such as hydrogen sulfide emitted by volcanoes, increased acidity in precipitation is generally attributed to atmospheric sulfur dioxide and nitrogen oxides released by the combustion of fossil fuels (Haines 1981).

Watersheds that have bedrock (such as granite) and soils low in soluble calcium are most sensitive to acid precipitation; such areas have little capacity to buffer the high hydrogen-ion concentrations delivered by rain and melting snow. Fisheries resources in regions with calcium-rich bedrock such as limestone are little affected by acid precipitation. Limestone (calcium carbonate) chemically combines with sulfuric or nitric acids to form a calcium salt and carbonic acid, which contributes significantly lower concentrations of hydrogen ions to water than do the stronger acids it has replaced. Thus limestone tends to buffer the pH of acid precipitation to levels that are at most only weakly acidic and are well within the tolerance limits of fishes.

Highly acidic conditions can cause concentrations of soluble forms of heavy metals, such as aluminum, to increase in surface waters. The combined toxicity of low pH and high concentrations of aluminum was believed responsible for mortality of brook trout in acidified waters of New York state (Cronan and Schofield 1979) and for larval blueback herring and striped bass in acidified tributaries of the Chesapeake Bay (Hall et al. 1985; Klauda and Palmer 1987). Although low pH and high concentrations of soluble aluminum act synergistically to disrupt the health of fishes, aluminum toxicity may be higher at intermediate than at more severe levels of acidity. For example, aluminum is more toxic to brook trout at a pH of 4.9 than at a pH of 4.4 (Siddens et al. 1986).

The pH of some lakes in the Adirondack Mountains of New York state has been chronically low for decades, causing the elimination of resident fish populations. In many other waters the pH through much of the year is at least minimally suited for survival of most fish species. However, episodic dosing that occurs during spring storms or during the melting of winter snow packs can temporarily cause a short-term "spike" of hydrogen ions into surface waters, increasing acidity to levels that are physiologically stressful or even lethal to fishes. Blueback herring and brook trout have exhibited higher rates of mortality and lower rates of growth when exposed to episodes of low pH and high concentrations of soluble aluminum than when the exposure to these conditions was continuous (Siddens et al. 1986; Klauda and Palmer 1987; Fiss and Carline 1993). Similarly, growth was retarded when brook trout were first exposed to sublethal levels of acidity but recovered to normal rates even when the fishes were exposed to the same low pH over extended periods of time (Tam and Payson 1986).

The Impacts of Acidified Waters on Fishes.   Low pH can cause stress and increase mortality of fishes in several ways: Efficiency of oxygen uptake may be reduced, fishes may be unable to regulate ion levels in the blood and tissues, and high concentrations of soluble heavy metals, such as aluminum, that are liberated in acidic waters may be toxic (Haines 1981).

Moderately high hydrogen ion concentrations may cause gill epidermal tissues to secrete excessive amounts of mucus. The resulting thick layer of mucus will reduce the rate of oxygen diffusion across the gill membranes into the bloodstream (Daye and Garside 1976). High hydrogen-ion concentrations can cause damage to gill tissues, potentially disrupting respiratory and excretory functions of the gills. At extremely high hydrogen-ion concentrations in water, the pH of the blood is lowered; this reduces the capacity of hemoglobin to absorb oxygen (Haines 1981).

Soluble aluminum produces similar effects to gill tissues, causing epithelial necrosis and initiating excessive production of mucus that can lead to suffocation.

Low pH in water can cause excessive loss of sodium, chlorine, and other ions from the bloodstream. Chloride cells in the gill epidermis normally prevent the diffusion of specific ions such as sodium and chlorine ions from the bloodstream to the surrounding water. However, in the presence of high hydrogen-ion concentrations, chloride cells are less effective in preventing ion loss (Haines 1981). Low pH also increases the permeability of gill tissues, causing excessive loss of other blood ions and increased uptake of hydrogen ions (Fromm 1980). To offset blood acidosis caused by the uptake of hydrogen ions, calcium and potassium may be mobilized from the skeleton and lost through the gills to surrounding water (Peterson and Martin-Robichaud 1986).

Reproductive failure may be the greatest cause of the decline of most fish populations exposed to chronically acidic waters (Booth et al. 1993). Egg and larval life stages of fishes are usually more susceptible to the effects of acidity than are adults. Also, adult females may fail to produce and release eggs properly, probably due to an inability to maintain appropriate levels of calcium in the blood serum (Haines 1981).

Mitigating Effects of Acidification.   The effects of low pH in small lakes and ponds with low flushing rates can be reduced effectively by periodic applications of either hydrated lime or agricultural limestone (Kretser and Colquhoun 1984). Liming will increase soluble calcium concentrations and the pH in acidified lakes (Marcus 1988). Because the solubility of most metals increases at lower pH, acidified lakes that have been treated with lime will also exhibit decreases in the levels of soluble forms of metals such as aluminum and zinc (Marcus 1988). Liming of an acidified lake in Ontario, Canada, led to marked increases in survival rates of lake trout eggs and juvenile life stages (Gunn et al. 1990). Booth et al. (1993) constructed spawning shoals consisting of limestone rocks 3 to 20 centimeters in diameter in a lake with long-term pH levels that were suited to survival of juvenile and adult lake trout, but with episodic acid spikes that caused high embryo mortality. The pH of interstitial water in the spawning shoals was higher than in natural spawning areas, and embryo survival increased substantially. Production of brook trout in an Adirondack lake after liming was similar to that in nonacidified lakes of the same region (Schofield et al. 1989); however, reacidification within 5 months after liming led to rapid declines in brook trout biomass and production due to increased mortality and reduced growth rates of surviving fishes. Because lime introduced into lakes may be flushed out slowly through outlet streams, treated lakes will ultimately reacidify, requiring periodic applications of lime to retain suitable conditions for enhanced fish production.

Neutralizing acidified waters is more difficult in streams than in ponds or small lakes. One-time applications of lime are rapidly flushed downstream in stream systems. Agencies have attempted to develop means of neutralizing low pH in streams without requiring the nearly constant presence of personnel for continuous application of lime. Workers in Pennsylvania investigated whether chunks of limestone or soda ash (Na_2CO_3) placed in wire baskets in streams would provide long-term neutralization of low pH. Limestone chunks elevated pH for about two weeks, and soda ash for up to several months. However, with time the surface of these materials became covered with silt or benthic algae such as diatoms, reducing the surface area that was in immediate contact with water. Thus, after these initial time periods, progressively slower rates of dissolution of these materials led to subsequent reductions in the pH of study streams (Arnold et al. 1988).

Other workers have developed a variety of automated stream dosers that release a slurry of dissolved lime on a predetermined schedule. Such systems can produce persistent neutralization of acidified waters, but they are subject to clogging or other malfunctions because they are left unattended (Zurbuch 1984; Janicki and Greening 1988). When difficulties with mechanical failure or clogging are resolved, two issues still limit the utility of automated dosers. The amount of lime needed to maintain suitable dosing may be difficult to supply in streams flowing through remote areas lacking easily reached access roads. Secondly, the cost of such systems can prohibit their use in all but a few carefully selected streams; such costs may far outweigh any economic benefits in fishing activity gained by neutralizing acidity. Thus the technology that would allow long-term neutralization of acidified streams in a cost-effective and dependable manner has not yet been fully developed.

HABITAT QUALITY AND THE FUTURE

As the population of the United States grows, the quality of aquatic habitats will continue to decline unless we become more effective stewards of natural resources. Effective habitat protection depends on the emphasis that government places on environmental protection versus economic development and other human needs. Frequently, as a society we treat the symptoms rather than the source of environmental problems. For example, we raise salmon in hatcheries and barge them downriver to try to circumvent the results of intensive water development. Similarly, we attempt to raise the pH of acidified waters by liming rather than reduce atmospheric acids at their source. Wild fisheries resources from headwater streams to continental shelves are vulnerable to or seriously affected by ongoing habitat problems. Protecting habitat quality will remain a major prerequisite to effective fisheries conservation and management in the future.

10
Manipulating Habitats and the Biotic Community

Throughout much of the 20th century state and federal fisheries agencies have developed a variety of programs to improve the productivity of fisheries by manipulating the ecosystems in which fishes live. The basic goal of such programs is to "funnel more of the . . . [aquatic system's] resources of energy and material" through particular species of important fishes (White 1975), or to improve the frequency of interactions between fishes and anglers and commercial harvesters. This chapter reviews some of the more frequently used manipulations, grouping them into the following categories:

1. Modification of the physical structure of aquatic habitats
2. Manipulation of the chemical environment by fertilization
3. Modification of the species makeup of the biotic community

MODIFYING THE STRUCTURE OF AQUATIC HABITATS

Habitat Manipulation in Streams

Most habitat-enhancement projects conducted within the stream channel focus on one or more of the following (modified from White 1975):

1. Increasing available cover for protection from potential predators or "social" (intraspecific) competitors
2. Increasing available shelter from rapidly flowing and turbulent currents
3. Facilitating removal of silt and clay particles to improve the quality of gravel substrates for food production and spawning

4.   Increasing the expanse of a stream that possesses optimal flow velocities and
     water temperatures for particular fish species

Pool habitats, which can provide protection from predators and competitors and shelter from fast currents, often occur sparsely in high-gradient, headwater streams. Thus many stream habitat-manipulation projects have focused on increasing the number and size of pools in stream sections. To meet the above objectives, fisheries managers place structures within the stream channel that modify flow direction, velocity, or turbulence. Boulders, logs, deflectors, and check dams are among the most commonly used habitat-improvement structures.

Deflectors are constructed to block a portion of the stream channel. These structures increase on-site velocity and turbulence by deflecting flow or concentrating it into a narrow portion of the channel. By increasing velocity or turbulence, deflectors cause silt and other small-particled materials in the substrate to be flushed from the immediate area, thus improving the quality of gravel and rock riffle areas. They also can deepen the water column through the narrowed channel and can create a small pool of quiet water on their downstream side (Everhart et al. 1975). Deflectors can be simple structures such as fallen trees or single logs, or they can be complex, large structures constructed with boulders and logs that are anchored to a stream bank or the streambed.

Check dams are low dams that increase water depth to no more than one to several feet above natural depths. These structures are often placed in stream stretches where the gradient is too high to allow natural development of pools. The turbulence created by water flowing over the low dam flushes silt and shifts gravel away from the dam, forming a pool (Figure 10.1). Shallow pools are also formed immediately upstream from the dam.

Figure 10.1   A fallen tree for cover (foreground), a log deflector (midground), and two log check dams (background) constructed to enhance juvenile coho salmon habitat in a stream in Mount Hood. (Reprinted by permission of *Fisheries*, Volume 18, Number 8, American Fisheries Society.)

Underpass deflectors, or digger logs, have also been used to modify stream habitat. Underpass deflectors consist of an anchored log that crosses the stream perpendicularly, blocking the flow of water from the surface down to several inches above the streambed. Although intended to flush out silt, underpass deflectors collect debris and essentially become dams. They also attract fewer sport fishes such as trout than do similar-sized logs lying angled into the stream or parallel to the flow of water (White 1975).

Shelter and cover can also be provided by relatively simple techniques, such as placing single, large boulders in fast–flowing areas of a stream (White 1975), felling trees into the stream so that they lie obliquely to the flow of water, or anchoring bundles of brush in water next to the bank. Flow deflects around boulders, creating a small area of relatively calm water on their downstream side. Fishes may use such sites for shelter. In many headwater streams, fallen trees with intact root wads lodged along the bank provide substantial pool habitat and cover (see Chapter 9). Although intended to simulate such natural woody debris, felled logs and brush piles provide only temporary cover. Because they are easily dislodged during high waters, they can be washed away, accumulating as debris piles in downstream areas and causing localized erosion and bank slumping.

Other methods have been used to improve specific habitat conditions in salmonid streams, but none as widely as the above. Agencies have attempted to improve spawning conditions by rehabilitating gravel substrates, using a variety of methods including bulldozers and portable pumps to dislodge silt that is subsequently flushed downstream. At other times, gravel has been added to stretches of stream with only limited natural gravel substrates. All of these methods have limited application in habitat manipulation (Reeves et al. 1991). Reduction of sediment loading by protecting vegetation in the riparian zone is the most effective means of preventing degradation of spawning gravels with silt.

Effectiveness in Enhancing Fish Productivity. Researchers have conducted few long-term studies to evaluate the effectiveness of stream habitat-manipulation projects (Reeves et al. 1991; Riley and Fausch 1995). Some studies that have measured short-term effects have reported an increase in the biomass and catch rates of fishes in modified streams. Saunders and Smith (1962) reported that the standing crop of trout in a stream on Prince Edward Island doubled after the stream was modified with extensive construction of check dams, deflectors, and anchored brush piles. Trout populations and/or fishery yields increased on 11 different streams in the eastern United States in which habitat-improvement programs had been conducted (White 1975). Similarly, the number of catchable trout in treated sections of a variety of Wisconsin and Colorado streams increased after installation of log dams and deflectors (Hunt 1988; Riley and Fausch 1995). Overwinter survival and smolt production were several times greater in manipulated than in unaltered stretches of a stream on Queen Charlotte Island, British Columbia (Reeves et al. 1991). Moore and Gregory (1988) used deflectors and semicircular rows of large rocks along stream banks to increase the amount of area consisting of low-velocity flows and backwaters, or "lateral habitats." These habitat-improvement structures increased the abundance of age-0 cutthroat trout, whereas abundances declined in other stream sections where lateral habitats were removed. Nickelson et al. (1992) compared the relative abundance of fishes in pools created by habitat improvement structures to the abundance found in natural pools. In the summer, abundances were the same for both types of pools; in the winter, abundances were higher in natural than in constructed pools. However, brush piles anchored in constructed pools increased abundance of juvenile coho salmon to levels found in natural pools in the winter (Nickelson et al. 1992).

In some instances, extensive habitat manipulation has failed to change the abundance or biomass of resident fishes or the fisheries they support. Construction of numerous half-log covers in a 700-meter length of McMichaels Creek, Pennsylvania, did not increase the number of large brown trout in the stream (Hartzler 1983). Extensive construction of deflectors in two warmwater streams in Ohio produced higher numbers and densities of fishes and greater diversity of species than in areas without deflectors. However, fisheries supported by smallmouth bass and other members of the sunfish family did not improve (Carline and Klosiewski 1985). When ineffective, habitat improvement projects appear to fail for one of two reasons: "1) the desired change in the habitat is not created, or 2) the desired change in habitat fails to produce the desired change in the fish population" (Nickelson et al. 1992).

The success of habitat-improvement projects in streams may vary according to the species of fishes present. Hartzler (1983) felt that the density of brook trout would be more likely to increase after initiation of habitat-improvement projects than would densities of brown and rainbow trout. After studying the response of brown trout to the addition of cover, Lowery (as reviewed in Hartzler 1983) felt that habitat manipulation probably would not dramatically increase the abundance of adults of this species. However, Riley and Fausch (1995) reported significant increases in abundance and biomass of both adult brown and brook trout after habitat manipulation.

Some workers have suggested that abundances of fishes and catch rates of associated fisheries increase in altered sections of streams largely because fishes from adjoining areas are attracted to these sites. Thus streamwide production has not necessarily changed; fishes simply redistribute due to the presence of altered stream habitats. If harvest increased but production did not, stream habitat alterations may deplete a fish population, not improve the fishery it supports on a long-term basis. Gowan et al. (1994) and Riley and Fausch (1995) demonstrated that fishes inhabiting altered habitats originated from other stream areas. However, fishes immigrated to altered sites from substantial distances, not from adjacent areas. Apparently, immigrant trout contacted and remained at altered sites while in the process of long-distance movements. In this instance, structures that created additional cover and shelter habitat probably increased streamwide production, because the fishes supported by such sites were those that had not previously settled into suitable habitat elsewhere in the stream.

Creating cover and shelter sites is more common in small than large streams (White 1975), in part because the cost of large-scale habitat-improvement projects is often prohibitive. Although these structures can influence changes in the nature of a fishery, the increase in fish and fishery productivity may be modest in relation to the cost of constructing and maintaining the habitat-improvement structures.

However, habitat-enhancement projects may be cost-effective means of increasing catch rates if alternatives also are costly, as with hatchery production and stocking. Rather than extensive habitat enhancement, in many instances land management agencies are employing land-management practices, such as protection of riparian zone vegetation, that provide natural habitat complexity through the addition of large woody debris (as mentioned in Chapter 9).

Creating Cover in Standing Freshwaters

Water-level fluctuations in many reservoirs inhibit the development of rooted aquatic vegetation and the shelter or cover it provides for numerous fish species that inhabit shoreline areas. Artificial structures that create cover habitat are often constructed in reservoirs where natural cover otherwise would not occur. Managers also create cover in natural lakes where natural cover is thought to be limited. Artificial cover in reservoirs and lakes may consist of timber left standing in submerged areas close to the shore or structures placed in the water and anchored in some manner to the substrate to prevent subsequent movement.

Standing Timber.   Inshore areas with standing timber in the water column will attract relatively dense concentrations of species such as the largemouth bass and black crappie, and they will attract anglers (Davis and Hughes 1971). By the third year after Bussey Lake, Louisiana, was opened to fishing, over 90 percent of all anglers fished in areas with standing timber (Davis and Hughes 1971). Angler harvest is often high in such habitats; harvest in timbered areas of Bull Shoals, Missouri, was over 3,000 pounds per acre, whereas that in areas void of timber averaged slightly more than 100 pounds per acre (Burress 1961).

Providing areas of standing timber can attract far greater numbers of anglers than are justified by angler success. A higher percentage of anglers catch and keep at least one fish, but the average number of fishes caught per hour of fishing may be lower in timbered areas (Davis and Hughes 1971; Brown 1986). Some fish species do congregate around timber, and this type of cover is easily seen by anglers searching for a place to fish. Thus high harvest levels in such areas is due to the concentration of both fishes and of anglers. However, individual anglers may not necessarily improve their catch rates or total harvests by fishing in areas with standing timber. It is not clear whether standing timber or any other types of artificial cover consistently increase abundance and productivity of sport fish populations, or whether the existing populations simply concentrate around such cover.

Creating Anchored Cover.   Twenty-eight of 45 state agencies responding to a survey conducted in the 1970s had artificial reef programs (Prince and Maughan 1978). A variety of materials including brush, automobile tires, beds of stakes, and waste concrete block or boulders have been used to create anchored cover (commonly called artificial reefs) in reservoirs and lakes.

Unlike brush reefs and beds of stakes, tire and stone reefs are essentially permanent structures that do not require rebuilding with time (Figure 10.2). However, many people may consider piles of tires that are visible from the water's surface unsightly. Submerging tire reefs to depths that are out of sight would address such

Figure 10.2   A submerged tire reef. (Photograph courtesy of Richard B. Stone.)

a problem but might reduce the effectiveness of cover in attracting fishing effort. If anglers prefer fishing at sites with cover, they need to be able to find such areas easily. Time spent searching for a place to fish instead of actually fishing might frustrate anglers, potentially leading to less satisfaction with the angling experience.

The Effectiveness of Anchored Reefs.   All types of artificial structures tend to attract more fishes than do open-water areas. About 10 times and 30 times as many fishes were attracted to tire and brush reefs, respectively, as were attracted to areas without cover in Barkley Lake, Kentucky (Brown 1986). Densities of largemouth bass, black crappie, bluegill, and channel catfish were five to eight times higher around artificial reefs than in natural shoreline areas of two Idaho warmwater reservoirs (Mabbott 1991). Although fishes clearly concentrate around artificial reef structures, studies have not conclusively indicated that productivity of fish populations is increased by such artificial habitats. Bluegill production increased in small ponds only when artificial structures covered at least 50 percent of the surface area of the bottom (Pardue 1973). Wege and Anderson (1979) felt that construction of artificial structures in large ponds did not affect total fish production. However, Prince and Maughan (1978) stated that production increased after artificial reefs were placed in Mountain Lake, Virginia.

Freshwater artificial reefs are most often constructed in smaller systems due to the time, effort, and expense that are necessary to create and maintain reefs big enough to be effective in large reservoirs or lakes. However, expansive reefs constructed of waste concrete, boulders, or other materials have been used in systems as large as the Great Lakes. Large reefs in Lake Erie are marked with surface buoys to allow anglers to find them easily. Smallmouth bass and other species do concentrate around these reefs, and anglers do catch them. As in many smaller systems, it is not clear whether these artificial habitats enhance population characteristics or simply concentrate fishes, exposing them to concentrated fishing effort and perhaps higher fishing mortality than might occur were the fishes not concentrated.

Artificial Cover in Coastal Ecosystems

Both artificial reefs and fish-aggregating devices (FADs) have been widely established in open waters of continental shelves to enhance recreational and commercial fisheries.

FADS.   A FAD consists of some type of umbrellalike, floating cover that is anchored to the ocean bottom. Such cover tends to attract schools of pelagic fishes that typically wander open ocean areas (R. Buckley 1989). Aggregating such schools of fishes in known areas can increase catches while significantly reducing the time and cost typically associated with searching for these species (R. Buckley 1989). Anchoring FADs can be difficult in natural areas where substrates are soft; forceful currents and storm surges can move FADs off of the site of anchoring. The majority of FAD sites on the Atlantic coast of the United States include some type of artificial reef system that provides a solid anchoring base as well as habitat for a benthic fish community (McGurrin 1989).

Fisheries scientists have generally assumed that FADs enhance the harvest within a fishery but do not increase biomass production that supports the fishery. Brock (1985) proposed that regional production of some species such as the yellowfin tuna may be enhanced by FADs. He noted that yellowfin tuna attracted to FADs may focus their feeding activity on different, and thus potentially untapped, foods from those fed upon by yellowfin not associated with FADs. However, increasing the efficiency of harvest may overcome any benefit gained in increased production. FADs may be stimulating overharvest of juvenile yellowfin tuna in nearshore areas that support heavy fishing pressure (Buckley 1989).

Marine Artificial Reefs.   Marine artificial reefs stimulate fishes to aggregate and can supplement biomass production to levels significantly higher than would otherwise be reached (Bohnsack and Sutherland 1985; Buckley 1989). Fishes use artificial reefs "for shelter, feeding, spawning and orientation" (Bohnsack and Sutherland 1985). Fishes colonize artificial reefs rapidly, reaching maximum population sizes within as short a time period as several months. Fish communities inhabiting these structures become stable within 1 to 5 years, although seasonal variations in population abundances and species composition may occur (Bohnsack and Sutherland 1985).

A variety of materials have been used to create artificial reefs in coastal waters. Many states generally have approved projects that use "materials of opportunity" (that is, whatever is available at the lowest cost; Buckley 1989; McGurrin 1989). Automobile bodies, scrap tires, waste concrete, building rubble, and sunken ships or barges frequently are used to establish artificial reefs in coastal waters (Stone 1978; Fisheries 1983). Some materials such as automobile bodies, although easily and cheaply attained, are not suited to reef construction as they tend to disintegrate within a few years after the reef is formed (Stone 1978).

Most artificial habitat in U.S. waters of the Gulf of Mexico has been created by the construction of over 3,000 petroleum platforms. A variety of fishes congregate around platforms, including recreationally important taxa such as snappers, groupers, speckled and sand trout, Atlantic croaker, and sheepshead, and predators such as barracuda, cobia, and schools of bluefish (Gallaway and Lewbel 1982). Platforms provide better fishing success than do adjacent soft-bottomed habitats (Gallaway and Lewbel 1982). These artificial habitats attract the majority of offshore recreational anglers and charter boat operators in Louisiana and Texas and provide two to three times higher fishing success than do natural reefs in California waters (Dugas et al. 1979; Harville 1983; Figure 10.3). Oil companies constructing platforms in public wa-

Figure 10.3 Recreational fishing around an oil platform in the Gulf of Mexico. (From R. Dugas, V. Guillory, and M. Fisher. 1979. Oil rigs and offshore sport fishing in Louisiana. *Fisheries* 4(6): 2–10. Reprinted by permission of The American Fisheries Society.)

ters are required to remove them once they are no longer in use. Once decommissioned, some oil platforms have been sunk at state or federally approved sites to create submerged reef habitat (Fisheries 1983). These cooperative efforts have allowed oil companies to remove platforms and to bolster their public image in the process.

Reefs have been constructed by nonpublic institutions more often as a means of disposing of solid waste materials than as a means of enhancing fisheries (Buckley 1989). Although some of these reefs are successful, many have failed to stimulate the productivity of fish resources and fisheries because of the materials used or the lack of emphasis on placement and construction in a manner most likely to meet fishery objectives.

Water-Level Management in Reservoirs

The initial period of high biological productivity in newly filled reservoirs often has resulted in higher levels of fisheries production than can be sustained through time (Kimmel and Groeger 1986; see the description of reservoir productivity in Chapter 2). Although a decline in initial productivity of fish populations in reservoirs is inevitable, the period over which highly productive fisheries are sustained can be prolonged by proper management. Reducing high harvest rates by the imposition of regulations such as minimum legal size limits and slot limits have improved fishing in reservoirs for species such as the largemouth bass over various periods of time (Redmond 1986; see Chapter 7). Fisheries agencies also have maintained high productivity levels by raising or lowering water levels within reservoirs on a seasonal basis.

Objectives of Water-Level Manipulations. Water-level manipulations have been used to improve the quality of reservoir fisheries for nearly half of a century. Objectives of such programs may include:

1. Affecting the reproductive success of fish species
2. Increasing biological productivity
3. Suppressing growth of rooted aquatic vegetation
4. Stimulating growth rates of predator species by modifying predator–prey interactions

Modification of water levels can influence the reproductive success of species that spawn in near-shore habitats. Reproductive success of nuisance species (such as the common carp) can be disrupted by lowering water levels during or immediately after the spawning season. The lower water levels exclude spawners from the inshore spawning habitats, cause high mortality of eggs deposited in subsequently exposed inshore substrates, and increase the mortality of young-of-the-year fishes that require inshore nursery habitats for feeding and shelter (Shields 1958; Groen and Schroeder 1978; Ploskey 1986). However, this method has not consistently reduced the abundance of dominant, unwanted fish populations to desired levels (Ploskey 1986).

Flooding of the shoreline area including expanses of terrestrial grasses during spawning has increased spawning success of recreationally important species such as the yellow perch and northern pike that scatter or lay eggs on submergent vegetation (Hassler 1970; Nelson and Walburg 1977; see Chapter 3). Flooding of shoreline areas for up to several months after the spawning season has improved growth and survival of young-of-the-year largemouth bass (Aggus and Elliott 1975; Rainwater and Houser 1975) and has increased year class strength of northern pike (Ploskey 1986) by increasing the amount and quality of nursery habitat.

Inundating shoreline areas in the spring also can increase nutrient input into a reservoir by initiating decomposition of herbaceous shoreline vegetation. This release of soluble nutrients can stimulate phytoplankton production (Ploskey 1986), which in turn may increase productivity at all trophic levels. It also improved water clarity in Kansas reservoirs by causing precipitation of clay particles suspended in the water column (Groen and Schroeder 1978).

Prolonged drawdowns have been used to decrease production of rooted aquatic vegetation. Such drawdowns cause desiccation or freezing of root systems of some species of aquatic macrophytes, thereby reducing densities of these species and the expanses of submerged substrates covered by them. Reduction in rooted aquatic vegetation may redirect soluble nutrients toward phytoplankton production, influencing zooplankton production and the production of fish species dependent on these biota for food (Bennett 1970).

Although some coverage of submerged substrates with rooted aquatic vegetation provides critical spawning and nursery habitat for a variety of fish species, too great an expanse of rooted aquatics can inhibit growth rates of predator fishes by disrupting their ability to capture prey. Savino and Stein (1982) experimentally showed that predator efficiency is significantly lower in areas with high densities of rooted aquatics. Thus suppressing growth of rooted vegetation and lowering water levels in spring and summer months may affect predator–prey interactions in fish communities. Spring and summer drawdowns concentrate forage fishes in open areas separate from dense aquatic vegetation; such changes in the distribution of forage species can result in increased efficiency of predation imposed on them, thus accelerating the growth rates of predator species (Heman et al. 1969; Ploskey 1986).

The Kansas Program.   The Kansas Fish and Game Commission conducts a program of seasonal water-level manipulations on a series of flood control reservoirs. The changes in water level support fisheries and wildlife management as well as control flooding. This program provides for (Groen and Schroeder 1978; Willis 1986) the following:

1. *Gradual increases in the water level in the spring to inundate shoreline herbaceous vegetation and rocky areas.* This is intended to enhance spawning success and to provide additional nursery habitat for newly hatched young-of-the-year fishes.

2. *Stable, high water levels in the late spring through early summer.* This allows shoreline vegetation to decompose, adding nutrients to the water and improving water clarity by precipitating colloidal clay particles from the water column.

3. *Lower water levels from midsummer to the fall.* Lowering of water levels allows revegetation of shoreline areas. It also concentrates forage fishes away from cover, making them more available to predators.

4. *Water levels in the fall sufficiently high to inundate a narrow band of herbaceous vegetation right at the shoreline.* This activity enhances fall habitat requirements of migratory waterfowl.

5. *Low water levels for the winter.* This reduces ice and wave damage to shoreline vegetation. It also provides maximum storage capacity for spring floods, preventing unplanned spring drawdowns (Figure 10.4).

Such management programs may call for seeding inshore areas with millet, rye, or other herbaceous species during drawdown in the late summer in order to ensure sufficient growth of vegetation in this area the following year.

The water-level management program in Kansas led to higher population abundances of walleye, white crappie, and white bass than those found in the same reservoirs before initiation of the management program (Willis 1986). However, summer drawdowns were not conducive to stimulating survival and production of largemouth bass. A water-level management program in Lake Eufaula, Oklahoma, which included stable, high water levels through the summer, increased recruitment of young largemouth bass and doubled that species' abundance (Wright 1991). The Kansas program is most suited to enhance fish populations of species whose young-of-the-year inhabit open-water areas by midsummer (Willis 1986).

MANIPULATING THE CHEMICAL ENVIRONMENT BY FERTILIZATION

Fertilization can significantly increase fish production in small ponded waters. Early studies at Auburn University (for example, Swingle and Smith 1939; Smith and Swingle 1940) indicated that accelerated rates of phytoplankton production resulting from the application of commercial fertilizers enhances biological production of taxa throughout aquatic food webs, including species of forage and sport

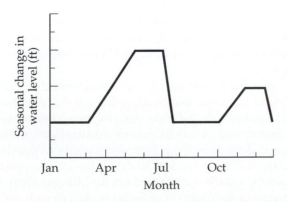

Figure 10.4 Seasonal water-level fluctuations in Kansas flood-control reservoirs. (From D. W. Willis. 1986. Review of water level management on Kansas reservoirs. Pages 110–114 in G. E. Hall and M. J. Van Den Avyle, editors. Reservoir fisheries management: strategies for the 80's. Reprinted by permission of The American Fisheries Society.)

fishes such as sunfishes and largemouth bass. However, not all ponds respond positively to fertilization. The levels of soluble nutrients in many pasture ponds are high enough due to organic waste input from livestock that additional fertilization has little impact on fish productivity (Lichtkoppler and Boyd 1977).

Types of Fertilizers

Early workers experimented with standard agricultural fertilizers containing soluble nitrogen, phosphorus, and potassium, believing that fish production would increase significantly more when combinations of all three of these nutrients were used than when any were applied singly. However, more recent work has shown that phosphorus alone can stimulate as much of an increase in productivity as when combined with nitrogen and potassium (Swingle et al. 1963; Dobbins and Boyd 1976; Boyd and Sowles 1978).

The type of phosphorus applied will significantly influence the level of increased fish production that is achieved. Liquid phosphates produce better results than granular forms. Much of the dissolution of granular phosphates takes place after the granules have fallen through the water column and settled on mud substrates. Most of the phosphate dissolved after settling onto substrates is absorbed by surface muds, preventing its uptake by phytoplankton (Boyd 1981a). Plankton absorb up to 100 percent of all liquid phosphate within 24 hours after its application (Boyd and Musig 1981). Thus liquid fertilizers increase the concentrations of soluble phosphate in the water column and the rate of phytoplankton production more effectively than granular forms (Boyd et al. 1981). However, heavy rainfall shortly after application tends to flush liquid phosphate out of systems. Even though much of the phosphate content of granular fertilizers tends to be absorbed by bottom muds, its slow rate of dissolution will have some residual effect on plankton production after heavy rainfall (Reeves and Harders 1983).

Effects of Fertilization

Production of sunfishes in ponds in the Southeast can increase from an average of about 100 pounds per acre per year without fertilization to as much as 370 pounds per acre per year after fertilization (Metzger and Boyd 1980). The level of increase in production depends on the type of phosphate used (that is, liquid versus granular forms) and the frequency and concentrations of applications (Boyd 1981b). Phosphate fertilizers must be applied many times each year, because after being absorbed by phytoplankton phosphates eventually end up in bottom substrates as organic detritus (Metzger and Boyd 1980).

Many fisheries managers have been cautious to recommend fertilization of ponds in northern states. After intensive fertilization, the organic detritus content of bottom substrates builds up substantially, increasing oxygen consumption of microbial decomposers (as mentioned in the section on cultural eutrophication in Chapter 9). During prolonged periods in the winter when northern ponds are covered with ice and snow, little of the oxygen lost to decomposers is replaced by either photosynthesis or dissolution at the water's surface. Resulting oxygen depletion can lead to winterkill, a partial or total die-off of fishes (Everhart et al. 1975).

Even when winterkill is not a factor, fertilization in northern states may not provide as great a benefit as further south. Fertilization experiments in Illinois ponds increased the standing crops of largemouth bass and bluegills by 23 percent (Hansen et al. 1960). Bluegills were larger in fertilized than unfertilized ponds, but bass were not. The total weight of bluegills harvested from fertilized ponds was nearly three times higher than that from unfertilized ponds, but the harvest of large-mouth bass was slightly lower. These researchers concluded that fertilization provided little benefit to largemouth bass fishing.

Although fertilization increases fish production in small ponds, it may not change productivity when locally applied within larger bodies of water. Farquhar (1988) fertilized selected coves in two reservoirs in Texas in order to determine whether cove fertilization would increase plankton production and survival of young or stocked fishes within large reservoir systems. Although dissolved nutrient concentrations increased within coves immediately after fertilization, they returned to prefertilization levels within a week. Phytoplankton and zooplankton production did not differ significantly between fertilized and unfertilized coves. Dilution due to high rates of water exchange in the reservoirs prohibited any increase in plankton production with the fertilization efforts (Farquhar 1988).

Fertilization may improve fish production in many instances, but its effect on fisheries may not be entirely positive. Heavy fertilization may reduce the rooted aquatic vegetation in the system (see the section on controlling aquatic vegetation in this chapter), but it may stimulate the growth of expansive mats of filamentous algae along shallow shorelines and at the water's surface. Dense growths of filamentous algae hindered angling activities in fertilized ponds in Illinois (Hansen et al. 1960).

MANIPULATING THE BIOTA OF AQUATIC COMMUNITIES

Fisheries agencies not only have modified the physical and chemical conditions of aquatic habitats to enhance the quality of fisheries but have also frequently modified the makeup of the biotic community by adding or attempting to eliminate particular animal or plant taxa. The stocking of hatchery fishes to create or enhance fisheries directly was discussed in Chapter 8. This section reviews three other biotic community manipulations:

1. The addition of animal taxa to provide a food (forage) base for fish populations supporting fisheries
2. The control or elimination of "undesirable" fish species
3. The reduction of cover created by dense beds of rooted aquatic vegetation

Creating a Forage Base

The productivity of fish populations and the quality of fisheries that they support can be limited by the absence of forage species that provide an optimal food base. Fisheries agencies have stocked a variety of animal taxa into aquatic systems to improve the transfer of energy from lower trophic levels within a food web into biomass production of recreationally important fish populations.

To be successful, forage introductions must (Ney 1981):

1. Provide an abundant food supply
2. Not experience wide fluctuations in abundance
3. Effectively transfer primary production to higher trophic levels
4. Be vulnerable to predation based on their size, behavior, and habitat preferences
5. Be innocuous

Establishing Invertebrate Forage.   Attempts to improve the food base of fishes by introducing species of insects or other aquatic invertebrates have generally been unsuccessful. Early introductions of aquatic insects generally failed to become established. Fisheries managers conducting such introductions apparently did not understand that highly mobile adult stages of many insects would have permitted the natural invasion and establishment of those taxa had the watershed provided optimal habitat conditions (Everhart et al. 1975).

The opossum shrimp, an invertebrate native to oligotrophic lakes in eastern North America, was widely introduced into western waters to establish a forage base after its introduction into Kootenay Lake, British Columbia, led to a substantial increase in growth and size at maturity for kokanee salmon in that system. Growth rates of target fishes have improved in some instances after the introduction of this species. However, in many instances the growth and abundance of fishes such as lake trout, bull trout, rainbow trout, and kokanee have declined once the opossum shrimp became established (Nesler and Bergersen 1991). In Lake Tahoe predation by an expanding population of opossum shrimp caused the collapse of native cladoceran populations, which previously had been a dominant component of the plankton community and a major food of kokanee. The change in the plankton community eventually led to the collapse of the kokanee stock (Northcote 1991). Introductions of opossum shrimp have caused similar declines in kokanee populations in other western lakes and reservoirs (Wydoski and Bennett 1981; Beatty and Clancey 1991; Bowles et al. 1991; Martinez and Bergersen 1991; Northcote 1991). In many coldwater systems, opossum shrimp remain in deep water during the day and migrate to the surface to feed on plankton at dusk. Such daily migrations make them poor forage for sight-feeding planktivorous fishes (Nesler and Bergersen 1991). Declines in planktivorous fishes such as kokanee and Arctic char have caused subsequent declines in abundance and growth of predatory species of salmonids (Northcote 1991).

Opossum shrimp can increase growth rates of fishes in systems with a strong thermocline in the summer when shrimp are restricted to deep coldwaters and cannot migrate to the surface to feed on and deplete large-bodied cladoceran populations; in highly eutrophic lakes where increases in plankton production partially counters predation by shrimp; and in shallow-water systems where shrimp can be found and fed upon during daylight hours by planktivorous fishes (Nesler and Bergersen 1991). However, the frequent decline in the quality of fisheries after introduction of this invertebrate has led fisheries agencies to take great caution when considering establishing this forage base in new waters.

Establishing Forage Species of Fishes.   The introduction of forage fishes has also produced inconsistent results. Threadfin shad have been introduced widely into reservoirs to provide a pelagic forage base for open-water predators such as the striped bass, white bass, and walleye (Noble 1981; Wydoski and Bennett 1981). Some fish predators that feed in inshore areas, such as the white crappie, have displayed increased growth after the introduction of threadfin shad; other littoral species such as the largemouth bass have not (Noble 1981). A closely related forage species, the gizzard shad, also has been extensively introduced into southern reservoirs. However, adult gizzard shad become too large to be eaten by most freshwater predatory fishes. Thus, unlike the smaller-bodied threadfin shad, a large proportion of the biomass of gizzard shad populations is not available as forage for most predator fishes. The abundance of gizzard shad can become so great that production of other forage species (such as the bluegill sunfish, which compete with them for available plankton foods), will decline (Noble 1981). This may lead to reduced rates of growth and smaller average sizes of species such as largemouth bass and crappies.

The rainbow smelt and alewife have been stocked into New England watersheds as forage for a variety of predator fishes. The rate of growth of landlocked Atlantic salmon and lake trout can increase significantly when rainbow smelt are present (Kircheis and Stanley 1981). Smelt are extensively used as food by these predators because they inhabit the same deep, coldwater habitats of lakes and reservoirs as the predators during the summer months. Other forage species that can tolerate warmer waters would not be forced into the deep hypolimnion waters to which lake trout and Atlantic salmon are restricted during that time of the year. However, rainbow smelt do not provide a perfect forage base for coldwater predators in New England. Landlocked smelt populations tend to exhibit extreme fluctuations in abundance; thus their effectiveness as a forage base is limited during years when their abundance is low. When abundant, rainbow smelt can compete for the same plankton food as that eaten by juvenile stages of salmonids and other recreationally important taxa. The alewife, stocked as forage in reservoirs and lakes throughout New England, can cause similar problems. Competition with alewives resulted in suppressed growth of kokanee salmon populations in Connecticut (Kircheis and Stanley 1981). The introduction of rainbow smelt and the invasion of alewives into Lake Michigan led to population collapses or extinctions of native fish species that had previously served as the forage base for predator fishes (Stewart et al. 1981).

The success of other forage fish introductions has been similarly inconsistent. The introduction of redside shiners into Paul Lake, British Columbia, provided additional forage for rainbow trout larger than 25 centimeters in length. However, growth and survival of juvenile rainbows in the same watershed were reduced, apparently because of competition with redside shiners for amphipod food resources (Wydoski and Bennett 1981).

Kokanee salmon are perhaps the most successful forage introduction into western lakes and reservoirs. Small kokanee provide prime forage for larger-bodied species of salmonids. For example, after the introduction of kokanee salmon into Priest Lake, Idaho, growth rates of lake trout and bull trout increased significantly (Wydoski and Bennett 1981). No reports of negative impacts of the introduction of kokanee have been published (Wydoski and Bennett 1981).

Potential impacts other than providing a food base are often not considered when fisheries agencies stock forage species into waters. "The concept that any species has no adverse interspecific impact is utopian; the 'open niche' [in any aquatic ecosystem] is a relative, not an absolute, term" (Ney 1981). Introduced forage species have caused the decline or disappearance of native fauna, both within the systems in which they were stocked and outside these systems when the introduced taxa expand their range of occurrence. Indeed, fisheries agencies have had to use fish toxicants, water-level manipulations, and the stocking of additional fish predators to control many forage introductions that have produced unwanted effects in fish communities (Noble 1981). Although introducing a forage base can significantly improve the productivity of fish populations and the recreational fisheries that they support, substantial evidence indicates that agencies must carefully consider undesirable impacts that may accrue before carrying out programs of forage fish introductions.

Controlling Undesirable Fish Species

Throughout the 20th century fisheries agencies have established programs to reduce the abundance of or to eliminate a variety of undesirable fish species from specific systems. Generally, state agencies have defined undesirable taxa of fishes as those species that are "not considered beneficial to humans" (Wiley and Wydoski 1993). Control programs have been established to eliminate the following:

1. Native nongame species that eat or compete with game fishes
2. Nonnative species that have caused a decline in native fish populations
3. Fish communities dominated by an overabundance of nongame species, followed by stocking of a fish assemblage that will support more productive fisheries
4. Resident fish communities in order to establish put-grow-and-take trout fisheries in marginal habitats (as discussed in Chapter 8)

Many "undesirable" species, such as the sea lamprey, common carp, and rainbow trout, have invaded or been introduced into watersheds well outside their native ranges, disrupting resident fish communities and causing declines in native species. Exotic introductions have had particular affect on native fish faunas of the arid southwest and the southeast, especially Florida (Shafland 1986; Minckley et al. 1991). Even when stocked within their native range, some species such as bluegills and gizzard shad may reach such high population densities that they alter normal patterns of energy flow through food webs, resulting in inadequate biomass production of desired species such as the largemouth bass (see Chapter 8 and the section on establishing forage species of fishes in this chapter). Many species deemed undesirable have been native suckers, minnows, catfishes, and other nongame taxa with little recreational or commercial importance that were believed to eat or compete with recreationally important species, thus reducing biomass production and survival of these resources. In some instances, the "undesirable" fishes are natives to the basin and the valued fishes are introduced through stocking. For example, Wiley and Wydoski (1993) state that "in Pyramid Lake, Nevada, cutthroat trout prey upon Tui chub and the chub is not a problem there. However . . . [elsewhere, native] minnows such as the Tui chub and Utah chub compete with stocked fingerling rainbow trout. Consequently, they are controlled by stocking strains of trout that are piscivorous, stocking warm or coolwater predators, or through chemical rehabilitation [that is, through the use of toxicants]."

Control Programs.   In the past fish control crews played an integral and active role in the management activities of most freshwater fisheries agencies (Noble 1980). A variety of techniques have been used to control or eliminate unwanted species, including seines, nets, traps, weirs, electrofishing methods, toxicants, and manipulating water levels. Fish toxicants have been used frequently in systems where complete draining of habitats is not possible. Bennett (1970) and Marking (1992) review commonly used fish toxicants, and Marking describes their current status under the licensing procedures of the EPA.

Rotenone, an extract from plants in the family *Leuminosae,* has been the most commonly used nonspecific fish toxicant. It has been used in small ponds to eliminate entire fish communities and in large ponds and reservoirs to reduce the densities of overabundant populations of nuisance species such as the common carp or panfish species such as the bluegill. Rotenone kills fish by blocking oxygen uptake, thus causing suffocation (Davies 1983). Its toxicity varies with species and size of fish as well as water temperature. Rotenone detoxifies rather rapidly in warmwaters exposed to direct sunlight but is persistent in colder waters; it is undetectable within 4 days after application in water temperatures of 24°C, yet it may persist for a month under ice during winter months (Marking 1992). It can be immediately detoxified by application of potassium permanganate (Davies 1983). Rotenone has several disadvantages:

1. It repels fishes without killing them if they are able to escape areas of treatment.
2. It does not kill fish eggs.
3. It kills other gill-breathing organisms, although it is only moderately toxic to humans (Everhart et al. 1975; Marking 1992).

Antimycin, an antibiotic produced by *Streptomyces* and marketed under the name Fintrol has also been widely used as a fish toxicant. Similar to rotenone, antimycin causes suffocation by inhibiting respiratory function. Unlike rotenone, antimycin kills fish eggs, it does not repel fishes, and fishes exposed to it cannot recover if they escape the area of application (Bennett 1970; Everhart et al. 1975). The rate at which it detoxifies in water is influenced by alkalinity, water temperature, and sunlight (Davies 1983). Like rotenone, it can be detoxified immediately by application of potassium permanganate (Davies 1983).

Eliminating Native Nongame Species That Eat or Compete with Game Fishes.   Both public-education and scientific-community publications have suggested that native nongame taxa severely affect biomass production and quality of gamefish populations. Such statements as "These species often crowd out more desirable game and pan fishes" (Bennett 1970) and "Almost all species of fish compete with trout for the available food supply of a pond" (J. W. Mullan and W. A. Tompkins, undated handbook of the Massachusetts Division of Fisheries and Game entitled "Trout pond management in Massachusetts"; circa late 1950s) occurred frequently in the fisheries management literature as recently as the 1970s. Terms attached to undesirable species reveal the regard that many agency personnel traditionally had for such taxa: "Rough fish," "coarse fish," "trash fish," and other uncomplimentary monikers have been applied frequently to these species.

Agency interest in controlling "undesirable" native fish species persisted long after fish ecologists began to question the effect that these taxa have on resident game fishes. As early as the 1960s, research by Keast (1965; 1966) and others countered conclusions that most or all species of fishes compete broadly for all invertebrate food resources. Indeed, Keast's thorough analytical approaches, and those of many researchers who followed, indicated that different species of fishes eat different invertebrate taxa and/or gather their food in different habitats or microhabitats; thus competition for food resources is minimized in many interactions between fish species. This is particularly true of competition believed to occur between game-fishes and species of suckers, minnows, freshwater sculpins, or other taxa that differ substantially in their feeding morphology and behavior from the gamefishes.

Removal of perceived predators and competitors has not significantly improved standing stocks or production in trout populations. In one long-term study, Flick and Webster (1975) did not detect any changes in the rates of body growth or mortality of brook trout during more than a decade of removal of nongame fishes. Similarly, 10 to 11 years after toxicants were applied to the Feather River, California, populations of nongame fishes had fully recovered and wild trout populations were similar to pretreatment years even though trout had been stocked annually since reclamation (Moyle et al. 1983).

Moyle (1977a; 1977b) addressed the impacts of both competition and predation in coldwater fish communities in his aptly titled papers, "In Defense of Sculpins" and "Are Coarse Fish a Curse?" He reviewed substantial evidence that nongame native fish species have little negative impact on production of trout populations in streams where habitats are not disrupted by human activities. Trout production in streams is much more influenced by factors such as the quality of spawning and nursery habitat, production of food types specifically selected by trout, and the abundance of appropriate feeding territories and shelter sites than by predation or competition pressure caused by other native, nonsalmonid species.

Fish Community Reclamation.   For some time, agencies have reclaimed or rehabilitated small reservoirs and ponds by eradicating entire fish communities and then stocking more desirable species into the reclaimed system. Many reclamation projects have been conducted to eliminate fish communities dominated by very large biomasses of unwanted species such as the common carp or by overabundant, slow-growing panfishes. Often, dominant, unwanted species or overabundant, stunted panfishes were the result of past introductions by agencies or anglers. Thus reclamation focused on correcting past introductions that did not provide desirable results.

Other community reclamation programs were conducted to create put-grow-and-take trout fisheries in waters that did not provide optimum habitat conditions (see Chapter 8). Introduced fishes rarely do well if stocked into suboptimal habitats because they are unable to gather food resources or utilize spawning habitats when interacting with fishes better adapted to prevailing conditions. Thus, in a sense, reclamation programs sometimes resulted in fisheries managers putting particular fish species where nature would not. Such programs also can eliminate resident fish species that would support recreational fisheries. For example, reclamation coupled with put-grow-and-take stocking in the Northeast often eliminated large-mouth bass and panfish populations in small, warmwater ponds. In recent years agencies generally have coupled fish community reclamation more often with stocking to produce self-sustaining fish communities rather than stocking to produce hatchery-maintained put-grow-and-take fisheries.

Northern Squawfish and Sea Lamprey Control.   The northern squawfish minnow and the sea lamprey have been deemed so undesirable that extensive efforts have been directed toward developing toxicants that are specific to these taxa. Chemical and nonchemical control programs focused on these species represent two of the largest efforts at controlling the abundance and occurrence of fish species in the United States.

The most intensive fish control program in the United States has been directed toward the sea lamprey. The sea lamprey has been blamed widely for the collapse of lake trout populations in the Laurentian Great Lakes of midwestern North America. The invasion of the sea lamprey into the upper Great Lakes (Lakes Huron, Michigan, and Superior) coincided with collapse of lake trout stocks that had supported regionally important commercial and recreational fisheries. The exact roles that commercial fishing versus expanding sea lamprey populations played in the collapse of native salmonids in this watershed are still being analyzed and argued (as in Coble et al. 1990). However, it is generally agreed that recovery of lake trout stocks and maintenance of the extremely successful put-grow-and-take Pacific salmon fisheries depend in part on controlling population abundances of the sea lamprey.

An effective lampricide, TFM, was developed after extensive testing of about 6,000 different chemicals (Everhart et al. 1975). TFM can be used to poison adult lamprey as they enter tributary streams during spawning migrations or to kill sea lamprey during the several years that they inhabit the streams as filter-feeding ammocoete larvae before metamorphosing into the predaceous adult form and migrating back into the lakes. Nonchemical means of control, such as constructing weirs or electric grids on spawning tributaries to disrupt spawning migrations of adult lampreys, have not been as effective as application of TFM. Thus control programs coordinated by the Great Lakes Fish Commission have emphasized use of this chemical to control postmetamorphosis recruitment. Control programs regularly treat tributaries of the Great Lakes that otherwise would support substantial sea lamprey recruitment.

Although treatment concentrations suitable for killing lamprey larvae are believed to be harmless to bony fishes, some fish kills have occurred during treatments of streams with TFM. The Great Lakes Fish Commission at one point in the 1980s had to remove specific streams from the treatment schedule in order to protect spawning migrations of pink salmon.

As found in many nuisance species control programs, sea lampreys have not been eliminated from any Great Lakes drainage that provides them with optimal spawning and nursery habitat. If streams are not retreated every several years, offspring production levels rapidly return to those characteristic of pretreatment times. The realistic objective of this management program is clearly toward long-term control of abundance, not the elimination of this introduced species.

The toxicant squoxin was developed to reduce the abundance of the native northern squawfish in watersheds of the northwestern United States, largely because of the predation pressure this species imposes on Pacific salmon during their downriver migration. Reducing northern squawfish populations is considered a critical element of the management of severely depleted salmonid stocks of the Columbia River basin even though these stocks are depleted largely as a result of a prolonged history of habitat degradation, disruption of migratory pathways by dams and flow regulation, and overharvest (see Chapter 12). Predation of salmon smolts by the squawfish occurs mostly in reservoir pools above dams (Rieman and Beamesderfer 1990), where downriver migration of smolts is markedly slowed from rates characteristic of the river basin prior to its impoundment (see Chapter 12). Although the severity of predation by the Columbia River squawfish is largely a result of human modifications to the river basin, controlling the abundance of this species has received unusual attention as a management goal of agencies in that region of the country. Agencies have used toxicants, explosives, and intensive trapping and netting to reduce northern squawfish populations, but none of these methods has become a routine, persistent management activity (Rieman and Beamesderfer 1990).

The emphasis on northern squawfish control led the state fisheries agencies in Washington and Oregon to establish a bounty program for Columbia and Snake River squawfish in 1990. The program offered a $3 bounty for every squawfish 11 inches or greater in length that was turned into designated check stations (Fisheries News 1991). As a result of this bounty program, over 20,000 squawfish were harvested above the John Day Dam in 1990. Well over 100,000 were harvested in an expanded program by 1991, and over 199,000 were harvested in 1995. Information on this program and updates on the number of squawfish turned in for the bounty is disseminated over a "squawfish hotline" operated by the Washington Department of Wildlife.

Bounty programs, once a common element of wildlife management, generally were phased out as management tools of wildlife agencies long ago. It is ironic that this method would be employed in the 1990s as part of an attempt to protect fisheries resources in severe decline due largely to other human-induced problems.

Recent Focus of Reclamation Programs.   Before 1970 most reclamation projects were intended to enhance production of game fishes, often nonnative to the treated water basin (Rinne and Turner 1991). Agencies used this method to increase recreational fishing opportunities and did not consider the broader issue of impacts on other fishes. For example, Becker (1975) stated that no consideration was given to rare or endangered fishes in reclamation projects in Wisconsin. Reclamation was probably responsible for the extinction of stocks of specific native fishes in the western United States (Rinne and Turner 1991).

In more recent years, fish community eradication programs are still considered a viable fisheries management technique by fisheries agencies. For example, when the EPA provided funding for approved state projects to restore water and habitat quality in degraded waterways, a few projects were funded that proposed eradication of existing fish communities and stocking of more desirable species (Keup 1979). However, as agencies have gained broader responsibilities as stewards of all resources, not just those with recreational and commercial importance, greater focus has been placed on what is lost as well as what is gained by fish community eradication.

Reclamation as Part of Management of Endangered Species.   Reclamation is being used in certain western watersheds to correct fisheries management mistakes of the past. By the late 1980s, 15 of 25 recovery plans for threatened and endangered fish species in the western United States included reclamation of watersheds (Rinne and Turner 1991). Ironically, the reclamation efforts are focused on purposely introduced fish species that now threaten the recovery of the endangered stocks of native fishes. Among the recovery plans are those developed for native trout species such as the Apache, Gila, cutthroat, and golden trouts, which have largely been displaced in stream systems by introduced brook or brown trout, or by hybrid fishes produced by matings between introduced rainbow and the native trouts. In these recovery plans, agencies use toxicants to eliminate the nonnative salmonids, with subsequent reintroduction of the native endangered species into the system. Even these recovery plans may produce less than desirable outcomes. The transplant of the endangered Gila trout into one reclaimed watershed in New Mexico was successful, although "a native population of Rio Grande suckers was eliminated in the process" (Rinne and Turner 1991).

Controlling Aquatic Vegetation

Rooted aquatic vegetation in the littoral zone of lakes and reservoirs provides shelter from predators for young fishes, improves water quality by stabilizing sediments, and enhances production of the invertebrate community (Bain and Boltz 1992). Durocher et al. (1984) found a direct correlation between the percentage of a reservoir's area that was covered with aquatic vegetation and both the standing crop and number of largemouth bass that were recruited to a harvestable size. The maximum coverage of substrate by aquatic vegetation in the study ponds was 20 percent of the total area. Wiley et al. (1984) estimated that intermediate plant densities and coverage produced optimum levels of largemouth bass production.

However, dense stands of aquatic vegetation covering large expanses of the water column might produce several undesirable effects. Such "weed beds" might bind up high proportions of available inorganic nutrients, causing reductions in phytoplankton production and the zooplankton that feed upon planktonic algae, thus reducing planktonic food resources available to the fish community (Bennett 1970). Stands of rooted vegetation that cover substantial proportions of the submerged substrates of small ponds and reservoirs can disrupt the efficiency of prey capture by predator fishes, potentially reducing production in the predator populations (Rottman 1977). Such stands also can cause significant declines in angling activity. Colle et al. (1987) measured an 85 percent decrease in angling effort in Orange Lake, Florida, after an exotic plant, hydrilla, had expanded from modest coverage to over 80 percent of the lake's substrates. Due to stunting, the number of harvestable-sized bluegill and redear sunfish declined, but that of largemouth bass and black crappie remained stable. Although catch per hour in the fishery remained high, angling activity declined due to the difficulty of fishing in dense hydrilla beds (Colle et al. 1987). Many nuisance stands of aquatic vegetation occur because of excessive nutrient input into an aquatic system or the introduction of exotic species such as hydrilla and Eurasian millfoil. Thus human activities often create the vegetation coverage that must be controlled.

Control of rooted aquatics can modify fish productivity. Complete removal of rooted aquatic vegetation in Lake Conroe, Texas, stimulated an increase in growth rates of young largemouth bass, black crappie, and white crappie (before removal, submergent vegetation covered 44 percent of the lake's 20,000 acres). Both species of crappies reached a harvestable size at an average of 1 to 2 years younger than before vegetation was removed (Maceina et al. 1991). Bettoli et al. (1992) attributed accelerated growth of largemouth bass to a significant reduction in the minimum size at which young bass were able to start feeding on forage fishes (140 millimeters—5.5 inches—in length before vegetation removal versus 60 millimeters—2.4 inches—in length after).

Because moderate aquatic vegetation coverage provides fishery benefits but major vegetation coverage does not, reservoir and pond habitat management often has included control of rooted aquatics. Some forms of aquatic vegetation can be controlled in reservoirs by regulating water levels (discussed elsewhere in this chapter). However, water-level manipulations are not possible in natural ponds and lakes or often are not compatible with water uses other than fishing in reservoirs. Thus mechanical, chemical, and biological methods have been used to control rooted aquatic vegetation in many systems.

Control Methods.   Mechanical removal has proven to be a time-consuming and costly method, and results are effective only for a short period of time. Some taxa of macrophytes develop large tubers or root systems that are very difficult to remove from the substrate. Thus dense stands of many aquatic plant taxa return in very short periods of time after mechanical removal.

Chemical control may take one of two forms: Fertilizers can be used to stimulate phytoplankton production, and herbicides can be employed to eliminate nuisance vegetation directly. Phytoplankton production can be increased substantially by the application of fertilizers (Smith and Swingle 1940; Swingle and Smith 1947). After fertilization, algal blooms become so dense that light penetration into the water column is not sufficient to support photosynthesis of rooted aquatic vegetation; ultimately, rooted aquatics are eliminated or restricted to shallower waters (Smith and Swingle 1940; 1942; Hansen et al. 1960). The increase in plankton production stimulates fish production significantly (see the section on increasing productivity through fertilization in this chapter) and the reduction in rooted aquatics improves angling conditions (Smith and Swingle 1940).

The application of herbicides was a favored means of vegetation control during the mid-1900s. Bennett (1970) stated that over 20 commercial herbicides were commonly used to control aquatic vegetation during that time period and listed those chemicals that were most frequently used to control specific types of aquatic vegetation. However, herbicides can be expensive to use, and their application in watersheds supporting multiple recreational and domestic activities may not be appropriate.

Biological Controls.   In the last several decades, agencies have directed substantial research toward developing effective biological controls. A variety of invertebrate and vertebrate species have been introduced into watersheds to control aquatic vegetation through grazing (Bennett 1970). Numerous fishes that consume rooted aquatics, including a variety of cichlids, characoids, and cyprinids, have been tested by researchers. Most species that were tested proved unsatisfactory for the following reasons (Shireman 1984):

1. Individuals did not consume sufficient quantities of vegetation to control rooted aquatic production.

2. Fishes were so prolific that they caused overcrowding and potential fish community disruptions.

3. Fishes were unable to persist in winter water temperatures typical of many areas of the United States.

The grass carp has received the broadest support as an effective biological control based on its ability to consume large amounts of a variety of plant taxa. However, the introduction of this species into watersheds has produced problems. Grass carp have had various effects on fish communities into which they have been introduced (Taylor et al. 1984). Some researchers have measured a decrease in biomass production of species such as the bluegill and largemouth bass after the introduction of grass carp, whereas others have reported that production of these species did not change significantly in the presence of introduced grass carp. Removal of vegetation by grass carp will make juvenile fishes more vulnerable to predation, thus reducing levels of recruitment in sport fish and panfish species.

By ingesting and digesting macrophytes, grass carp convert the tissues of rooted aquatic vegetation into soluble and available nutrients, the release of which can result in dense phytoplankton blooms. The transfer of nutrients from rooted aquatics to phytoplankton blooms "could be more damaging to water quality than dense stands of macrophytes" (Hestand and Carter 1978). Grass carp are selective feeders, often grazing heavily on certain taxa of rooted aquatic plants while virtually ignoring other taxa. Thus, when grass carp are stocked at low densities, they may reduce the abundance of some plant species, releasing nutrients into the water column that can also support increased production of other macrophytes that might have been the intended target of grass carp introduction (Shireman 1984).

Grass carp ultimately may provide no better than an all-or-none approach toward control of nuisance macrophytes (Shireman 1984). Establishing stocking densities of grass carp that will reduce vegetation stands and maintain them at some optimal lower density has proven difficult, if not impossible. If sufficient densities of grass carp are stocked to produce significant reductions, total eradication will probably occur; if significantly lower densities of fishes are stocked, then measurable reductions may not be achieved. Management objectives typically focus on control rather than eradication of rooted aquatics, because complete removal can result in the loss of feeding, spawning, and shelter habitats for some fish species (Taylor et al. 1984). Total eradication has led to substantial changes in fish communities, some of which may not be desirable. Bettoli et al. (1993) recorded declines in biomass of black and white crappies, an increase in average size but a decrease in catch rates of largemouth bass, and a shift in species that dominated the recreational fisheries in Lake Conroe, Texas, after all vegetation was eliminated by grass carp. To date, grass carp have provided fisheries managers with a useful but limited tool to enhance recreational fisheries by controlling expanses of rooted aquatics.

Although some agencies have actively promoted the use of fishes to control aquatic vegetation, others have moved more cautiously toward the introduction of these fish taxa. Past experiences with the negative effects produced by the introduction of exotic species (see Chapter 8) have led many agencies to restrict or prohibit the introduction of herbivorous fishes into public or private waterways.

THE ROLE OF MANIPULATING AQUATIC SYSTEMS

Fisheries agencies have developed a variety of methods to manipulate aquatic systems. Some of the methods, such as fertilization and water-level management, clearly influence yield or harvest of fisheries by increasing biomass production of fish populations. Others, such as construction of artificial structures within systems, at a minimum can influence the frequency of interaction between fishes and humans by increasing catch rates of anglers in recreational fisheries, and may stimulate increases in biomass production at least in some instances. Still others, such as the elimination of native nongame fishes through reclamation or the introduction of at least some forage species, have failed due to inaccurate assumptions concerning the biological interactions that such actions are intended to influence. Many manipulations have proven to be effective on a small-scale basis (within selected reaches of stream systems, in specific reservoirs or lakes, or in specific coastal regions). Regardless of how effective various methods have proven to be, programs manipulating systems will be effective most often when integrated into broad-scale strategies for fisheries conservation. Providing suitable biotic and abiotic environments and preventing overharvest must remain the major goals of efforts of fisheries agencies.

11

Conservation of Endangered and Threatened Resources

The United States harbors one of the most diverse temperate freshwater fish faunas of the world, and hundreds of separate, reproductively isolated stocks of anadromous fishes spawn in its river systems (Nehlsen et al. 1991; Warren and Burr 1994). In addition, the freshwater mussel family *Unionidae* reaches its greatest diversity on this continent (Williams et al. 1993). Numerous species, subspecies, and stocks of these faunas are imperiled with extinction, as are all six species of sea turtles and a number of marine mammals that inhabit U.S. waters (NOAA 1993). Within the latter half of the 20th century, the growing public awareness of the countrywide and worldwide threatened extinction of a substantial number of animal and plant taxa has led to passage of federal and state laws intended to reduce the rate of extinction and to facilitate recovery of animal and plant taxa that might be lost without human intervention.

This chapter reviews the extent and causes of imperilment of aquatic fauna in the United States and conservation programs that have been developed to facilitate recovery of these taxa.

PRINCIPAL ELEMENTS OF IMPERILED SPECIES CONSERVATION

Here the term *imperiled* includes all taxa—species, subspecies, and reproductively isolated stocks in the instance of anadromous fisheries—that are seriously threatened with extinction by a variety of human activities. The federal Endangered

Species Act of 1973 (ESA; see Chapter 6) established guidelines for defining the level of imperilment that resources face. Under these guidelines, a taxon that is in danger of extinction throughout all or a significant portion of its geographic range is designated *endangered,* and one that is likely to become endangered in the foreseeable future is *threatened* with extinction. These designations provide taxa with some level of federal protection from harvest and loss of critical habitat. In some instances, it is difficult to determine differences in protection afforded fishes classified as endangered versus others classified as threatened (Johnson 1987). States generally use the same designations when defining the status of fauna imperiled within their borders.

The Process of Endangered Species Conservation

Federal and state programs for the conservation of endangered taxa typically include the following elements (Bean 1986):

1. A listing procedure with criteria for selecting taxa that require protection
2. The development of a recovery plan to facilitate recovery of the resource
3. Restrictions on the capture and taking of protected taxa
4. Authority and funding to acquire and protect habitat deemed critical to protect listed taxa

In addition, the federal ESA places restrictions on activities that must be authorized, funded, or conducted by federal government agencies, if those activities would further endanger listed taxa (Section 7 of the ESA; see Chapter 6). This provision has created strong conflict between protecting endangered resources and using or developing the "landscape" (for example, the snail darter at Tellico dam, and Devil's Hole pupfish controversies; see Chapter 6). However, very few proposed projects that require federal review of impacts potentially jeopardize the existence of federally protected species. For example, in 1984 only about 0.25 percent of the federal actions subject to review for potential conflict with endangered species mandates were found likely to jeopardize listed species (Bean 1986).

IMPERILED AND EXTINCT AQUATIC RESOURCES OF THE UNITED STATES

Recent Extinctions

It is very difficult to sample aquatic systems effectively enough to verify that historically rare taxa are extinct, especially if the taxa are small bodied or secretive in their behavior, or if they are found in habitats that are difficult to sample. Miller et al. (1989) provided examples of supposedly extinct fish species that have been rediscovered, such as the Owens pupfish and the Shoshone pupfish in California. Because of this, lists of recent extinctions are often conservative, with researchers including only those taxa for which there is substantial evidence of extinction. Even so, scientists have concluded that numerous aquatic organisms have become extinct in this century.

Human activities have led to the extinction of 27 species and 13 subspecies of freshwater fishes in North America in the 20th century. Thirty of these were native to the United States (Table 11.1). Many of these taxa were restricted to extremely

TABLE 11.1 A list of the freshwater fish species of the United States that have become extinct during the 20th century (Miller et al. 1989).

	Factors Responsible for Extinction				
Species	Physical habitat alteration	Introduced species	Chemical alteration or pollution	Hybridization	Overharvesting
Miller Lake lamprey			×		
Longjaw cisco		×		×	×
Deepwater cisco		×		×	×
Lake Ontario kiyi		×	×		×
Blackfin cisco		×		×	×
Yellowfin cutthroat trout		×		×	
Alvord cutthroat trout		×		×	
Silver trout		×		×	×
Maravillas red shiner		×			
Independence Valley tui chub		×			
Thicktail chub	×	×			
Pahranagat spinedace		×			
Phantom shiner	×	×		×	
Rio Grande bluntnose shiner	×	×		×	
Clear Lake splittail	×	×	×		
Las Vegas dace	×				
Grass Valley speckled dace	×	×			
June sucker	×	×		×	
Snake River sucker	×			×	
Harelip sucker	×		×		
Tecopa pupfish	×	×		×	
Monkey Spring pupfish	×	×			
Raycraft Ranch poolfish	×	×			
Pahrump Ranch poolfish	×	×			
Ash Meadows poolfish	×	×			
Whiteline topminnow	×				
Amistad gambusia	×			×	
San Marcos gambusia	×	×	×	×	
Blue pike	×	×	×	×	×
Utah Lake sculpin	×	×	×		

small geographic ranges, such as two forms of the Pahrump poolfish—the Raycraft Ranch poolfish, which was restricted to one desert spring, and the Pahrump Ranch poolfish, which was restricted to two adjacent springs on a single ranch property (these forms were described as subspecies of the Pahrump poolfish by R. R. Miller in 1948; Miller et al. 1989). Such restricted distributions made these taxa very vulnerable to extinction once humans disrupted the conditions of the aquatic systems in which they lived. Other taxa occurred in much larger geographic areas. For example, the longjaw cisco, deepwater cisco, blackfin cisco, and blue pike (a subspecies of the walleye) occurred in 2 or 3 of the five Great Lakes. However widely distributed these taxa were, they were still restricted to a small number of aquatic systems.

Although no anadromous species spawning in U.S. river basins has been driven to extinction since European settlement of North America, numerous stocks, or gene pools, of anadromous species have. More than 100 West Coast stocks of Pacific salmon are extinct (Nehlsen et al. 1991). Similarly, stocks of the anadromous Atlantic salmon have been lost on the East Coast, although some such as the Connecticut River and Merrimack River stocks were extinct well before the 20th century. At one time, spawning stocks of the striped bass migrated into many river systems along the Atlantic coast. Today, spawning stocks are largely restricted to the Chesapeake Bay, Hudson River, and Roanoke River (NOAA 1993). Because individuals in anadromous species home—migrate over long distances to reproduce in the reach of a river basin where they started life—numerous separate stocks or populations of particular species may spawn in a river system. Thus gene pools uniquely adapted to conditions of specific regions of river basins are lost when such stocks become extinct.

Freshwater mussels also have suffered high rates of extinction. The American Fisheries Society Endangered Species Committee projected that 18 species and 3 subspecies of mussels may be extinct. The committee chose to label these taxa as "potentially extinct" rather than extinct due to the difficulty of determining that the last individuals of such long-lived, secretive species no longer exist. Fourteen of the 24 species and subspecies in the genus *Epioblasma* were included in this list. The 21 taxa represent about 7 percent of all of the species and subspecies of freshwater mussels native to North America (Williams et al. 1993).

Imperiled Taxa

In 1990 the Nature Conservancy estimated that 20 percent of the freshwater fishes in the United States (approximately 160 species), 55 percent of freshwater mussels, and 36 percent of crayfishes were imperiled (Warren and Burr 1994). The Endangered Species Committee of the American Fisheries Society (AFS) designated 148 species or subspecies of freshwater fishes native to the United States as either endangered or threatened (Williams et al. 1989). Nehlsen et al. (1991) identified 214 native spawning stocks of Pacific salmon in California, Oregon, Washington, and Idaho that were at some risk of extinction. As with extinctions discussed above, many endangered or threatened fish taxa are endemic to restricted geographic regions. Some of these taxa are reduced to extremely small numbers of individuals.

TABLE 11.2 **Endangered or threatened species of sea turtles and marine mammals found in United States waters.**[a]

Species	Range	Status
Loggerhead turtle	Atlantic and Pacific Oceans	T
Green sea turtle	Atlantic and Pacific	T[b]
Leatherback sea turtle	Atlantic and Pacific	E
Hawksbill sea turtle	Atlantic and Pacific	E
Kemp's Ridley sea turtle	Atlantic	E
Olive Ridley sea turtle	Pacific	T
Caribbean monk seal	Gulf of Mexico	E
Guadalupe fur seal	California to Mexico	T
Hawaiian monk seal	Hawaii	E
Stellars sea lion	Pacific Coast of N. America	T
Blue whale	Worldwide oceanic	E
Bowhead whale	Northern latitudes	E
Finback whale	Worldwide oceanic	E
Gray whale	North Pacific	E
Humpback whale	Worldwide oceanic	E
Right whale	Worldwide oceanic	E
Sei whale	Worldwide oceanic	E
Sperm whale	Worldwide oceanic	E

[a]Marine mammals from taxonomic groups whose management is under the responsibility of NMFS.
[b]Listed as endangered in Florida, and threatened in the rest of the U. S. Atlantic and Pacific.

All six species of sea turtles that inhabit U.S. territorial waters have been identified as endangered or threatened under the ESA (Table 11.2), as have eight species of whales, the Stellar sea lion, and several species of seals. Some turtle stocks are extremely depleted. For example, only about 700 to 800 female Kemp's ridley turtles currently nest each year along Mexico's Atlantic coast, whereas over 40,000 were counted on one day on a single beach in that region in 1947 (NOAA 1993). Some marine mammals are similarly reduced to extremely low numbers. The Atlantic right whale population is currently more than 95 percent depleted from its estimated abundance prior to the 1800s. The Pacific right whale is believed to be near extinction. Only a few sightings of this species have been reported in the last 25 years (NOAA 1993).

The AFS Endangered Species Committee listed 120 freshwater mussels that are endangered or threatened (Williams et al. 1993). When added to the 21 taxa that may be extinct, the total represents nearly 50 percent of all extant North American freshwater mussels.

Patterns of Imperiled Taxa. The greatest number of endangered and threatened freshwater fish taxa listed by the AFS come from only a few families. The most diverse family of North American freshwater fishes, the minnow family *Cyprinidae* and darters of the perch family *Percidae*, have 46 and 30 imperiled species, respectively. The *Salmonidae* (trouts and whitefishes), *Catostomidae* (suckers), *Ictaluridae* (catfishes), and *Cyprindontidae* (killifishes and related species) all have 10 or more species that the AFS has listed as endangered or threatened.

A high percentage of the total number of species of several genera or families are imperiled. For example, of the 13 species of the pupfish genus *Cyprinodon* that occur in the United States (AFS 1991), 9 species are imperiled. Five of the eight species of North American sturgeons(Figure 11.1)—the shortnose sturgeon, lake sturgeon, pallid sturgeon, Alabama sturgeon, and a subspecies of the Atlantic sturgeon, the Gulf sturgeon—are listed by Williams et al. (1989) as threatened or endangered, as are three of the six species of cavefishes (*Amblyopsidae*).

Imperiled fauna are disproportionately distributed among genera of the freshwater mussel family *Unionidae*. Twenty-three of the 24 taxa in the genus *Epioblasma* are extinct or endangered; the other taxon is threatened (Williams et al. 1993). Twenty-three of 31 taxa in *Pleurobema* are deemed endangered or threatened, as are 8 of 11 in the genus *Alasmidonta*, 6 of 7 in *Medionidus*, and 10 of 20 in *Quadrula* (Figure 11.2).

The greatest number of imperiled freshwater taxa are concentrated within the southwestern and southeastern regions of continental United States. The Southeast is a region of high species diversity; of the 10 states with the highest number of freshwater fish species, 7 occur in the Southeast. In contrast, the southwestern states of Arizona, Utah, and Nevada are among the four states with the lowest number of native species (Warren and Burr 1994). Both of these regions are areas of high endemism, with numerous taxa existing in extremely small geographic ranges. For example, endemic species constitute most of the native fish fauna in drainage basins such as the Colorado River, Bonneville, Lahontan, and Death Valley; 67 to 78 percent of the native fish fauna in these systems are endemic (Warren and Burr 1994). Thus, although the number of imperiled species is unrelated to the diversity of species native to particular regions, it is directly related to the proportion of the native fauna that are endemic to restricted geographic areas.

Forty-one percent of the native fish fauna of California are extinct, officially listed as threatened or endangered, or deserving of consideration for listing due to their depleted and fragile population status (Moyle and Williams 1990). These authors found that the majority of imperiled freshwater fishes in California were:

Figure 11.1 The lake sturgeon. (Photograph courtesy of Tom McHugh/Steinhart Aquarium; Photo Researchers, Inc.)

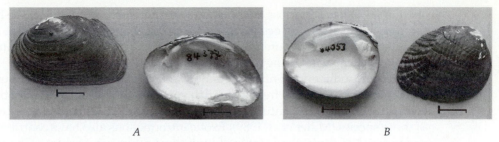

Figure 11.2   The endangered James spinymussel (*Pleurobema collina*: A) and birdwing pearlymussel (*Lemiox rimosus*: B). (Photographs courtesy of Richard J. Neves.)

1. Endemic to very small areas within one basin

2. Found in isolated springs, warmwater rivers, or big rivers

3. Part of a fish assemblage of less than five species

The first two conditions are common to many imperiled fish taxa throughout the United States.

Freshwater mussel taxa are largely restricted to eastern North America and the Mississippi River drainage basin. The greatest species diversity, the highest rate of endemism, and the greatest number of imperiled taxa of this fauna occur in the southeastern United States (Neves 1993).

Factors Contributing to Extinction or Imperilment.   The decline and loss of aquatic organisms generally is a result of several or more contributing factors, including the following:

1. Deterioration or loss of habitat

2. Introduction of nonindigenous or exotic species

3. Chemical alteration or pollution of water

4. Disease

5. Overharvest

Habitat deterioration and loss have contributed to the status of 73 percent of the North American freshwater fishes considered imperiled by Williams et al. (1989), and have contributed to the demise of 63 percent of the U.S. freshwater fish taxa driven to extinction in this century—19 of 30 extinct species, including 4 which became extinct solely due to habitat destruction (Table 11.1). Habitat deterioration has been the greatest factor contributing to the status of most imperiled and possibly extinct freshwater mussels (Neves 1993) and has been critical to the endangerment and loss of many anadromous fish stocks (see Chapter 12).

Habitat change or loss has such wide impact in part because aquatic systems are subject to habitat deterioration produced by a wide variety of human activities (see Chapter 9). Other than in Alaska, most surface waters in the continental United States are significantly modified from their natural condition. It is not surprising that habitat deterioration has threatened the existence of so many fishes and shellfishes, especially those endemic taxa that are very vulnerable to the habitat alteration or loss that may occur within restricted geographic boundaries. Sea turtle populations have also suffered from habitat loss due to development and human activities on coastal nesting and nursery habitats.

The introduction of nonindigenous and exotic species has contributed to the extinction of 24 of the 30 freshwater fish species and subspecies lost in the United States during the 20th century. This was the only cause of extinction listed for two of those species (Miller et al. 1989). Exotic introductions also have contributed to the endangered or threatened status of numerous other fish taxa (see Chapter 8). Although historically the introduction of nonnative species has had little impact on freshwater mussels, the invasion of the zebra mussel into the Great Lakes and Mississippi River drainage basins poses "an ominous threat" to numerous rare mussel species (Neves 1993). Once introduced into the western basin of Lake Erie, possibly in ballast released from shipping vessels, this Eurasian native caused substantial damage to the native mussel fauna in just a few years; unionid bivalves underwent dramatic declines in offshore waters after infestation by the zebra mussel (Schloesser and Nalepa 1994).

The introduction of new species generally has not been a widespread problem for anadromous salmonid stocks. However, expansive stocking of nonnative genetic strains of the same species that occur naturally in river basins has led to introgressive hybridization and the loss of wild gene pools (see Chapter 8).

Overfishing generally has not been a major contributor to the extinction or imperilment of fishes and freshwater mussels. Fishing contributed to the extinction of only 6 of the 30 freshwater fish species listed in Miller et al. (1989; Table 11.1). None of these taxa was driven to extinction by fishing alone. Only 3 percent of the North American freshwater fish taxa considered endangered or threatened or meriting special concern by Williams et al. (1989) are jeopardized by overharvesting (*special concern* is a category used by these authors for taxa that need further study because their status is uncertain or that can become threatened or endangered with minor disturbances in their habitat).

Harvest has not been a major factor in the imperilment of freshwater mussels. However, a commercial market for mussel shells that are ground into small particles and used as "seed" to stimulate pearl production in cultured Pacific oysters is growing. Thus greater interest in harvest may place another stress on the continued health of numerous imperiled mussel taxa in the future (Williams et al. 1993).

Overexploitation has been a major cause of imperilment of sea turtles and marine mammals, although both groups of marine organisms are now protected in U.S. waters. Bycatch mortality of sea turtles and marine mammals in a variety of fisheries, including shrimp trawl fisheries, groundfish gill-net fisheries, drift gill-net fisheries, and purse seine fisheries, has further exacerbated the conditions of imperiled sea turtles and marine mammals and has resulted in restrictions on the use of certain gears. Thousands of sea turtles have been incidentally captured in shrimp trawls annually. However, the recent use of turtle-excluding devices in shrimp trawls has significantly reduced turtle mortality in this fishery (see Chapter 7). Whales also have been affected by nonconsumptive uses. Incidence of injuries and deaths due to collisions with large vessels is a growing concern in protection of imperiled whale populations. Human interference and contact through activities such as whale-watching excursions have apparently caused humpback whale females to abandon traditional calving and calf-rearing grounds in nearshore waters of Hawaii (NOAA 1991).

Disparities Between Federal Listings and the List of Candidates Proposed by the American Fisheries Society. The number of aquatic taxa listed under the ESA as federally endangered or threatened has increased slowly since passage of that legislation. Thirty-two species of fishes native to the United States were listed as endangered and 12 more as threatened by the time the ESA was reauthorized in 1982 (Johnson and Rinne 1982). The 1982 reauthorization included a streamlining of the listing procedures, which was intended to accelerate the previously burdensome, slow listing process (Williams 1994). By 1987, 73 species were listed either as endangered or threatened (Johnson 1987). Ninety-seven species and subspecies of freshwater fishes, 4 stocks of anadromous Pacific salmon, and 56 species and subspecies of freshwater mussels were listed by September 1994.

A backlog of taxa remain federally unlisted even though they are potential candidates for protection. This backlog of unlisted candidate species largely is due to two factors (Kohm 1991):

1. The process of determining status and listing species is time-consuming.
2. Some federal administrations have been reluctant to list new species. For example, no new species were listed during the first year of the Reagan administration.

By the late 1980s over 3,000 species were "stuck in the listing pipeline" (Kohm 1991). By September 1994 the 97 taxa of freshwater fishes that were federally listed as endangered or threatened constituted only about 60 to 65 percent of the species and subspecies considered imperiled by the AFS or Nature Conservancy (Williams et al. 1989; Warren and Burr 1994). The 56 listed freshwater mussel taxa represented only about 40 percent of those meriting listing according to Williams et al. (1993). In addition, the four stocks of Pacific salmon listed—the Sacramento River winter run, Snake River spring–summer run, Snake River fall run of chinook salmon, and Snake River sockeye salmon stock—represent an extremely small proportion of salmonid stocks that are imperiled.

The major disparity between the number of anadromous taxa that are potentially entitled to protection and the small number that are listed is due in part to the emphasis placed on species and subspecies, not reproductively isolated stocks, in the listing process of the ESA. In theory, hundreds of stocks of Pacific salmon species could become extinct without any single species properly being classified in danger of extinction throughout its geographic range of occurrence. Until recently, stocks were not listed as endangered or threatened under guidelines of the ESA. Also, Pacific salmon stocks migrate into river basins that are intensively used by humans. Many people in the Pacific Northwest view protection and recovery of endangered Pacific salmon stocks as directly conflicting with the regionally important economic activity supported by these basins, such as that generated by logging of watersheds, by hydropower facilities, by diversion of water for irrigation and urban uses, and by uses of waterways for shipping. In such instances, conflicting social priorities can make the listing process an arduous, prolonged task.

State Listings.   Most states have actively pursued listing and protecting imperiled species within their boundaries with the support of the federal ESA, which provides funding for the development of cooperative programs with states that have suitable endangered species programs. By 1987, 270 species of freshwater fishes were listed as endangered or threatened by at least one state (Johnson 1987). At that time eight taxa listed as federally endangered or threatened received special concern but not protected status by states—such as the Little Kern golden trout and Paiute cutthroat trout so listed by California and the Sonora and Yaqui chubs by Arizona—whereas three federally protected species, the beautiful shiner, Alabama cavefish (Figure 11.3), and watercress darter, were not listed by any state (Johnson 1987). Some taxa are state listed because they are rare within state boundaries although not vulnerable to extinction throughout their entire range. Four states (Alabama, Alaska, Hawaii, and Louisiana) did not list any fishes with protected (endangered or threatened) or special concern status at the time of Johnson's review. By 1991 all states but Alabama and Louisiana had established cooperative agreements with the U.S. Fish and Wildlife Service to initiate recovery efforts (Ernst 1991).

PROGRAMS OF ENDANGERED SPECIES CONSERVATION

The ultimate goal of the ESA is to get taxa off the endangered and threatened list, not place them on it (Bean 1986). Once listed, taxa are provided with protection detailed in federal and state endangered species legislation in order to deter further

Figure 11.3   The Alabama cavefish, a federally endangered species that is not protected by state endangered species laws. (Photograph courtesy of Dr. Richard Mayden, University of Alabama, Tuscaloosa.)

endangerment. The recovery plan is the tool used to restore taxa to stable, self-sustaining populations. The creation of recovery plans for federally listed species is the responsibility of the Secretary of the Department of the Interior or of the Department of Commerce. Such plans are developed by the U.S. Fish and Wildlife Service or the National Marine Fisheries Service, or their designees. Recovery plans serve as guides for the conservation of listed taxa. Government agencies are not legally required by the ESA to ensure that recovery plan recommendations are fully accomplished. In some instances lawsuits by environmental, conservation, or other citizen groups have been necessary to ensure that recovery plans are implemented (Carlson and Muth 1993).

The lack of recovery plans has retarded the recovery process for some taxa. The listing of a taxon does not guarantee the immediate development of a recovery plan, let alone its implementation. In 1989 recovery plans had been established for only 61 percent (47 of 77) of the fish taxa federally listed as endangered or threatened. Accomplishing the recovery of an endangered taxon often is a lengthy and uncertain process. Not surprisingly, relatively few federally listed animal and plant taxa have been removed (delisted) or upgraded from endangered to threatened status.

A variety of strategies may be incorporated into recovery plans, including the following (Carlson and Muth 1993):

1. Protecting critical habitat conditions
2. Propagating for reintroduction into historic basins, for introduction into new waters, or in extreme instances as a means of temporarily preventing extinction that otherwise would be immediate and inevitable
3. Establishing refuges
4. Controlling or eliminating introduced, nonnative species

Protecting and Modifying Habitat

Habitat deterioration and loss is such a common factor leading to the imperilment of freshwater taxa that habitat issues represent a major focus of efforts directed toward recovery of most listed taxa. As Rinne and Turner (1991) noted, "as go the habitats, so go the species." Land acquisition is the direct means that the ESA provides to ensure that critical existing habitat remains intact for federally listed species. Only a few states have significant land-acquisition programs associated with their efforts to conserve nongame and endangered species (Vickerman 1989). Often actions other than land acquisition are required as part of recovery efforts.

Numerous western freshwater taxa are imperiled at least in part due to altered flow conditions or lowered water levels. Providing instream flows of proper timing, magnitude, and duration is considered critical to the recovery of the Colorado River squawfish and other listed species of the upper Colorado River basin (Tyus 1991; Wydoski and Hamill 1991). The development of the Quail Creek water storage project on the Virgin River, Utah, required the provision of minimum flow levels in the river to support the federally endangered woundfin minnow (Deacon 1988). However, the required minimum flow standard, originally established due to "downstream water rights" (see Chapter 6) and deemed suitable to meet the needs of the woundfin, was not met throughout much of the summer in 1985 and 1986 (Deacon 1988). Establishing instream flows in streams of the arid West can be particularly problematic, because previously appropriated water rights may exceed the maximum usage that would allow the flow to be maintained at levels critical to protected species. The recovery program for endangered Colorado River fishes includes provisions to acquire and appropriate water rights and to convert these rights into instream flow for fishes (Wydoski and Hamill 1991).

Stream "improvement" activities have been conducted in numerous western streams to enhance the population condition of native trout. The success of these programs has been inconsistent and the results have been questioned (Rinne and Turner 1991).

Genetically uncontaminated stocks of rare native trout in many regions of the West are restricted largely to stream habitats with natural barriers such as waterfalls or dry reaches of streams that separate headwaters from downstream areas (Rinne and Turner 1991). Where natural barriers do not occur, artificial structures have been constructed in streams to prevent invasion of other fish taxa that might compete with, feed on, or hybridize with the protected taxa, further imperiling their existence. Efforts to eliminate nonnative species from specific habitats of endangered species often are combined with construction of obstructions to prevent reinvasion (see the section on removal of nonnative species in this chapter).

Other types of habitat modifications have been included in recovery programs of specific endangered taxa. Some of these have been minor. When water levels in Ash Meadows, Nevada, threatened to drop below a natural shelf providing critical spawning and feeding habitat for the endangered Devil's Hole pupfish, an artificial fiberglass shelf was suspended below the surface and artificially lighted to stimulate production of attached algae. This effort proved unsuccessful, requiring that water levels be maintained above natural shelf areas in the pool to prevent loss of this species (Carlson and Muth 1993). Cattails and other inshore emergent vegetation were periodically removed from refugia holding populations of the endangered Owens pupfish to prevent the loss of open, shallow-water habitat (Minckley et al. 1991).

A few habitat modifications have been expansive in cost and effort. The cui-ui, endemic to the Pyramid Lake–Truckee River system in Nevada, faced extinction after water withdrawal lowered the level of Lake Pyramid to the point where mature cui-ui could not migrate up the Truckee River to spawn. Over 90 percent of the spawning aggregation that gathered at the mouth of the river in 1983 were from the 1969 year class, whereas most of the rest were from the 1950 year class. Thus low water levels and the lack of suitable migration conditions had prevented the production of surviving offspring in all but one year between 1951 and 1982. The Marble Bluff dam has four fish ladders and a 5-kilometer-long channel, which were constructed to allow migrating cui ui to enter the mouth of the Truckee River and bypass a sand delta that obstructed further upstream migration (Scoppetone and Vinyard 1991). Juvenile fishes appeared in this system during the 1980s. However, spawning migrations and offspring production that occurred in the early to mid-1980s were supported by wetter than normal years (Scoppetone and Vinyard 1991). In 1986 a 100-year flood stimulated a large migration, which led to a tripling in the abundance of adult fishes by 1990 (Emlen et al. 1993). After 1987 the combination of drought and water withdrawal in the Truckee essentially prevented spawning runs (Emlen et al. 1993). Even with the Marble Bluff facility, only a small number of adults ascend the river in dry years. Drought and increasing salinity in Pyramid Lake still threaten the continued existence of this endemic species (Scoppetone and Vinyard 1991).

Similarly, the Colorado River Recovery Program, a $60 million 15-year project, is focused on recovery of the endangered Colorado River squawfish, bonytail chub, humpback chub, and razorback sucker (Figure 11.4). This plan includes creation of artificial backwaters, jetties, fish ladders, and other structures to enhance the habitat conditions for these species (Wydoski and Hamill 1991).

Propagation for Stocking into Historic or New Basins

Most endangered species recovery programs include provisions for establishing new populations of endangered or threatened species, either within or outside their original geographic range of distribution (Williams 1991). Under guidelines of the ESA, protected taxa can be stocked into new basins to create experimental populations that legally are not deemed essential to survival of the taxon. They are provided less protection in those waters than in systems where they still occur naturally. Thus the introduction of a listed taxon into a new area would not automatically prevent development or use of the water or surrounding landscape. Without this provision, introduction of a protected species into new waters would be extremely difficult politically.

Although propagation methods are well understood for some fish taxa, such as many salmonids, they must be developed for others. Agencies often develop propagation techniques at their own fish hatchery facilities. The Dexter National Fish Hatchery in New Mexico has served as a center for propagating imperiled southwestern fishes for two decades (Carlson and Muth 1993). States may develop

Figure 11.4 The razorback sucker, an endangered species of the upper Colorado River basin. (Photograph courtesy of Tom McHugh/ Steinhart Aquarium; Photo Researchers, Inc.)

cooperative arrangements for the propagation of some taxa. For example, Ohio's Division of Wildlife established a joint program for propagation of the western banded killifish, a state endangered species, with the curators of the Columbus, Ohio, Zoological Garden (Ross 1990).

Shannon et al. (1993) believe that propagation will be necessary to prevent the extinction of many freshwater mussel species. However, techniques for successful rearing of mussels through larval stages are not yet developed. Reintroduction of mussels depends on transplantation of juveniles and adults. To date, many attempts at transplantations have failed.

Establishing Refuges

Refuges can be established in small, fishless natural systems or in artificially produced and maintained bodies of water that are designed for one to several species of target organisms, but not to maintain and conserve the biotic communities in which these species occur naturally.

Some refuges are large and serve multiple species of imperiled fishes. The Dexter National Fish Hatchery in New Mexico, which held populations of 24 species and subspecies of federally protected fishes from 1974 to 1989, was established to protect genetic stocks of imperiled fishes that might become extinct in nature, and to propagate large enough numbers of selected species to allow stocking into previously inhabited systems (Johnson and Jensen 1991). Many refuges are small natural or constructed pools that serve a single taxon. For example, the Gila topminnow was released into over 100 isolated pools in Arizona, most of which were developed originally for livestock watering.

Although refuges have only recently gained widespread use in recovery of imperiled fish taxa, a few were established decades ago. The impoundment of Hank and Yank's Spring, Arizona, into a concrete box followed by introduction of the Sonora chub in 1949 created one of the most stable habitats for this species in the United States (Williams 1991). Even with the creation of this stable artificial habitat, periodic restocking has been required to maintain the population in this refuge. One of the most persistent single-species refuges, the Devil's Hole pupfish refugium below Hoover Dam, held a population of this species for 14 years without restocking (Williams 1991).

Generally, refuges are not intended to provide a permanent means of recovery for endangered species, but a means of protecting and dispersing rare gene pools. Distributing previously localized gene pools into multiple waters over wider geographic regions lessens the potential for natural or human-caused habitat disruptions to cause extinction of that taxon (Williams 1991). Several elements limit the importance of refuges to some recovery programs. The construction and maintenance cost of many refuges can be very high. Many populations in small refuges have displayed dramatic shifts in their genetic makeup from that representative of the wild populations from which they were taken. Low numbers of individuals are susceptible to genetic drift (see Chapter 8), and the artificial conditions of many refuges can strongly select for genetic changes. For example, Devil's Hole pupfish populations housed in artificial refuges grow to larger sizes than those in natural settings, and they exhibit "misshapen bodies" (Williams 1991).

Naturally occurring refuges have been more successful than artificial ones in sustaining rare fishes through time. However, when refuges are not on publicly owned and specifically protected lands, numerous factors such as the introduction of nonnative species, collapse of water sources due to diversions or pumping of underground sources, and habitat disruption can threaten the long-term utility of these systems.

Removal of Nonnative Species

Propagation and reintroduction of endangered species cannot be successful unless the causes of decline of that taxon are corrected. Numerous western fishes are endangered in part due to the introduction of nonnative fauna. Resulting competition or predation upon the native fishes by the introduced taxa or introgressive hybridization between the introduced and native fishes can lead to the loss of the native form (see Chapter 8). Where such interactions have been a major factor leading to endangerment, elimination of nonnative fauna may be necessary to accomplish recovery (see Chapter 10).

The removal of nonnative fishes by use of toxicants or in some instances by capture with seines or other collecting gear has been part of recovery programs for many endangered western fishes. Removal of nonnative fishes has been an integral activity in the management of imperiled western trout, such as the Gila trout in New Mexico and Apache trout in Arizona. Introduced brook, brown, and rainbow trout and hybrids between the Apache trout and rainbow trout were successfully removed from 8 of 13 streams in Arizona treated with toxicants (Rinne and Turner 1991). Apache trout were reestablished in seven of those eight systems by 1991. Hybrid or nonnative trout still occurred in the other five streams after the use of toxicants. All streams in which nonnative trout were successfully eliminated were simple habitats, lacking backwaters and other areas where toxicants might not be effective and possessing barriers that prevented reinvasion by fishes from downstream areas. The five streams in which removal failed were more complex systems that included marshes, springs, or other habitats that protected portions of target populations from effective exposure to toxicants. Similarly, during attempts to improve the condition of a population of the Gila trout in Iron Creek, New Mexico, extensive marshy habitats and small springs associated with that creek system prevented effective removal of the introduced brown trout (Rinne and Turner 1991).

Removal also has been used in recovery efforts for nontrout endangered species, with various degrees of success. Toxicants were applied to selected reaches of the Owens River basin to eliminate introduced largemouth bass, rainbow trout, and brown trout as part of recovery efforts for the Owens River pupfish. Prior to removal efforts, fish barriers were constructed to prevent reinvasion by these nonnative species. Repeated treatments with toxicants were required due to natural reinvasion by mosquitofish and to unauthorized introductions of largemouth bass and possibly mosquitofish into the refuges (Minckley et al. 1991). Similarly, toxicants and the construction of barrier weirs did not eliminate mosquitofish from two of three pools at Bylas Springs, Arizona, that contained populations of the Gila topminnow. At least one of the spring pools retained some mosquitofish that were not killed by the toxicants, and reinvasion occurred when flood waters breeched weirs that had been constructed at the outlet to each spring. After introgressive hybridization between the nonindigenous sheepshead and the endangered native Leon Springs pupfish had threatened the survival of the native species, repeated poisoning was necessary to eliminate the sheepshead minnow, while persistent netting was necessary to examine and remove surviving hybrids (Minckley et al. 1991).

"WHY BOTHER?" AND "ARE WE DOING IT RIGHT?"

The value and relevance of protecting imperiled taxa has been analyzed and argued for decades. Scientists have provided several basic premises for preventing the extinction of living things (Carlson and Muth 1993):

1. Taxa provide ecological benefits to humans.
2. Some animals and plants provide direct economic benefit or may potentially do so in the future.
3. Taxa may provide aesthetic benefits to humans.

Ehrlich and Ehrlich (1981) and others have maintained that diversity of organisms in natural systems determines the stability and persistence of the biota of those systems. Mass extinctions would jeopardize the stability of biotic communities upon which humans depend for their own existence. Thus humans must be concerned with human-caused accelerated rates of extinctions because of the long-term threat to the quality and condition of the lives of their descendants. However, Rolston (1991) points out that potentially valid arguments supporting prevention of mass extinctions may not be easily extended to the protection of all imperiled species individually. Many people would argue that the possible loss of a species of fish that may exist in only a single desert pool or stream does not warrant protection from development of water resources in ways that are beneficial to humans.

Some imperiled fishes provide obvious economic benefits to humans, such as the numerous stocks of Pacific salmonids. However, the protection necessary to facilitate recovery of these resources may have economic costs that exceed economic benefit derived from their existence. Many imperiled aquatic taxa confer no measurable economic benefits.

Perceived aesthetic benefits of fisheries resources vary greatly among humans. Some aesthetic benefits are measurable. For example, imperiled salmonids may be valued by anglers. However, because aquatic organisms live in a different and in many ways secretive medium (water), most species are not as intimate a part of the outdoor experiences of many people as are some taxonomic groups such as birds. Whales, coral reef fish communities of the Florida Keys, and migrating anadromous fishes do provide an equivalent experience to that enjoyed by so many bird-watching enthusiasts. However, most fishes and other aquatic resources are "out of sight and largely out of mind" (Rolston 1991).

In addition to direct economic or aesthetic benefits, a growing segment of society supports the conservation of endangered fauna and flora on moral grounds (Carlson and Muth 1993). Many people believe that allowing the worldwide rate of extinction, which is projected to exceed dozens of species per hour by the end of the 20th century (Myers 1985), to continue unabated is unconscionable.

Are Endangered Species Merely an Expression of the Real Problem?

Success in endangered species conservation has been quite modest, at best. A few highly visible organisms have been delisted—for example, the American alligator and brown pelican—or have had their condition significantly improved by conservation actions guided by endangered species legislation—the bald eagle is under review for possible delisting or upgrading in listing status. Some have been upgraded due to new information. For example, the snail darter has been upgraded from endangered to threatened not because its existence has been significantly enhanced by recovery programs but because it was found in additional areas once it was listed as endangered. However, after more than two decades since passage of the ESA, the vast majority of threatened and endangered species remain imperiled.

Ultimately, perhaps the greatest contribution made by an emphasis on endangered species conservation in the latter half of the 20th century will be the role that this has played in the evolution of concern for ecosystem management and biodiversity conservation (see Chapter 12). The evolution of efforts in endangered species conservation has coincided with a growing awareness that responsible stewardship includes conserving not only recreationally and commercially important species but the health of all natural systems and the biotic communities that inhabit them. Efforts in endangered species conservation were spawned in the 1960s by widespread concern over the possible extinction of "charismatic megafauna," or large-bodied mammals and birds (Cairns and Lackey 1992). By the 1980s the focus of that concern was expanding to the conservation of entire biotic communities (called biodiversity conservation), not just specific high-visibility taxa that live within those communities. State programs that focus on management and conservation of nongame resources have been established largely because of the growing awareness of the intrinsic importance of all life forms and of the health of the biotic communities in which they live. Indeed, the list of taxa imperiled with extinction is always growing due to alteration or deterioration of the quality of the habitats those taxa live within, or modification of the biotic communities of which they are members. Thus the presence of endangered species often can be viewed as a signal that the persistence of natural systems is being threatened. In most instances, if endangered fisheries resources successfully recover as a result of endangered species programs, it would be because the systems in which they live are protected. Just as the goal of the ESA is to remove taxa from its protected lists, ultimately a goal of fisheries conservation should be to protect the quality of aquatic systems in order to reduce the number of potential candidates for listing in the future (Williams 1991).

12

Case Studies and Future Opportunities

Chapters 7 through 10 reviewed basic conservation and management actions or strategies that fisheries resource agencies may take. Obviously, conservation of specific resources may require coordinating fisheries regulation, habitat protection, or improvement and stocking in a comprehensive management strategy. This chapter reviews four intensively managed fisheries resources: Atlantic coastal striped bass, Pacific anadromous salmonids, Lake Erie walleye, and New England continental shelf groundfishes. Each of these fisheries resources supports regionally important fisheries that have been managed in a complex biological, sociological, and political setting. Each of these resources has suffered through substantial declines, all due to fishing mortality and all but New England's groundfishes due to serious habitat perturbations. Conservation efforts focused on these resources have required interagency, interstate, and/or international cooperation. In addition, strategies that state agencies have used to manage trout stream fisheries are reviewed. The management setting for many trout stream fisheries is much less complex than that of others listed above. These resources are reserved for recreational use only. Habitat issues complicate successful conservation and management of particular trout fisheries. However, this review focuses on the dichotomous manner in which agencies have attempted to meet high fishery demand for resources that generally display limited production potential.

The management histories of numerous fisheries resources may be well suited for this chapter. Those listed above were selected because their management ranges to date from successful to unsuccessful. As a group they illustrate the importance of effectively balancing biological needs of a resource with economic and sociological needs of its users. The successful management of striped bass, Lake Erie walleye, and many trout stream fisheries has occurred in part due to changes in the standard management process (for example, passage of the Striped Bass Act in 1984) or particularly conservative control of harvest to ensure resource recovery or protection. Thus their management focuses on long-term, sustainable interactions between resources and their users. Management of Pacific anadromous salmonids and New England groundfishes has been influenced by economic and sociological issues that have superseded the biological considerations in the management process. To date, management of Pacific salmonids represents attempts to treat the symptom of declining resources rather than address the causes of the declines, whereas that of New England groundfishes has favored immediate economic health of the fishing industry rather than a sustainable, long-term relationship between the resource and its users.

STRIPED BASS—MANAGEMENT TOWARD RECOVERY

The Atlantic coastal migratory stock of anadromous striped bass has long supported important recreational and commercial fisheries. Due largely to long-term overfishing and deterioration of spawning habitat, abundance of Atlantic coastal striped bass declined precipitously through the 1970s, which resulted in a depleted resource and collapsed associated fisheries by the early 1980s. Significant changes in conservation strategies facilitated a remarkably rapid recovery of this resource. The recovery was accomplished by the following:

1. Adopting a coastwide strategy of harvest restriction and severely reducing or curtailing commercial harvest to support recreational uses of the resource
2. Adopting a management philosophy that would allow rapid responses to changes in the condition of the coastal migratory stock
3. Initiating an expansive monitoring program to estimate stock condition and the intensity and impact of harvest more accurately so that changes in management more effectively improved or protected the condition of the resource

Success of this conservation program was made possible by passage of federal legislation that radically changed the manner in which multiple states cooperated in joint management. The following material summarizes a more detailed review of this management program presented in NOAA (1993).

The coastal migratory stock of Atlantic striped bass is composed of fishes from two populations, one that spawns in the Chesapeake Bay and another in the Hudson River. Juvenile and adult fishes from these populations migrate seasonally along the Atlantic coast, although individuals from the Hudson River are more restricted in their coastal distribution than are Chesapeake fishes. The coastal migratory stock has supported recreational and commercial harvests for a long time, even though periodic declines in abundance, probably due in part to overfishing, have occurred since the 1700s. During the mid-20th century, commercial harvests of striped bass increased significantly (Figure 12.1), while recreational interest in this resource grew. Yearly fishing mortality rates of juvenile and adult fishes in the coastal migratory stock ranged from 24 to 59 percent throughout the 1970s, even though commercial landings began to drop precipitously during that decade.

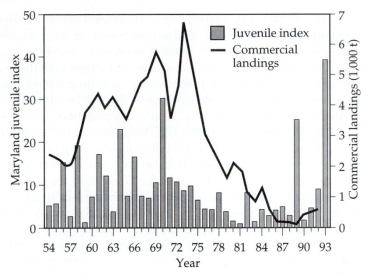

Figure 12.1   Yearly commercial harvests of juveniles and adults, and indices of abundance of young-of-the-year striped bass from Maryland waters of the Chesapeake Bay. The decline in commercial harvests starting in the mid-1970s occurred due to depletion of the stock and in more recent years to restrictive harvest regulations. The index of abundance of young-of-the-year fishes represents the number captured per seine haul in a standardized yearly seining survey (reprinted from NOAA 1993).

Spawning of the Chesapeake Bay population, which historically had produced strong year classes of offspring about every 2 to 4 years, produced low to very modest numbers of offspring for more than a decade after the record year class produced in 1970. This prolonged period of reproductive failure, likely caused by a progressively depleted resource and deteriorating environmental conditions in Chesapeake Bay, caused a severe collapse of the coastal stock and the recreational and commercial fisheries it supported. Commercial harvests in 1979 were only about 20 percent as high as they had been just 6 years previously (Figure 12.1).

Interstate Management of Coastal Striped Bass

By 1981 the Atlantic States Marine Fisheries Commission (ASMFC; see Chapter 6) had adopted an interstate fishery management plan for striped bass that included a minimum legal size limit of 14 inches in estuaries and 24 inches along the coast, and a closure of spawning areas to fishing during the spawning season. As the stock continued to decline and more stringent restrictions were clearly needed, Congress passed the Striped Bass Act of 1984 (see Chapter 6), which required all states to follow further management plans created by the ASMFC or face a moratorium on the harvest of striped bass in their waters (previously, states could decide whether or not they would participate in any management plan developed by the ASMFC). Such moratoria were threatened several times and implemented once by the federal government before all states complied with the ASMFC recovery plan.

The ASMFC established a minimum legal size limit that would be increased annually to prevent harvest of all striped bass from the modestly abundant 1982 year class until 95 percent of the females from this cohort had matured and spawned (see Chapter 7). Harvest of fishes older and larger than those in the 1982 year class was allowed because a total moratorium on fishing would have been difficult to enforce, and because there were so few of these fishes in the stock that their removal would not seriously jeopardize future reproduction.

By 1989 females from the 1982 year class were producing most of the eggs in Chesapeake Bay. The 1989 production of offspring in Maryland waters of Chesapeake was the most successful since 1970 and the second highest recorded since 1954 (Figure 12.1). Recruitment in the Hudson River population, which had been similarly protected by the harvest restrictions even though it had not been as seriously depleted as the larger Chesapeake population, also reached record levels in the late 1980s.

After 1989 several strong year classes of Chesapeake fishes were produced, including the highest ever recorded for that population in 1993. Harvest restrictions continued to protect the rapidly increasing number of young fishes during this time period. Based on criteria established when the management plan was adopted, the ASMFC declared that the coastal migratory stock was fully recovered in 1994. This triggered a planned change in the management regime from restricting fishing mortality to a level that facilitated population growth, to one designed to maintain the recovered stock at a desirable level of abundance. As part of this process, The ASMFC developed Amendment 5 to the striped bass management plan. Amendment 5 allowed states to reduce the minimum legal size limit from 36 to 28 inches, to increase the daily catch limit for anglers from one to two fishes, and to increase commercial quotas. Interestingly, not all recreational users were convinced that regulations should be changed in this manner. As a result of public hearings held in March 1995, the Massachusetts Marine Fisheries Commission reduced recreational size limits only from 36 to 34 inches with a daily catch limit of one fish, and increased the commercial quota from 238,000 to 750,000 pounds (Amendment 5 of the ASMFC allowed the Massachusetts commercial quota to rise to 1,000,000 pounds). These conservative changes in harvest regulations resulted from testimony from anglers who were concerned that the full changes recommended in Amendment 5 might stimulate another decline of the striped bass coastal stock. Although striped bass have been declared restored, "fisheries managers may have to wait a few more years to restore the angling public's confidence" (McKiernan 1995).

Passage of the Striped Bass Act was a pivotal event leading to this stock's recovery, because the decision to fully protect the 1982 year class so restricted harvest that "it seems likely that few states would have fully complied" with the management plan had they not been required to do so (NOAA 1993). The decision of the ASMFC to restrict coastwide commercial harvest, and of some states to totally prohibit commercial harvest, was instrumental in effectively controlling fishing mortality once minimum legal size limits were eased. Thus, although the proximate stimulus facilitating recovery was the gradually increasing minimum legal size limit that protected one year class of fishes, far-reaching principles were behind the success of this conservation effort, including the following:

1. The federal challenge to the authority of states to unilaterally manage the portion of an interstate resource housed within their respective boundaries

2. The decision to favor recreational uses of this resource while severely curtailing commercial harvest, which reduced total fishing pressure and maximized the success of one user group while preventing another historical user group from full access to the resource

PACIFIC ANADROMOUS SALMONIDS—THE COMPLEXITY OF FISHERIES CONSERVATION AND MANAGEMENT

Conservation of Pacific salmon (chinook, sockeye, coho, pink, and chum salmon and the steelhead trout) that spawn in river basins of California, Oregon, Washington, and Idaho is an incredibly complex problem for fisheries agencies in that region. These economically valuable and aesthetically valued resources are challenged by a diverse array of factors that have caused their decline and threatened their existence. Threats include a number of activities whose collective economic and political value to the region dwarf the considerable economic value of the salmon resources themselves. Thus the social and political framework within which the conservation of Pacific salmonids has had to operate is a particularly difficult and frustrating one. Indeed, Pacific salmon have received a lot of attention, including the passage of federal legislation and a series of federal court decisions that have detailed the country's responsibility in conserving them. Large amounts of money have been expended on their behalf. However, numerous populations in West Coast river basins have not been sufficiently protected by these management attempts. Pacific salmon resources are seriously depleted in rivers from California to Washington because their management has been driven more toward treating the symptoms of population decline rather than addressing the causes of the widespread decline that these resources have suffered.

Major factors that affect Pacific salmonid resources include:

1. Harvest
2. Deterioration of spawning and nursery habitat due to a variety of land- and water-use patterns, including logging, grazing, and diversion of water for agricultural irrigation and urban uses
3. Loss of spawning grounds and the substantially increased mortality of migrating adults and smolts due to hydropower and flood-control dams

Harvest

Although Alaskan harvests of Pacific salmonids remain robust, harvests in other southern West Coast states (Washington, Oregon, and California) have dropped precipitously. Ocean catches of coho salmon declined by 95 percent between 1976 and 1993 and chinook by 80 percent between 1988 and 1992 (Stevens 1994). Such declines do not reflect a reduction in fishing interest or effort but in the condition of the stocks supporting the fisheries.

Because Pacific salmon are harvested as adults within coastal waters and river basins during their spawning migrations, and until recently outside U.S. territorial waters by high seas drift net fisheries, regulating harvest is a complicated process. International agreement was required to end high seas directed fishing and incidental harvest in a drift net squid fishery (NOAA 1993). Multiple agencies and other entities may be involved in cooperatively developing harvest regulations for some river systems. For example, harvest of Pacific salmon in the Columbia River basin is overseen by the U.S. Fish and Wildlife Service; state fisheries agencies of Washington, Oregon, and Idaho; and the Columbia River Inter-Tribal Fish Commission (CRITFC, composed of members from the Nez Perce, Umatilla, Warm Springs, and Yakima tribes; Marsh and Johnson 1985). The state of Washington works with 21 treaty tribes to allocate allowable harvests of anadromous salmonids in its coastal rivers (Moring 1993; see Chapter 6). Such interagency groups not only must establish allowable harvest levels that will help sustain the salmonid stocks but must recommend allocations to commercial, recreational, and tribal interests.

Harvest limits have become increasingly restrictive as salmon stocks have declined. In April 1994 the Pacific Fisheries Management Council, which oversees coastal harvest of anadromous salmonids, voted to prohibit commercial and recreational fishing for salmon from Cape Falcon, Oregon, to the Canadian border, and to restrict harvest south of that region; the council proposed a tribal harvest of 16,400 salmon in the prohibited waters (*The New York Times*, 10 April 1994). Commercial harvesters, who felt that they were bearing too much of the burden of aiding the recovery of many stocks depleted by habitat deterioration and loss, objected strongly to these proposals.

Fishing has not been a major factor in the decline of many anadromous salmonid stocks in that region. Habitat loss is the major threat to 90 percent of the imperiled salmon stocks in the Pacific Northwest (Lawson 1993). Effectively addressing the deterioration and loss of spawning and nursery habitat, and the disruption of migrations into and from freshwater systems, has proven to be significantly more complex and frustrating than has regulating harvest.

Habitat Problems

Watershed use and development and diversion of water from river channels have destroyed large expanses of freshwater habitat in numerous river systems. In California widespread loss of stream habitat caused by logging, agricultural, and urban uses has led to the decline or disappearance of coho salmon stocks in all 582 streams of that state that once supported spawning populations of that species (Brown et al. 1994). Logging on public lands has been particularly harmful, and dams that prevent migration or water diversions that have caused spawning grounds to dry up have led to losses of stocks in some basins (such streambed drying has also occurred in reaches of the Umatilla River in Oregon and Yakima River in Washington; Daniel 1993).

For years, lands under the stewardship of the U.S. Forest Service were managed largely for timber production. However, growing concern over the declining condition of western river basins led to a gradual change in the management practices of the Forest Service. At times, this change has been stimulated by pressure (including lawsuits) brought by conservation, environmental, and fishing industry groups. In 1993 the Clinton administration developed forest management guidelines for federally managed public lands in the Pacific Northwest (Stevens 1994). One set of guidelines, often referred to as PACFISH, was specifically intended to protect anadromous fish spawning and nursery habitat on public lands in that region. These forest land management guidelines generally have been supported by conservation groups and, in at least some instances, opposed by the timber industry and some congressional representatives. For example, amendments included in the Senate appropriations bill to fund the Department of the Interior prevented implementation of PACFISH guidelines on the Tongass National Forest in Alaska for federal fiscal year 1994 (T. Williams 1994). However, PACFISH guidelines are being considered for use in some other western areas. For example, in August 1994 the NMFS supported timber harvests in the Nez Perce National Forest (Idaho) and issued an incidental "take" permit (under the Endangered Species Act) on the condition that a 300-foot-wide buffer, a no-cut area on either side of the stream channel, be established in certain areas proposed for logging.

However slow the progress in establishing effective land-use strategies for protecting anadromous salmonid resources is, resolving the impacts of dams on these stocks is proving to be equally frustrating.

Dams

Many western river basins have been dammed to produce electricity, to control flooding, and to store and divert water for agricultural and urban uses. Such intensive water use has produced unfavorable conditions for anadromous salmonid stocks. Some stocks were driven to extinction when dams prevented their migration to spawning grounds. Chinook salmon were eliminated from the Sacramento River after construction of a dam in that system. Dams have cut off nearly one-third of the Columbia River basin that had been accessible previously to migrating fishes (Wood 1993; see Chapter 9).

Upstream Migration Past Dams.   Fish ladders have been constructed on many dams to allow upstream migration of adult fishes. However, even the most effective passage facilities still cause some mortality of migrating salmon. Eight dams with fish passage facilities occur on the Columbia River–Snake River system (Figure 12.2); about 5 to 10 percent of adult salmon reaching each dam fail to pass upstream (Wood 1993). Thus about 30 to as many as 50 percent of the fishes attempting to access the upper Snake River fail to do so because of dams. Mortality that occurs during downstream migration of smolts—juvenile fishes migrating to the sea—can be even greater.

Downstream Migration Past Dams.   Smolt migration to the ocean generally occurs during high river flows of spring months. High springtime flows typically are stored in reservoirs above dams to control flooding, or for diversion or generation of electricity

Figure 12.2   Dams on the mainstem of the Columbia River–Snake River basin. Dams from Bonneville to Lower Granite on the Snake River have fish passage facilities for upstream migration of
adult anadromous salmonids. (From M. H. Gessel, D. A. Williams, R. F. Brege Krema, and D. R.
Chambers. 1991. Juvenile salmonid guidance at the Bonneville Dam second power house, Columbia
River, 1983–1989. *North American Journal of Fisheries Management* 11:400–412. Reprinted by permission
of The American Fisheries Society.)

later in the year. In the spring, the greatest amount of water released at hydropower
dams is that which flows into turbine intakes and past turbines. Thus, migrating smolts
may suffer high rates of mortality at dams because they follow this flow.

A number of bypass systems have been developed that attempt to divert downstream migrating smolts away from flow entering turbine intake pipes. These bypass systems rely on the release of significant amounts of water through channels,
canals, or in some instances simply over the top of the dam. Smolts are unable to find
these alternative routes past the dam unless enough water is released to attract them.

Estimates of mortality of migrating smolts at dams have varied from about 10 to 33 percent (Raymond 1979). The most recent estimate for the dams on the Columbia and Snake rivers, 13.5 percent per dam (Northwest Power Planning Council, 1994, unpublished report) would produce a cumulative mortality of about 70 percent for smolts migrating from the upper Snake River (Gessel et al. 1991). Total mortality may be even higher in dry years when little water flushes down the river basin; up to 90 percent of migrating smolts might die annually during drought periods. Additional deaths occur within reservoirs upstream from dams due to increases in the amount of time that is necessary to migrate through the system (Gessel et al. 1991).

Increased Migration Time. In highly regulated basins such as the Columbia River, the long reaches of pooled water in reservoirs substantially slows the rate of downriver migration of smolts. Travel time through such systems can be significantly increased, particularly during years of low river flow (Berggren and Filardo 1993). Delayed migration might disrupt the physiological changes that need to occur to allow smolts to enter the marine environment. Delayed migration can also increase mortality caused by riverine predators, such as the introduced walleye and smallmouth bass and the native northern squawfish in the Columbia basin. Eleven percent of all Columbia River smolts entering the John Day reservoir via the upstream McNary dam are eaten by squawfish; most of this mortality occurs in the immediate vicinity of dams (Reiman et al. 1991). Because of this increased predation mortality, the squawfish has been the focus of a predator control program in the Columbia basin (see Chapter 10). Interestingly, its reputation as a predator may not be fully deserved. Gadowski and Hall-Griswold (1992) found that squawfish tend to select dead rather than living smolt prey in experimental studies. Because much of the mortality attributed to squawfish predation occurs in the immediate vicinity of dams, it is probable that some of the smolts that are eaten are already seriously injured or dead after passage through turbines.

Mitigation of Impacts Caused by Dams. Management programs that have addressed dam-induced mortality and population decline have included development of

1. Water-release schedules that enhance conditions for migration and passage at dams (see Chapter 9)
2. An expansive hatchery system to supplement depleted wild stocks (see Chapter 8)
3. A program to capture wild and hatchery-derived smolts and barge them past most of the dams blocking their migration (see Chapter 9)
4. Population-control programs directed at the northern squawfish to reduce mortality of migrating smolts caused by predators (see Chapter 10)

The first two have been widely applied to anadromous salmonid stocks; all four have been an integral part of management within the Columbia River basin.

Enhancing Flows. Only one of the above actions, the development and implementation of water-release schedules, focuses directly on mitigating mortality caused by dams in river basins. Water-release schedules are intended to significantly increase the flow of water through bypass channels or over dams so that more smolts use these routes instead of intakes to turbines. In the Columbia River, releases are also designed to create greater flow through extensive reservoir systems, guiding smolts better in their downstream migration and reducing the time needed to complete the journey.

In large basins such as the Columbia River, water releases to enhance salmon migration may conflict with a variety of uses of water that collectively are extremely important to the economic condition of the region. Dams are constructed because water serves so many valuable needs. Thus, when water releases are studied or proposed, objections may arise from many sources.

Obviously, releases prevent the same water from being used to generate electricity, which has significant economic implications. Recent proposals for water releases and reservoir drawdowns indicate the other issues related to such actions. Proposed reservoir drawdowns on the San Joaquin River to support restoration of salmon stocks caused widespread objection by the farming community that receives diverted waters of that basin for irrigation. Recent proposed springtime drawdowns in the Columbia–Snake River reservoirs have met with objections from the following sources:

1. The utility industry was concerned with reduced electricity generation. However, the loss is not a substantial proportion of the energy produced by hydropower facilities within the basin. If used to turn turbines, the water releases needed to promote downstream smolt migration effectively would produce about the same amount of electricity as a medium-size coal-burning plant (Wood 1993).

2. The agricultural community was concerned about the loss of water that might have been used for irrigation and the increased cost of shipping harvested grain by railway or highway if spring releases reduce the likelihood that crop products can be barged downriver later in the year.

3. In at least one instance, a state fisheries agency has objected. In July 1994 the state of Montana filed a notice of potential lawsuit to prevent the drawdown of Montana reservoirs to provide downriver springtime flows in the Columbia basin. Montana feared that such drawdowns might jeopardize resident populations of bull trout and cutthroat trout.

The federal government created the Northwest Power Planning Council (NWPPC; see Chapter 6) to coordinate water use in the Columbia basin in a manner that enhances anadromous salmonid resources while protecting the economically important industries supported by water of the basin. The central emphasis of the council is the development of a water budget that supports flow levels within the basin to enhance spring migration of smolts (Wood 1993). Implementation of the water budget has proven to be extremely difficult. An adversarial relationship has developed between the Columbia Basin Fish and Wildlife Authority (CBWFA; made up of 13 Indian tribes and 7 state and federal fish and wildlife agencies) and operators of the hydropower system within the basin. The CBWFA has argued that allocation of water among power generation, flood control commitments, and reservoir refill does not allow the water budget to be adequately addressed in drought years, whereas the Bonneville Power Administration (a federal agency) has supported industry's objections to some plans proposed by CBWFA for enhanced flows (Wood 1993).

Hatchery Supplementation.   Hatchery production has been viewed as a means of restoring depleted stocks and mitigating the loss of wild fishes caused by water development projects and land-use practices; between 1980 and 1989, 18 federal fish hatcheries produced over 703 million Pacific salmon and steelhead trout that were stocked into basins of the Pacific Northwest (USFWS 1989). However, relatively few stocks have been significantly enhanced or restored by this massive hatchery effort. Hatcheries historically have operated in ways that maximize biomass production and stocking rates; such agendas may harm rather than enhance native stocks (see Chapter 8). In many instances, nonnative strains of salmonids have been stocked into basins housing declining native strains of the same species. Even when native strains have been used, hatchery spawning practices have not been designed to conserve the genetic variability of the wild stocks in basins being stocked. In some instances, stocking of large numbers of hatchery fishes has apparently accelerated rather than prevented the decline and loss of some native stocks (as described in Chapter 8). In addition, the proportion of hatchery-reared fishes that survive after stocking has persistently declined in long-term supplemental stocking programs.

Although substantial emphasis has been placed on hatchery production and stocking, there is little cause to feel that much success has been achieved. Recent improvements in artificial spawning, rearing, and stocking strategies may increase success in restoring abundance of wild fish stocks while conserving the integrity and variability of their gene pool (such as the Yakima River basin stocking strategy developed by the Northwest Power Planning Council; see Chapter 8). However, unless mortalities caused by damming and habitat deterioration are successfully addressed, hatcheries alone will provide little success in the conservation of wild salmonid resources (Marsh and Johnson 1985).

Barging and Predator Control.   Of all management activities applied to Pacific salmonids, barging and predator control programs are the best examples of responding to the symptoms rather than the causes of impacts.

Barging, which obviously establishes another artificial element in the life cycle of these salmonids, essentially serves as an alternative to increasing flows during migration. Supported by the utility industry and other water users in the Columbia River basin, barging does increase survival somewhat over the extremely low rates experienced by smolts migrating in the absence of enhanced flows (Daniels 1993). However, crowding on barges exposes fishes to diseases and to physiological stress that might increase mortality during the barging or after release. In addition, adult anadromous fishes are able to return to spawn in streams in which they had hatched by following odors within river basins that they imprint upon as juvenile fishes during their migration to the sea (Moyle and Cech 1988). The impact of barging on imprinting and ultimately on the ability of salmon to return to their natal streams as adults is not well understood. Thus the decrease in turbine-related mortality that may occur as a result of barging juvenile salmon past dams may not be adequate to stimulate long-term recovery of wild salmonid resources.

The Columbia River northern squawfish control program was established to reduce rates of predation suffered by salmon smolts during downriver migration (see Chapter 10). Because much of the squawfish predation imposed on salmon smolts in the Columbia River basin occurs within the vicinity of dams, scientists believe that squawfish predation rates are higher than levels characteristic of the basin before construction of multiple dams and filling of their reservoirs. In addition, this mortality represents only a portion of the total smolt mortality caused by human modifications of the basin and watershed. Predator control is an old approach to resource management: Eliminate natural predators in order to reduce the impact of high human-induced mortality. Although squawfish bounties are less expensive than alternatives such as flow augmentation, it is questionable whether the recovery and stability of Columbia River anadromous salmonids will hinge heavily on controlling a native minnow species by offering a bounty for its capture.

The Effect of Climate on Salmon Survival and Production

Generally, Pacific salmon harvest levels have been regulated to allow sufficient levels of "escapement"—the number of migrating adults that escape capture in a fishery in order to reach spawning grounds and reproduce—to produce some desired level of offspring production (Beamish and Bouillon 1993). However, the abundance of spawning salmon generally has been poorly correlated with abundance of the surviving offspring that they produce (as described in Chapter 3). Beamish and Bouillon (1993) and Beamish (1993) demonstrated that long-term changes in climatic conditions in the North Pacific Ocean parallel trends in ocean survival and production of salmon and other fishes. The intensity and duration of the Aleutian Low, an extensive low-pressure weather system that develops over the North Pacific Ocean in early winter and dissipates by the following summer, influences patterns of upwelling of nutrients to the surface waters, in turn influencing levels of plankton production and ultimately fish production throughout wide areas of the North Pacific Ocean (Beamish 1993). For example, during the most intensive lows, such as those occurring in the late 1970s, ocean productivity of a variety of fisheries resources, including Pacific salmon, was unusually high, supporting increased levels of ocean harvest of these resources (Beamish 1993).

Interestingly, the hypothesis that levels of salmon production in the ocean are correlated with gradual fluctuations in climatic conditions conflicts with hypotheses that blame persistent declines in the abundance of salmon stocks on increasing levels of hatchery production. Whereas some researchers have reasoned that such declines are in part due to a growing dependence on ill-adapted hatchery fishes, others believe that recent declines are at least partly due to changes in the Aleutian Low. Although some causes of decline are measurable, such as turbine mortality and reproductive failure due to deteriorated spawning and nursery grounds, scientists do not yet understand the relative role that factors such as cycles of ocean survival and production and reduced survivability of hatchery fishes are playing in the persistent decline of Pacific salmon resources.

Where is Management of Pacific Salmonids on the West Coast Going?   To date, the results of management are not promising. Although substantial effort and money have been directed toward Pacific salmon management on the West Coast of the United States, actions have dealt more with masking the impact of land- and water-use activities than with resolving those impacts in a manner that protects the condition of wild resources. In one sense, public opinion favors effective conservation of these resources. One survey has indicated that 75 percent of Washington residents are willing to pay higher electric bills, and over half are willing to pay higher taxes, to support restoration of salmon runs (*Northwest Energy News* 1994). People also favored management that focuses on maintenance of wild fishes rather than on fisheries permanently supported by hatchery production. This group listed restoration of salmon as a higher priority than other commercial uses of the Columbia River (*Northwest Energy News* 1994).

Regardless of such support for conservation of anadromous salmonids, effective mitigation of impacts produced by land- and water-use activities will be very difficult within the economic and social setting in which these resources are found. Logging, agricultural production, shipping, grazing, and particularly energy production collectively are too economically important to be easily restricted and modified to protect resources even as valuable as anadromous salmonids. For example, a 1993 report to Congress from the General Accounting Office (GAO) provided cost estimates for increasing springtime flow to protect Columbia–Snake River salmonids. These estimates, which varied according to the type of flow augmentation desired, ranged to substantially more than $220 million per year (GAO 1993). In December 1994 the NWPPC developed a reservoir drawdown plan to further supplement flow for smolt migrations in the Columbia basin. This plan proposed a release of water from the Lower Granite Dam in 1995, with additional drawdowns from other dams in subsequent years. The "lost" water would increase the cost of electricity to consumers by 4 percent over the following several years.

Local and regional communities depend heavily on intensive use of land and water resources in the West. Hatchery production, predator control programs, and barging may be the most acceptable management alternatives unless traditional values and philosophies concerning resource use are significantly altered.

OHIO'S MANAGEMENT OF LAKE ERIE WALLEYE

Walleye have supported important recreational and commercial fisheries in Lake Erie for much of the 20th century. After record U.S. and Canadian harvests in the late 1950s (up to 15 million pounds yearly), walleye resources collapsed due to severe deterioration of spawning and nursery habitat and to excessive commercial harvest. Some other historically important fisheries resources of the Great Lakes, such as lake trout, collapsed due to the combination of overfishing, habitat deterioration, and the introduction of exotic species and have not recovered. However, within two decades of its collapse, Lake Erie's walleye stock had recovered sufficiently to support very active recreational and commercial fisheries without significant resource declines since the early 1980s. The recovery was facilitated by substantial improvement in habitat conditions, particularly in the lake's highly enriched western basin, and was ultimately accomplished by international and interstate agreements to effectively regulate harvest, including the exclusion of commercial harvests in waters of two of the management areas (territorial waters of Ohio and Michigan). This section reviews the evolution of the management process for this resource by the state of Ohio, whose territorial waters house about half of the total walleye habitat found in Lake Erie. Unless otherwise cited, information presented comes from Ohio's Lake Erie Walleye Strategic Plan (ODNR 1994).

Collapse of the Resource and Its Commercial Fisheries

The collapse of Lake Erie walleye stock in the late 1950s has been attributed to habitat deterioration and overfishing. Land-use practices in the watersheds surrounding Lake Erie had been causing changes in habitat parameters for most of the 20th century, particularly in the lake's western basin (Figure 12.3), which contains a substantial proportion of the walleye habitat of the lake. By the middle of the century, high rates of silt loading from agricultural areas, increased nutrification from both urban and agricultural sources, and deforestation had led to the following conditions (Hartman 1972):

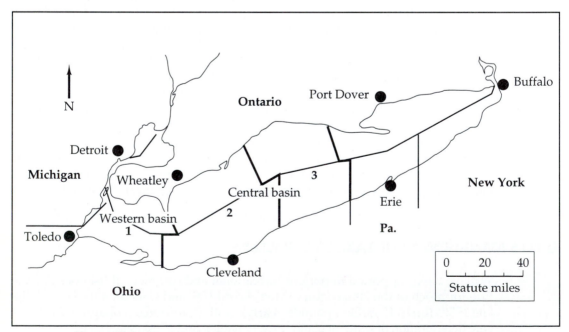

Figure 12.3　The Lake Erie basin, with territorial boundaries and Ohio walleye management areas (1-3) delineated. (Adapted from Lake Erie walleye task group report 1994.)

1. Turbidity in the lake increased significantly.

2. Silt accumulated in the sediments of the western basin, which might have degraded shallow-water spawning areas for fishes such as walleye.

3. The influx of nutrients and the buildup of organic materials in the lake's sediments accelerated. Increases in phytoplankton production caused by intensive cultural eutrophication led to gradual accumulation of substantial concentrations of organic material in bottom sediments. Decomposition of these organic materials led to oxygen depletion, particularly in the western basin. This basin, shallow enough that it stratifies only briefly in most summers, formed thermoclines for long enough periods in the summers of 1953 and 1955 to cause oxygen depletion from microbial decomposition of the enormous organic load that had built up.

4. A change was observed in the benthic community of the western basin. Previously abundant mayfly nymphs, which served as an important food base for fishes in the western basin, did not survive the periods of oxygen depletion of the early to mid-1950s. The abundance of other large-bodied benthic invertebrates, such as caddisflies and amphipods, also declined during this time period (Carr and Hiltunen 1965). The formerly dominant large-bodied invertebrates were replaced by a bottom fauna dominated by small oligochaetes and midge larvae. This change in the benthic invertebrate community apparently affected foraging of some western-basin fishes and was probably responsible for a decline in growth rates exhibited by yellow perch in the western basin that began in the late 1960s (Hayward and Margraf 1987).

Walleye harvests peaked at over 15 million pounds in the 1950s, in part because commercial gill netters converted from cotton to nylon nets and generally improved electronic gear (fish finders, ship-to-ship radios) on board their fishing boats; these changes significantly increased their capture efficiency (Hartman 1972). Starting in the early 1950s, a series of weak year classes of walleye were produced, probably due to a progressively degrading habitat and a declining stock. Both harvest levels and the stock declined steadily through the 1960s (Hartman 1972). This decline stimulated a series of management decisions by the Ohio Department of Natural Resources, culminating in the prohibition of all commercial gill netting in Ohio waters and allocation of all allowable harvest in Ohio waters to recreational users.

Managing Harvest After the Decline

The first major change in management strategy established after the population decline was not directly related to stock recovery. In 1970 concern over mercury contamination in Lake Erie walleye led to a lakewide cessation of commercial fisheries directed toward this species (under agreement of fisheries agencies in Ohio, Michigan, and the Province of Ontario). Reduced fishing mortality and slowly improving habitat conditions led to an increase in walleye abundance during the 1970s. As mercury levels in fishes diminished, Ontario reopened its commercial fishery. However, in 1972 Ohio placed a 5-year moratorium on commercial harvest of walleye in its waters, and in 1973 Michigan designated this species as a sport fish, permanently prohibiting commercial harvest. By 1975 the Great Lakes Fish Commission (see Chapter 6) established an interagency Lake Erie Committee, which set annual harvest levels for walleye. Portions of each allowable annual harvest were allocated to Ohio, Michigan, and Ontario according to the area of walleye habitat of the western and central basin of Erie that is housed within their territorial waters (51.4, 5.3, and 43.3 percent, respectively). Michigan and Ohio chose to allocate their entire quotas to recreational fishing, whereas Ontario allocated most of its quota to commercial harvest.

Ohio regulated harvest with daily catch limits, initially allowing 10 per license holder per day but reducing the limit to 6 per day in 1979 because its recreational harvests exceeded its annual quota. Even the reduced catch limit proved insufficient to regulate harvest, as yearly quotas were exceeded by an average of 60 percent between 1976 and 1989 (Figure 12.4). Ohio's harvest equaled or was lower than its quota only five times in those 14 years. During the same time, Michigan exceeded its quota three times (1987 to 1989) and Ontario twice (1977 to 1978; Lake Erie Task Group 1994). Ohio's overharvest was due in large part to the difficulty in accurately predicting effort in a recreational fishery managed with daily catch limits (see Chapter 7), particularly when the fishery was rapidly expanding due to a recovering stock. Ontario, which had allocated most of its yearly quota to commercial harvest, had much greater opportunity to annually ensure that harvests did not exceed quotas. Effort and harvest levels are more easily monitored and controlled in commercial fisheries with relatively modest numbers of participants. The reduction of daily catch limits from 10 to 6, and later to 5, was considered a good faith effort by Ohio to control its harvests. In spite of Ohio's persistent overharvest, walleye abundance within the lake increased steadily through the 1980s. Also, the number of walleye harvested per hour of angling in Ohio waters of the western basin, where the greatest fishing effort occurred within the lake, remained stable.

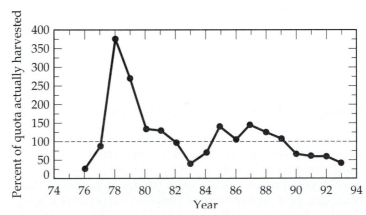

Figure 12.4   The percentage of the yearly allocation of allowable walleye landings in Ohio waters that was actually harvested. The allocation was exceeded in all years exhibiting harvest levels higher than 100 percent. (Data from Lake Erie walleye task group report 1994.)

Stock recovery proceeded through the 1980s due in part to the conservative approach taken by the Lake Erie Committee in establishing allowable quotas. Originally this committee chose to allow a 10 percent harvest of the adult stock annually, which was significantly lower than that which might have been sustained without causing stock declines. This conservative approach was adopted in part to offset possible imprecision in data bases used to calculate stock abundance. The rate of allowable harvest increased to 20 percent in 1980 and 28.5 percent in 1982 as walleye abundance grew, and it became apparent that harvest could be increased without harming the stock. In addition, the abundance of the walleye stock was underestimated through much of the mid-1970s through mid-1980s. Thus conservative harvest levels applied to underestimates of true stock abundance essentially removed fishing as a significant mortality factor for Lake Erie walleye. This facilitated rapid recovery of the stock.

The result of quota management in the Lake Erie walleye fishery certainly differs from some fisheries managed by quotas, particularly when commercial harvest constitutes a major use of a resource. In such conservation arenas, economic and social pressures can drive quotas toward the absolute maximum that a resource might be able to sustain. Under such pressures, uncertainties or inaccuracies of data bases will more likely facilitate decline rather than recovery and maintenance of fish stocks.

Ohio's Buyout Program

In 1983 Ohio chose to prohibit gill net fishing in its waters permanently. This measure did not influence directed harvest of walleye, because commercial harvest of this species had not been allowed in Ohio waters since 1968. However, gill netters harvesting yellow perch were causing significant levels of walleye bycatch mortality. The ban on gill netting was established to eliminate walleye bycatch mortality.

Rather than simply putting gill netters out of business, the gill net closure was accomplished by a buyout program established by Ohio's Department of Natural Resources. This agency bought back all 39 commercial licenses held by gill netters at that time. Each license holder was paid 94 cents per pound for the average harvest of the best 3 years of fishing that the licensee recorded between 1977 and 1982; the rate was based on ex-vessel values for yellow perch. Similar federal buyout programs have been proposed as a means of ending financial ruin faced by harvesters fishing for collapsed coastal fisheries resources, such as groundfish resources of the northeastern United States (see the section on the failure to control harvest of New England groundfishes in this chapter). However, such programs have not often been implemented to date due in part to their great expense.

Recent Changes in Ohio's Fishery

Lake Erie's walleye stock continued to increase in abundance through the 1980s, resulting in a lakewide record harvest of about 20 million pounds in 1988–1989 (ODNR 1994). During the 1980s the growing walleye stock expanded into the central basin (management units 2 and 3), substantially increasing abundance and angling effort there. Angling effort was 18 times greater in Ohio waters of the central basin in 1988 than it was in 1977, whereas in the western basin effort was only 2.7

times higher in 1988 than in 1977. Ohio's fishing effort declined in the western basin from 1988 to 1993, whereas it remained reasonably stable during that time in the central basin (Figure 12.5). Catch rates (number of walleye harvested per hour of

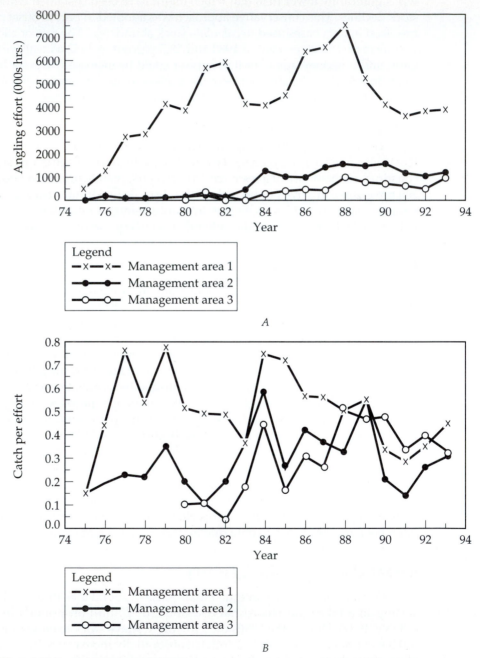

Figure 12.5 (*A*) The total angling effort measured in thousands of angling hours and (*B*) catch for each hour of angling in Ohio waters of management areas 1 (×), 2 (●), and 3 (O). (Data from Lake Erie walleye task group report 1994.)

angling) in both basins were lower in the 1990s than the peak years of the mid-1980s. Unlike earlier years, total harvest in Ohio waters was lower than that state's yearly quota from 1990 through 1993.

The decline in fishing effort and catch rates in the western basin in recent years has not been due to declining stocks or highly restrictive regulations. Walleye abundance has been stable through the 1990s, although at a somewhat lower level than the peak levels of the mid-1980s. Reduced effort and harvest in the western basin may have been related to changes in water clarity in the 1990s. Water clarity improved substantially during those years, possibly due to reductions in nutrient input into the western basin and to the impact of the introduced zebra mussel. The population of this exotic species, first noted in the lake in the late 1980s, expanded explosively, covering much of the suitable substrate throughout the western basin in just several years. Its capacity to filter a large proportion of the plankton in the water column of the western basin apparently led to an increase in water clarity of that area. In clearwater lakes, walleye tend to forage at twilight rather than throughout the day as they do in more turbid waters (Scott and Crossman 1973). Thus the improved water clarity in the western basin may have made walleye less available to daytime anglers than they had been in earlier years. In the central basin, reduced light penetration into its deeper waters allows walleye to feed more actively during the day when the greatest amount of angling is occurring.

Projections through the year 2001 predict that anglers will continue to spend from 5 to 8 million hours angling yearly for walleye in Ohio waters, harvesting about 10 percent of the adult stock yearly. With this fishing pressure, the adult stock should remain at 30 to 70 million fish. Angler catch rates rather than restrictive catch limits will continue to determine harvest levels, as they have since 1990.

In one sense, the successful recovery of Lake Erie walleye seems almost accidental; the recovery occurred although Ohio anglers persistently exceeded harvest allocations. In fact, it is not surprising that overharvest occurred, because it is very difficult to regulate recreational harvest effectively by using daily catch limits. However, the conservative approach taken to establish allowable harvest levels protected the stock not only from inaccurate population projections but from real overharvest by the recreational fishery. The decision to close commercial gill netting permanently by buying back all permits was an unusual management strategy but not surprising in that management setting. The expanding recreational fishery focused on the recovering walleye stock was creating important economic activity for coastal communities of Lake Erie. Angling for walleye is not a casual activity; anglers expend substantial dollars on angling supplies and equipment, marinas and other service providers, and charters in order to fish for walleye in the shallow shoals of the lake. Tourist-based expenditures in many coastal communities of Lake Erie are significantly boosted by walleye recreational fishing. In recent decades, agencies in inland areas of the country have managed increasing numbers of fisheries resources for recreational use when growing conflicts between recreational and commercial uses have occurred. Buying out 39 harvesters to ensure that the entire allowable harvest supports recreational fishing was a politically acceptable decision, because recreational users had established such strong influence on the management of this resource.

NEW ENGLAND GROUNDFISHERIES: FAILURE
TO CONTROL OVERFISHING

For several months in the autumn of 1994, northeastern newspaper headlines detailed the extremely poor condition of New England's groundfish resources and fisheries: "How did it get this bad?" (*Cape Cod Times,* October 30, 1994); "Empty nets, sinking hopes: Too many boats, too few fish" (*Portland Press Herald,* September 18, 1994); and "New England fishing areas shut down: U.S. ban includes Georges Bank" (*Boston Globe,* December 8, 1994). In November 1994 the New England Fishery Management Council (NEFMC) voted to recommend emergency measures to halt further collapse of groundfish resources (haddock, cod, species of hake, and flatfishes) off the New England coast and to develop long-term management strategies that included severe restrictions on harvest and large area closures of traditional fishing grounds on Georges Bank and elsewhere. This vote was taken less than a year after Amendment 5 of the New England Multispecies Groundfish Fishery Management Plan was implemented. Amendment 5 had been viewed widely by the fishing industry as harshly restrictive and destructive for commercial harvesters. However, less than a year after its implementation the NEFMC deemed it inadequate to foster conservation and recovery of severely depleted stocks of haddock, yellowtail flounder, cod, and other commercially important fish species.

On December 7, 1994, the Commerce Department accepted the recommended emergency measures of the NEFMC, which included the following:

1. Shutting down fishing in three areas (approximately 6,600 square miles of fishing grounds) to protect spawning and nursery habitats (Figure 12.6).

2. Increasing the minimum mesh size in codends of trawls to 6 inches for trawlers fishing on Stellwagen Bank and Jeffreys Ledge.

3. Prohibiting the use of mesh smaller than 6 inches in other areas unless the bycatch of regulated groundfish species is less than 5 percent of the total harvest. Possession of these regulated species will be prohibited by boats using less than 6-inch mesh.

The fishing industry responded with various levels of intensity, from those showing concern for their livelihood to those displaying outrage that the government would prohibit their capacity to fish in this way. Politicians representing districts with commercial fishing communities split in their response, some angrily declaring that the conservation measures were unfair or unnecessary, others lamenting that the regulations were necessary but unfortunate. Clearly, commercial fishing off New England's coastline was going to change dramatically, and the industry's activity would be diminished for an extended period of time into the future.

What led to the severe collapse of New England's groundfish resources? Groundfish resources of New England had been harvested at levels that could not be sustained for several decades prior to the 1994 management decisions. Declining fisheries resources under the management aegis of ICNAF (International Convention for Northwest Atlantic Fisheries) provided strong impetus for passage of

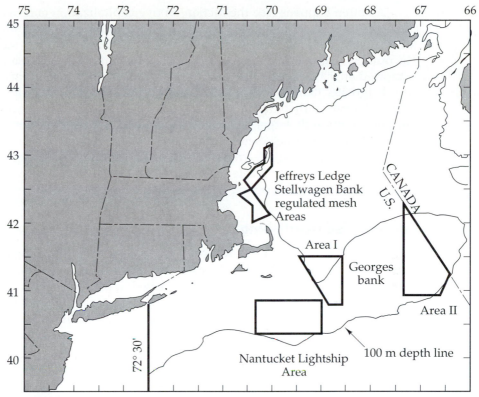

Figure 12.6 Closed areas and regulated mesh areas established in 1994 emergency measures of the New England Fishery Management Council. Fishery closures occurred in Areas I and II of Georges Bank and in the Nantucket Lightship Area. A year-round 6-inch square mesh requirement for the co-dend of trawls was established for regulated mesh areas. (Reprinted from New England Fishery Management Council *News*, November 2, 1994.)

the Magnuson Act in 1976 (see Chapter 6). Rather than stimulating effective conservation of New England's groundfish resources, this act indirectly supported unprecedented growth of New England's domestic groundfish fishing fleet. Government-sponsored studies conducted at the time of passage of the Magnuson Act indicated that resources no longer harvested by foreign distant water fleets would support substantially larger domestic fleets, creating a nationwide increase of up to 43,000 new jobs while improving the U.S. trade balance with increased export of fish products (Royce 1989). Initially, the federal government facilitated this growth through the Fisheries Obligation Guarantee program (which provided government-guaranteed loans to build or upgrade fishing vessels) and the Capital Construction Fund (which allowed boat owners to defer tax payments on boats; Garber and Canfield 1994a). These programs ended in 1979. Passage of the American Fisheries Promotion Act of 1980 (see Chapter 6) further stimulated growth of fishing fleets, providing grants for industrial development and funds that allowed boat and facilities owners to avoid defaulting on private loans.

As often happens (see the section on overcapitalization and overfishing in Chapter 4), the groundfish fleet rapidly expanded well beyond the number of vessels that could be economically sustained by the fisheries resources supporting it. The number of groundfish vessels fishing New England waters increased from 570 to more than 900 between 1976 and 1981 (Campbell 1989). The number of larger vessels (greater than 80 feet in length and 105 tons) in this fleet tripled between 1976 and 1986 (Garber and Canfield 1994a). These larger, more powerful vessels can pull significantly larger nets, can fish further distances from home ports, and can fish during more inclement weather conditions than smaller boats. In addition, vessel owners upgraded electronic fish-detection equipment that significantly increased their capacity to locate fishes. The groundfish fleet greatly increased its fishing effort during this period (Figure 12.7).

Uncontrolled Overfishing

Shortly after its inception, the NEFMC instituted a system of annual quotas for highly sought species. Several problems arose from this first management approach of the NEFMC. Some harvesters falsely reported lower catches than they brought to dock, which prevented accurate assessment of how rapidly quotas for particular species were reached. When reported catches reached quota levels, instead of closing the fishery the NEFMC increased the quota in order to allow harvesters to continue fishing. Such action was the result of council sympathy for harvesters who complained that closures would bankrupt their operations (Doeringer et al. 1986). By the early 1980s the council replaced its quota system with a regulatory policy that minimized intervention with harvester activities. Minimum mesh-size regulations to allow juvenile fishes to escape trawls, and seasonal closures to protect spawning grounds of haddock and yellowtail flounder served as the major emphasis of regulations into the 1990s. In spite of increased warnings from scientists that groundfishes were being overharvested (Table 12.1), this management system was maintained through this time period largely as a result of pressure from harvesters. When the council modified its regulation of the groundfish fishery, it made incremental, modest changes in existing regulations rather than instituting some type of

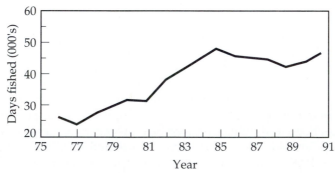

Figure 12.7   Total yearly fishing effort in the northeastern groundfish fishery, measured as thousands of days fishing, since passage of the Magnuson Act. (Redrawn from NOAA 1992.)

TABLE 12.1   Comments on the condition of groundfish resources in the United
States waters of the Northwest Atlantic from fisheries scientists of the
Northeast Fishery Research Center (NMFS:NOAA).[a]

1983 (NOAA 1984)

Haddock: "stock biomass will continue to decline given anticipated levels of fishing mortality"

Yellowtail flounder: "stock biomass should continue to decline in both areas if fishing mortality remains high"

1985 (NOAA 1985)

Haddock: "stock biomass dropped to very low levels in 1984 and is expected to decline even further"

Yellowtail: "commercial and research vessel survey indices of abundance have declined sharply"

1987 (NOAA 1987)

Haddock: "the stock will probably decline further in 1987-88"

Yellowtail: "fishing mortality remains substantially above F_{max}" (the maximum fishing mortality that can be sustained for extended time periods by a stock without significant declines in abundance)

Cod: "Unless F [instantaneous rate of fishing mortality] is reduced further, both yield and stock biomass will continue to decline"

1989 (NOAA 1989)

Cod: "United States otter trawl effort was the second highest ever". . ."at the present level of fishing effort, any gains in stock size will be short-lived"

Haddock: "The 1988 autumn survey index of 5.57 [an index of relative abundance] is only 20 percent of the 1979 value"

Yellowtail: "the population is severely overexploited"

1991 (NOAA 1991)

Cod: "To halt the declining trend in SSB [spawning stock biomass; Chapter 3], fishing mortality needs to be markedly reduced"

Haddock: "The low abundance of incoming year classes suggests that the survey index [abundance] will remain at record or near record low levels"

Yellowtail: "rebuilding of the stock will require a major reduction in fishing mortality and several good years of improved recruitment"

1992 (NOAA 1992)

Haddock: "abundance and biomass are at all-time lows"

Yellowtail: "the stock is still at a very low level and is comprised of few age groups"

[a]All quotes are taken from annual volumes of "Status of fishery resources off the Northeastern United States," an annual series of NOAA/U. S. Department of Commerce Technical Memoranda prepared by the Northeast Fisheries Science Center of NMFS.

quota system or controlling the size and fishing effort of the growing groundfish fleet. At the same time, the federal government was subsidizing growth of the groundfish fleet's harvesting capacity.

In 1993 the NEFMC voted the fifth amendment to its multispecies fishery management plan. Amendment 5 established guidelines for controlling effort of the groundfish fishing fleet. Among its restrictions, Amendment 5 called for an annual 10 percent reduction in fishing effort from that allowed the previous year, thus reducing total effort of the fleet to about 50 percent by 1999. This would be accomplished by controlling the number of days each vessel in the fishery would be allowed to fish and reducing that number each year accordingly. The fishing industry responded angrily to Amendment 5, fighting to change it or delay its implementation. One group of Maine fishermen sued to have the regulations stopped, while others in Gloucester, Massachusetts, violently demonstrated, dumping the processed fish aboard two refrigerator trucks onto a pier and overturning two pickups (Garber and Canfield 1994b). Although Amendment 5 was considered too harsh by many in the industry, others outside of the industry considered it to be too little action taken too late.

By the fall of 1994 fisheries scientists from the Northeast Fisheries Science Center reported to the NEFMC that groundfish stocks were in worse condition than the previous year when Amendment 5 was implemented. The council, deciding that the effort reductions were too gradual and modest to prevent further resource collapse, voted to establish emergency measures to immediately reduce harvest. Further, the council intended to develop permanent regulations to replace the emergency measures, which would "be far more stringent than those currently in place and no doubt [would] have serious economic impacts on the fishing industry" (P. Coates, chairman of the NEFMC's Groundfish Committee, as quoted in a NEFMC News Release, 2 November 1994). The Secretary of the Department of Commerce approved the emergency measures in November, stimulating an outcry from the industry that was widely printed in the region's newspapers. Scientists, whose warnings of declining stocks had been challenged for years by the industry, were now blamed for not providing timely, accurate information. The Management Council was blamed for slow, incremental changes in fisheries that required stronger control. Some elements of the industry blamed others for overfishing (see Chapter 4).

Why Did Management Fail?

One might assume that protection of New England's groundfish resources should be easily achieved. Although the deterioration of coastal wetlands has had an impact on continental shelf fisheries resources, New England's groundfish stocks have not been as seriously harmed by numerous land-use and water-use activities as have many freshwater and anadromous resources (such as Pacific salmonids; discussed elsewhere in this chapter). Thus management of these fisheries has not been complicated by the deterioration and loss of habitat. Sustainable fishery yields and healthy resources should be attainable if harvest is effectively regulated. Managing harvest requires predicting abundance, recruitment, and production rates of the fish stocks being harvested and precisely measuring fishing pressure, harvest levels, and the effect of removing specific numbers and biomasses of fishes from populations. It is difficult to accurately estimate such population characteristics of resources and harvest characteristics of fishing fleets. However, the collapse of New England's groundfishes and the fisheries they support has not occurred due to an inability to understand the impact of harvest on those resources.

Conservation of New England's groundfish resources has failed to date because the New England Fishery Management Council has responded more effectively to the immediate economic health of the fishing industry and coastal fishing communities than to the long-term condition of the groundfish resources. From its early years, the NEFMC conducted a policy of minimizing restrictions on harvesters. The protection it did try to provide to resources was circumvented by other government policy. For example, any possible protection provided by mesh-size regulations was overcome by the federal government-supported upgrading of the fishing fleet. Loan and mortgage protection programs provided the harvester with a means of improving his or her income by becoming more efficient in the face of declining resource availability. Declining catches per day of fishing merely reinforced the harvester's resolve to improve efficiency and effort (as discussed in Chapter 4). Increases in efficiency without regulated limits on effort or participation simply drove stocks into further decline. Closures protecting spawning areas of species such as haddock and yellowtail flounder became ineffective as reproductive effort and success declined due to growing depletion of adult stocks.

Unlike the conservative approach used to manage Lake Erie walleye, management decisions in the New England groundfish fishery consistently favored the harvester rather than the resource. This is in part due to the following:

1. It is difficult to restrict commercial harvesters when their livelihood is at stake. The Magnuson Act specifically requires management for optimum yield, that is, the fishery yield (or harvest) that provides the greatest benefit to the United States. Such management must consider economic and sociological issues as well as the biological condition of a resource. Commercial fishing activities often lead to overfishing, resource decline, and reduced profits, the "irrational fishing" of Royce (1987; see Chapter 4). Management too often favored the short-term economic health of the New England fishing industry instead of the long-term benefit gained from a sustainable interaction between users and a resource.

2. The majority of voting members of the New England Fishery Management Council have been commercial harvesters or have had direct ties to the industry. The Magnuson Act has allowed a user group to be responsible for controlling its own activities, which are tied to its economic well-being.

Under these circumstances, it is not surprising that New England's groundfish fisheries have not been conserved. Although restrictive management actions are now being taken, an industry is in jeopardy and there is no promise as to when or how well the resources will recover under this management regime. Many believe that the recovery of these and many other coastal fisheries resources will depend on a restructuring of the management system established by the Magnuson Act rather than a refocus of the present management system toward more effective conservation.

TROUT STREAM MANAGEMENT: MEETING HIGH ANGLING DEMAND

Trout species support widely and avidly pursued recreational fisheries. The majority of anglers who fish for trout prefer stream over lake fisheries (Griffith 1993). Generally, trout resources attract very high rates of angling pressure. For example, a reach of the intensively fished Yellowstone River in Yellowstone National Park sustained more than 1,300 hours of angling per hectare of water (> 520 hours per surface acre) during the brief, 3.5-month open fishing season in 1981 (Schill et al. 1986). Trout stream fisheries with longer open seasons may attract up to 2,500 hours of angling per hectare (about 1,000 hours per acre) per year. Much of this fishing pressure is due to a persistent increase in trout fishing activity in recent years; angling effort has doubled in some of the most popular trout streams during the last decade (Griffith 1993).

Interestingly, production and standing stocks of trout in coldwater streams are generally low compared with other systems. For example, although highly variable from system to system, trout biomass in 313 streams in the western United States that had suitable trout habitat averaged about 54 kilograms per hectare (about 48 pounds per acre; Platts and McHenry 1988), whereas ponds with coldwater trout habitat average about 250 kilograms of trout per hectare (about 220 pounds per acre; Flickinger and Bulow 1993); productive warmwater systems may support similar or greater biomasses of recreationally important species when compared with coldwater ponds. Thus, when managing trout stream fisheries, agencies have attempted to meet high angler demand with a limited supply of natural resources. To further complicate the apparent disparity between supply and demand, some species of trout are easily captured (Figure 12.8). Easily captured fishes in populations exhibiting limited production potential are very susceptible to overfishing.

Conservation and management programs have taken a dichotomous approach toward matching supply with demand in trout stream fisheries. On one hand, trout resources have been protected with widespread application of highly restrictive

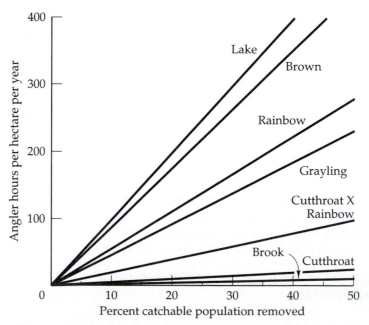

Figure 12.8 The vulnerability of various species of trout to angling, measured as the number of angling hours that are required to catch specific proportions of populations of each species. (From J. S. Griffith. 1993. Coldwater streams. Pages 405–425 in C. C. Kohler and W. A. Hubert, editors. Inland fisheries management in North America. Reprinted by permission of The American Fisheries Society.)

harvest regulations, including frequent use of mandatory catch-and-release regulations. In this instance, agencies have mandated that demand become nonconsumptive, that is, that the angling experience fulfills the angling demand by itself without including harvest and consumption of what is caught. Anglers must be satisfied with nonconsumptive experiences such as fishing in a pristine setting for wild fishes that are presumably difficult to catch.

Agencies also have chosen to create expansive put-and-take trout fisheries, essentially providing almost unlimited supply to meet the fishing demand. For example, fisheries agencies reared and stocked into public waters more than 53 million catchable trout in 1973 and more than 78 million catchable trout in 1983 (Behnke 1990). In order to meet predicted increases in demand for opportunities to fish for trout, state hatchery production facilities have grown to the point where their operation and management have become the largest single part of the fisheries conservation and management budget (Johnson et al. 1995). In this instance, agencies are managing a recreational activity, not conserving a wild resource. Many trout are stocked into streams and ponds incapable of supporting wild trout populations, whereas others are released into systems holding wild populations to enhance the opportunity of anglers to harvest fishes. Put-and-take trout are not expected to contribute to reproduction or long-term production of wild trout populations (see Chapter 8). Some anglers do not enjoy aspects of put-and-take fisheries, including potentially crowded fishing conditions and fishes that are very easily captured. However, many anglers are pleased with the high catch and harvest rates characteristic of effectively managed put-and-take fisheries.

Other recreational resources are managed with restrictive regulations, and stocking is the major element of successful put-grow-and-take fisheries associated with a variety of resources including striped bass, walleye, and Pacific salmon in the Great Lakes. However, perhaps no other fisheries are managed by extensive use of such extremely differing alternatives: No harvest of wild fishes versus liberal harvest of fishes domestically reared solely for the intent of being harvested. Segments of the recreational fishing community enthusiastically support and expect the opportunity to participate in either or both of these types of fisheries.

Growing Movement Toward Conservation of Wild Resources

Hatchery production and stocking has been a central element of fisheries management since federal and state agencies were first developed. Throughout the latter half of the 19th and much of the first half of the 20th century, hatcheries became the major response to growing fisheries problems, in part because scientific management, including regulating harvest to levels compatible with the production capacity of wild populations, had not yet developed (see Chapter 8). As put-and-take trout management programs developed, justification for expansion was based on the success of these programs in matching the fishing "needs" of a growing segment of the angling public. State agencies expanded hatchery facilities and the emphasis on hatchery production for put-and-take fisheries partly because a growing proportion of the angling public appreciated the ease of capture and the high harvest rates that are associated with such programs. The success of put-and-take programs stimulated the view that expansion was appropriate and necessary to meet a projected growing angling demand.

The first organized public objection to the emphasis on put-and-take management of trout fisheries occurred in 1959 with the founding of Trout Unlimited, which was created for the purpose of influencing the state of Michigan to deemphasize put-and-take fisheries and focus more on management of wild trout fisheries (Johnson et al. 1995). Even before studies were published indicating that stocking put-and-take trout may be detrimental to wild populations (Vincent 1984, 1987; see Chapter 8), state agencies were beginning to reexamine the growing emphasis on put-and-take stocking to meet angling demand. Although some states developed a greater focus on effective management of wild resources, the production of catchable trout in public hatcheries increased by over 50 percent in the 25 years after Trout Unlimited was founded (Johnson et al. 1995). Some agencies were concerned about the reaction of the angling public to a shift from hatchery-based, liberal harvest fisheries to restrictive harvest of wild fishes. Would anglers consider catching trout a quality experience if they were unable to keep them? In addition, would the catch rates in catch-and-release fisheries be high enough to substitute for the high catch rates typical of put-and-take fisheries?

Interestingly, angler acceptance of catch-and-release fisheries can be quite high, due not only to a desire to fish in a pristine, natural setting but also to the potentially high catch rates that can be attained in catch-and-release fisheries. Anderson and Nehring (1984) studied a stream system in Colorado that had one reach converted from a put-and-take/liberal daily catch limit fishery to a mandatory catch-and-release fishery supported solely by a wild trout population. After the change, the catch-and-release section supported higher catch rates than had been attained under previous stocking regimes, with the capture rate of trophy-size fishes (greater than 15 inches, or 38 centimeters) about 28 times greater than that achieved under stocking regimes. Most anglers fishing in that stream system preferred the catch-and-release fishery, including 99 percent of those fishing in the catch-and-release zone and 72 percent of those who fished primarily in other reaches still supporting put-and-take stocking.

Behnke (1990) estimated that about 20 percent of anglers wishing to fish for trout would prefer waters supporting wild trout populations that were protected with highly restrictive catch limits or catch-and-release regulations. Such an estimate reflects that a proportion of trout anglers find nonconsumptive aspects of their angling experience more important than keeping fishes. It also demonstrates that a large proportion of trout anglers may desire the harvest opportunity associated with hatchery-supported fisheries. Although supporting greater emphasis on sustainable wild trout fisheries, Behnke (1990) felt that "a major impediment . . . is the reliance on put-and-take catchable trout stocking to supply the demand for angling." Few studies have been conducted to investigate whether a reduction in production of hatchery fishes, concurrent with greater reliance on restrictive harvest of wild fishes, can sustain the quality of fisheries and the enthusiasm and interest of anglers. Based on cost-benefit analyses, Johnson et al. (1995) demonstrated that the rate of stocking in several heavily fished streams in Colorado was substantially greater than that necessary to fulfill the demand of anglers fishing in those systems. They projected that similar discrepancies may exist in other states that rely heavily on put-and-take hatchery programs.

Although historically trout management focused on ever-increasing production of hatchery-reared fishes to sustain interest in trout fisheries, managers are becoming more and more aware of the limited advantages that hatcheries provide in some management settings (as discussed in Chapter 8). The future of trout fishing in coldwater stream systems may shift toward greater emphasis on restricting harvest and protecting the quality of habitats for wild trout resources. In the future, it is possible that hatcheries will supplement rather than dominate the trout management arena.

THE FUTURE OF FISHERIES CONSERVATION AND MANAGEMENT

In this book I have provided numerous examples of the fisheries conservation and management process. I have included a historical perspective to outline what has been learned from more than a century of conservation and management and to describe the current condition of fisheries resources and the benefits gained from them. I have specifically focused on management failures as well as successes because much of the fisheries conservation and management process has been conducted on a trial-and-error basis. Thus the frequency of success achieved in programs has been based in part on learning from experience.

Historically, much of the emphasis of fisheries conservation and management has been species based; that is, species that are deemed important have been afforded substantial attention and management effort. The traditional approach of conservation and management has focused on protecting the benefit that society receives from some resource by manipulating either human behavior and activities or ecosystems in a manner that sustains or increases the benefit. Some species-related management approaches have attempted to maintain or create a relationship between humans and a resource without the resource being sustained in its natural setting, for example, the expansive hatchery effort that serves as a central focus of management of Pacific salmon stocks or put-and-take stocking of trout. On occasion, species-focused management has proven detrimental to other resources and biotic communities, as in the declines suffered by cutthroat and other trout species after introduction of rainbow trout.

If we view the condition of our fisheries resources as a measure of how successful we have been in conservation and management, our record is at best spotty. Although angling is one of the most actively pursued of outdoor activities, the condition of many of the resources that support this recreation is not good. Habitat deterioration and loss have led to the decline in the quality of many resources, while contaminants make a growing number unsafe for human consumption. The condition of many commercially exploited resources is poor, often due to our inability to control our use of them. Although we rear and stock huge numbers of fishes into our waterways, the conservation and management objectives associated with some of these programs are often unfulfilled. Fisheries agencies are often dependent on other agencies to protect aquatic habitats. Effective stewardship of fisheries resources is affected by the priority given to resource conservation by executive and legislative branches of government and by their interpretation of how the government should fulfill its legal responsibility to protect the quality of our natural resources.

To achieve greater success in the future, a growing number of fisheries scientists propose shifting focus from species-based management to conservation of the health and viability of whole biotic communities and habitats that hold them.

Ecosystem Management

Ecosystem management includes protecting or restoring the capacity of natural habitats to support the living resources that inhabit them (Sparks 1995). Ecosystem management emphasizes protecting those aspects of an ecosystem that are most important in influencing its capacity to support a natural, productive biotic community. For example, rates of flow and of sediment passage and deposition strongly influence the structure of habitats, biological production, and biotic interactions in rivers flowing through broad floodplains. Ecosystem management of riverine fisheries resources would focus on maintaining or restoring those factors that produce appropriate flow and sedimentation patterns (Sparks 1995).

For some time, agencies have understood the need to protect habitat quality. However, much habitat management has focused on production of specific taxa, not of entire biotic communities. In some instances, we have understood the habitat-related problems that threaten the sustainability of many wild resources but have opted to treat the symptom of declining fish populations with hatchery production, the introduction of new species, or other management activities that attempt to replace resources being lost rather than protect them.

Species-based management will continue to be an integral part of the fisheries conservation and management arena in the future, in some instances because of public desires (such as the maintenance of put-and-take or put-grow-and-take fisheries) and in others because it is the most appropriate way to regulate a fishery. If the major threat to the sustainability of a specific fishery is harvest related but not habitat related, as is true of many continental shelf resources as well as some freshwater resources, species-based management may meet the objective of sustaining the fishery. However, in the many instances where habitat deterioration and loss or changes in species assemblages found in aquatic systems threaten the sustainability of biotic communities, ecosystem conservation and management should play a greater role in the fisheries conservation and management process. As noted on numerous occasions in this text, such management will be complex and difficult due to the considerable use and abuse endured by the nation's waterways.

Regardless of the difficulties associated with fisheries conservation and management, we will still wish to use aquatic resources in the future, and the pressures placed on those resources and their habitats will continue to grow. Whether one becomes the technician or scientist involved in determining the condition of our resources or the impacts of fishing and habitat deterioration on them, the administrator who develops conservation and management strategy, the manager who ensures that programs are carried out, or the communicator who interacts with and educates interest groups and the general public concerning issues associated with fisheries resource conservation and management, much still needs to be learned and accomplished. If anything, those choosing a career in fisheries conservation and management in the future will never be bored because of lack of challenges.

Appendix:
Scientific Names of Species and Subspecies of Fishes and Invertebrates Listed in the Text

Species are listed alphabetically by the first letter in the first word of their common names.

FISHES

Acadian redfish	*Sebastes fasciatus*
Alabama cavefish	*Speoplatyrhinus poulsoni*
Alabama sturgeon	*Scaphirhynchus suttkusi*
albacore	*Thunnus alalunga*
alewife	*Alosa pseudoharengus*
Alvord cutthroat trout	*Oncorhynchus clarki ssp.*
American eel	*Anguilla rostrata*
American plaice	*Hippoglossoides platessoides*
American shad	*Alosa sapidissima*
Amistad gambusia	*Gambusia amistadensis*
Apache trout	*Oncorhynchus apache*
Arctic char	*Salvelinus alpinus*
Arctic grayling	*Thymallus arcticus*
Ash meadows poolfish	*Empetrichthys merriami*
Atka mackerel	*Pleurogrammus monopterygius*
Atlantic cod	*Gadus morhua*
Atlantic croaker	*Micropogonias undulatus*

Atlantic herring	*Clupea harengus*
Atlantic mackerel	*Scomber scombrus*
Atlantic menhaden	*Brevoortia tyrannus*
Atlantic salmon	*Salmo salar*
Atlantic silverside	*Menidia menidia*
Atlantic sturgeon	*Acipenser oxyrhynchus*
Atlantic wolffish	*Anarchichas lupus*
barracuda	*Sphyraena barracuda*
beautiful shiner	*Cyprinella formosa*
bigeye tuna	*Thunnus obesus*
black crappie	*Pomoxis nigromaculatus*
black rockfish	*Sebastes melanops*
black sea bass	*Centropristis striata*
blackfin cisco	*Coregonus nigripinnis*
blue catfish	*Ictalurus furcatus*
blue marlin	*Makaira nigricans*
blue pike	*Stizostedion vitreum glaucum*
blue rockfish	*Sebastes mystinus*
blueback herring	*Alosa aestivalis*
bluefin tuna	*Thunnus thynnus*
bluefish	*Pomatomus saltatrix*
bluegill	*Lepomis macrochirus*
bocaccio	*Sebastes paucispinis*
bonefish	*Albula vulpes*
bonnethead shark	*Sphyrna tiburo*
bonytail chub or bonytail	*Gila elegans*
brook trout	*Salvelinus fontinalis*
brown trout	*Salmo trutta*
bull shark	*Carcharhinus leucas*
bull trout	*Salvelinus confluentus*
burbot	*Lota lota*
butterfish	*Peprilus triacanthus*
California corbina	*Menticirrhus undulatus*
California halibut	*Paralichthys californicus*
capelin	*Mallotus villosus*
channel catfish	*Ictalurus punctatus*
chinook salmon	*Oncorhynchus tshawytscha*
chub mackerel	*Scomber japonicus*
chum salmon	*Oncorhynchus keta*
Clear Lake splittail	*Pogonichthys ciscoides*
cobia	*Rachycentron canadum*
coho salmon	*Oncorhynchus kisutch*
Colorado River squawfish	*Ptychocheilus lucius*
common carp	*Cyprinus carpio*

creek chub *Semotilus atromaculatus*
cui-ui *Chasmistes cujus*
cutthroat trout *Oncorhynchus clarki*

deepwater cisco *Coregonus johannae*
Devils Hole pupfish *Cyprinodon diabolis*
Dolly Varden *Salvelinus malma*
dolphin fish *Coryphaena hippurus*
Dover sole *Microstomus pacificus*

ehu (squirrelfish snapper) *Etelis carbunculus*
English sole *Pleuronectes vetulus*
eulochon *Thaleichthys pacificus*

fallfish *Semotilus corporalis*
finetooth shark *Carcharhinus isodon*
flathead catfish *Pylodictis olivaris*
fourhorn sculpin *Myoxocephalus quadricornis*
freshwater drum *Aplodinotus grunniens*

gag *Mycteroperca microlepis*
garibaldi *Hypsypops rubicundus*
Gila topminnow *Poeciliopsis occidentalis*
Gila trout *Oncorhynchus gilae*
gizzard shad *Dorosoma cepedianum*
golden shiner *Notemigonus crysoleucas*
golden trout *Oncorhynchus aguabonita*
goldfish *Carassius auratus*
goosefish *Lophius americanus*
grass carp *Ctenopharyngodon idella*
Grass Valley speckled dace *Rhinichthys osculus reliquus*
grasby *Epinephelus cruentatus*
great hammerhead shark *Sphyrna mokarran*
green sunfish *Lepomis cyanellus*
Guadelupe bass *Micropterus treculi*
Gulf sturgeon *Acipenser oxyrhynchus desotoi*

haddock *Melanogrammus aeglifinus*
harelip sucker *Lagochila lacera*
humpback chub *Gila cypha*

ide *Leusciscus idus*
(Icelandic) cod *Gadus morhua*
Independence Valley tui chub *Gila bicolor isolata*

jack mackerel *Trachurus symmetricus*

jewfish	*Epinephelus itajara*
June sucker	*Chasmistes liorus*
kelp bass	*Paralabrax clathratus*
king mackerel	*Scomberomorus cavalla*
kiyi (Lake Ontario)	*Coregonus kiyi*
kokanee	*Oncorhynchus nerka*
lake cisco	*Coregonus artedi*
lake herring (cisco)	*see lake cisco above*
Lake Ontario kiyi	*see kiyi above*
lake sturgeon	*Acipenser fulvescens*
lake trout	*Salvelinus namaycush*
lake whitefish	*Coregonus clupeaformis*
largemouth bass	*Micropterus salmoides*
Las Vegas dace	*Rhinichthys deaconi*
lemon shark	*Negaprion brevirostris*
Leon Springs pupfish	*Cyprinodon bovinus*
longbill spearfish	*Tetrapturus pfluegeri*
longfin mako shark	*Isurus paucus*
longfin smelt	*Spirinchus thaleichthys*
longhorn sculpin	*Myoxocephalus octodecemspinosus*
longjaw cisco	*Coregonus zenithicus* (listed as *c. alpenae* in Miller et al. 1989)
Maravillas red shiner	*Cyprinella lutrensis blairi*
Miller Lake lamprey	*Lampetra minima*
Moapa dace	*Moapa coriacea*
Monkey Spring pupfish	*Cyprinodon sp.*
mosquitofish (western)	*Cambusia affinis*
mountain whitefish	*Prosopium williamsoni*
muskellunge	*Esox masquinongy*
North Sea plaice	*Pleuronectes platessa*
northern anchovy	*Engraulis mordax*
northern pike	*Esoc lucius*
northern squawfish	*Ptychocheilus oregonensis*
ocean pout	*Macrozoarces americanus*
ocean sunfish	*Mola mola*
onaga (longtail snapper)	*Etielis coruscans*
opakapaka (pink snapper)	*Pristipomoides auricilla*
Owens pupfish	*Cyprinodon radiosus*
Pacific bonito	*Sarda chiliensis*
Pacific cod	*Gadus macrocephalus*

Pacific hake	*Merluccius productus*
Pacific halibut	*Hippoglossus stenolepsis*
Pacific herring	*Clupea pallasi*
Pacific ocean perch	*Sebastes alutus*
Pacific sardine	*Sardinops sagax*
Pacific whiting (hake)	*see Pacific hake above*
Pahranagat spinedace	*Lepidomeda altivelis*
Pahrump Ranch poolfish	*Empetrichthys latos pahrump*
pallid sturgeon	*Scaphirhynchus albus*
Peruvian anchoveta	*Engraulis ringens*
phantom shiner	*Notropis orca*
pink salmon	*Oncorhynchus gorbuscha*
pollock	*Pollachius virens*
porbeagle shark	*Lamna nasus*
pumpkinseed	*Lepomis gibbosus*
rainbow smelt	*Osmerus mordax*
rainbow trout	*Oncorhynchus mykiss*
Raycraft Ranch poolfish	*Empetrichthys latos concavus*
razorback sucker	*Xyrauchen texanus*
red drum	*Sciaenops ocellatus*
red hake	*Urophycis chuss*
redear sunfish	*Lepomis microlophus*
redfin pickerel	*Esox americanus americanus*
redside shiner	*Richardsonius balteatus*
Rio Grande bluntnose shiner	*Notropis simus simus*
Rio Grande sucker	*Catostomus plebeius*
rock bass	*Ambloplites rupestris*
rock sole	*Pleuronectes bilineatus*
round whitefish	*Prosopium cylindraceum*
sablefish	*Anoplopoma fimbria*
sailfish	*Istiophorus platypterus*
San Marcos gambusia	*Gambusia georgei*
sand bass (barred)	*Paralabrax nebulifer*
sand bass (spotted)	*Paralabrax maculatofasciatus*
sand lance	*Ammodytes americanus*
sand seatrout	*Cynoscion arenarius*
sauger	*Stizostedion canadense*
scup	*Stenotomus chrysops*
sea lamprey	*Petromyzon marinus*
sea raven	*Hemitripterus americanus*
sharpnose shark (Atlantic)	*Rhizoprionodon terraenovae*
sharpnose shark (Pacific)	*Rhizoprionodon longurio*
sheepshead	*Archosargus probatocephalus*
sheepshead minnow	*Cyprinodon variegatus*

shortfin mako shark	*Isurus oxyrinchus*
shorthorn sculpin	*Myoxocephalus scorpius*
shortnose sturgeon	*Acipenser brevirostrum*
Shoshone pupfish	*Cyprinodon nevadensis shoshone*
silver hake	*Merluccius bilinearis*
silver seatrout	*Cynoscion nothus*
silver trout	*Salvelinus fontinalis agassizi*
	(listed as *S. agassizi* in Miller et al. 1989)
skipjack tuna	*Katsuwonus pelamis*
smallmouth bass	*Micropterus dolomieui*
snail darter	*Percina tanasi*
Snake River sucker	*Chasmistes muriei*
snook	*Centropomus undecimalis*
sockeye salmon	*Oncorhynchus nerka*
Sonora chub	*Gila ditaenia*
Spanish mackerel	*Scomberomorus maculatus*
speckled (spotted) seatrout	*Cynoscion nebulosus*
spiny dogfish	*Squalus acanthias*
spot	*Leiostomus xanthurus*
spotted bass	*Micropterus punctulatus*
spotted seatrout	*Cynoscion nebulosus*
starry flounder	*Platichthys stellatus*
steelhead trout	*Oncorhynchus mykiss*
striped bass	*Morone saxatilis*
summer flounder	*Paralichthys denatus*
surf smelt	*Hypomesus pretiosus*
swordfish	*Xiphias gladius*
tarpon	*Megalops atlanticus*
tautog	*Tautoga onitis*
Tecopa pupfish	*Cyprinodon nevadensis calidae*
tench	*Tinca tinca*
thicktail chub	*Gila crassicauda*
threadfin shad	*Dorosoma petenense*
thresher shark	*Alopias vulpinus*
tiger shark	*Galeocerdo cuvier*
tui chub	*Gila bicolor*
ulua (giant trevally)	*Caranx ignobilis*
Utah chub	*Gila atraria*
Utah Lake sculpin	*Cottus echinatus*
vermillion snapper	*Rhomboplites aurorubens*
walleye	*Stizostedion vitreum*
walleye pollock	*Theragra chalcogramma*

watercress darter	*Etheostoma nuchale*
weakfish	*Cynoscion regalis*
western banded killifish	*Fundulus diaphanus menona*
white bass	*Morone chrysops*
white catfish	*Ameiurus catus*
white crappie	*Pomoxis annularis*
white croaker	*Genyonemus lineatus*
white marlin	*Tetrapterus albidus*
white perch	*Morone americana*
white shark	*Carcharodon carcharias*
white sturgeon	*Acipenser transmontanus*
white sucker	*Catostomus commersoni*
whiteline topminnow	*Fundulus albolineatus*
widow rockfish	*Sebastes entomelas*
windowpane (flounder)	*Scophthalmus aquosus*
winter flounder	*Pleuronectes americanus*
witch flounder	*Glyptocephalus cynoglossus*
woundfin (minnow)	*Plagopterus argentissimus*
Yaqui chub	*Gila purpurea*
yellow perch	*Perca flavescens*
yellowfin cutthroat trout	*Oncorhynchus clarki macdonaldi*
yellowfin sole	*Pleuronectes asper*
yellowfin tuna	*Thunnus albacares*
yellowtail flounder	*Pleuronectes ferrugineus*

INVERTEBRATES

abalone	*Haliotis spp.*
American lobster	*Homarus americanus*
Atlantic calico scallop	*Argopecten gibbus*
Atlantic hard calm (northern quahog)	*Mercenaria mercenaria*
Atlantic (eastern) oyster	*Crassostrea virginica*
bay scallop	*Argopecten irradians*
birdwing pearlymussel	*Lemiox rimosus*
blue crab	*Callinectes sapidus*
blue king crab	*Paralithodes platypus*
blue mussel	*Mytilus edulis*
brown (golden) king crab	*Lithodes aequispina*
brown shrimp	*Penaeus aztecus*
calico scallop	*see Atlantic calico scallop*
Dungeness crab	*Cancer magister*

James spinymussel   *Pleurobema collina*

long-finned squid   *Loligo pealei*

northern shrimp   *Pandalus borealis*

ocean quahog   *Arctica islandica*

Pacific hard clams   *familly Veneridae*
Pacific oyster   *Crassostrea gigas*
Pacific razor clam   *Siliqua patula*
pink shrimp   *Penaeus duorarum*
pismo clam   *Tivela stultorum*

red king crab   *Paralithodes camtschatica*

sea scallop   *Placopecten magellanicus*
sea urchins   *Strongylocentrotus spp.*
short-finned squid   *Ilex illecebrosus*
slipper lobster (Atlantic/Caribbean) *Scyllarides nodifer*
slipper lobster (Pacific)   *Scyllarides squamosus*
softshell clam   *Mya arenaria*
spiny lobster (Atlantic/Caribbean)   *Panulirus argus*
spiny lobster (Pacific)   *Panulirus marginatus*
stone crab   *Menippe mercenaria*
surf clam   *Spisula solidissima*

tanner (snow) crabs   *Chionoecetes bairdi*
                                     Chinoecetes opilio

white shrimp   *Penaeus setiferus*

Glossary

active capture gear—gear that is pulled through the water to actively entrap, entangle, or hook fishes.

alevin—early life stage of salmon and trout that, although possessing some traits of premetamorphosed larval fishes, exhibits many morphological and behavioral features that are characteristic of juvenile fishes.

allochthonous materials—organic matter, often leaf materials, that originates from outside of the water column; for example, tree leaf litter falling or washing into a stream channel.

ammocoete—premetamorphosis, larval life stage of lampreys.

anadromous fishes—species that reproduce in freshwater but spend a substantial proportion of their life in the ocean.

anglers—people who fish for personal, aesthetic experiences rather than solely for income or subsistence.

anoxia—the depletion of dissolved oxygen in water to levels insufficient to support fishes and sediment-dwelling invertebrates typical of that habitat.

autotrophic organisms—those organisms that can produce their own nutrition (food), such as photosynthetic plants.

backwaters—surface waters associated with, but separate from, the main channel of a river or the major basin of a lake.

banks—offshore elevations, or shallow areas, of the continental shelf that are separated from each other and the mainland by broad, deep channels.

benthic organisms—organisms existing on or within the substrates of aquatic systems.

biomass production—the amount of new biomass added to a population through body growth.

bottom fishes—fishes that reside at the bottom of the water column, spending much of their time at or near the surface of the substrate.

brackish—marine and estuarine waters with salinity between 0.5 and 30 parts per thousand due to the mixing of seawater and freshwater.

BRD (bycatch reduction device)—mechanism placed in shrimp trawls to reduce the incidental capture of fishes, particularly juvenile life stages.

broadcast spawning—release of eggs and sperm into the water column. Once fertilized, buoyant eggs of broadcast spawning fishes drift with currents while suspended in the water column.

buffers—substances that neutralize acids or bases, maintaining concentrations of free hydrogen ions (thus pH) at near-neutral levels.

bycatch—organisms that are captured incidentally during fishing activities directed toward other taxa (also called *incidental catch*).

carrying capacity—the biomass or abundance of a particular species that an ecosystem can support through some specific period of time.

catadromous species—fishes that spend most of their life in freshwater systems but migrate to the ocean to spawn.

channel, river—the substrates that are regularly flooded during natural flows of a river system.

closure—regulation that prohibits fishing activity in specified fisheries. Closures are specific to a particular time period and/or a particular place.

codend—the collecting bag at the posterior end of a trawl.

commercial harvesters—harvesters who receive a significant proportion of their income from the sale of their catch.

competition—interaction between two or more organisms or types of organisms attempting to use the same resource that is limited in availability.

conservation—the wise use of living, wild (naturally-produced) resources that allows humans to gain some sustainable benefit from them.

continental shelf—the submerged extension of continental masses, extending seaward from the shoreline to the 200-meter depth contour.

creel limits—regulations that restrict the number of fishes a single participant can harvest daily in a recreational fishery.

cultural eutrophication—nutrient enrichment that occurs due to activities of humans. Common sources of nutrient enrichment include effluents from municipal sewage treatment facilities and fertilizer run-off.

delta—an expanse of submergent and emergent substrates that build up along coastal areas of oceans or lakes due to the deposition of materials at the mouths of river systems.

demersal—fishes and other aquatic organisms that live near the bottom of the water column.

density-dependent—characteristics of fish populations that are subject to the influence of population density or abundance. For example, body growth of fishes is believed to be inversely proportional to population abundance.

deposition—the settling of suspended materials from the water column of flowing water. This settling usually occurs as the rate of flow of the water declines.

dike—a high barrier built along the banks of river systems to prevent flooding of floodplains during high-water periods of the year (also *levee*).

directed catch—the portion of any catch that is composed of the taxa specifically being pursued during the fishing activity.

dredge—rigid framed fishing gear that is dragged over substrates to harvest bivalve shellfish.

endemic—fully restricted to a small geographic area or to single or a small number of watersheds.

epifauna—invertebrates living on the surface of substrates of aquatic systems.

epiphytes—microscopic plants that grow on the leaf surface of macroscopic plants.

epilimnion—warm, well-mixed surface waters occurring above the thermocline during periods of stratification. See *thermocline* and *stratification.*

erosion—the removal of particles from substrates due to the abrasive action of flowing water.

estuary—transitional habitat that occurs where freshwater flows into seawater in coastal areas of continents. Estuaries typically form in bays that are somewhat protected from coastal erosional processes due to the buildup of barrier beaches at the bay's mouth.

eutrophic—aquatic systems with high concentrations of nutrients and high levels of biological productivity.

eutrophication—the process of nutrient enrichment of aquatic systems.

exotic fishes—species occurring in U.S. waters that are not native to the North American continent.

exvessel prices—the price paid to commercial harvesters for their catch when it is delivered to the dock.

fecundity—the number of eggs produced by a female.

fetch—the distance over which winds can blow across the water's surface in standing water systems.

fingerling—small, juvenile fish, usually in its first year of life; a general size category, not a specific life stage.

fishery—the total interactions among an aquatic resource, its aquatic environment, and humans. Fisheries resources may either be consumptively used (that is, harvested) or they may provide nonconsumptive benefits to humans.

flat, tidal—unvegetated substrates occurring in estuaries between the low and high tide lines.

floodplain—terrain surrounding a river channel that is inundated during periods of flooding.

flushing rate—the rate of water exchange in an aquatic system. Lakes with high flushing rates are those that have relatively high rates of inflow and outflow per unit time in comparison to the volume of water that is held by the lake.

food web—all of the organisms in a particular ecosystem involved in the transfer of energy from primary producers (photosynthesizing organisms) through all levels of consumers.

forage base—the supply of food resources available to a particular species.

gangions—the series of hooks and leaders attached to longline gear.

gear efficiency—a measure of the proportion of a group of fishes encountering harvesting gear that are captured by it.

gene pool—the total genetic makeup of all members of a population.

genotype—specific genetic makeup of an individual or group.

gill net—passive capture net that entangles fishes by the head or opercles (gill covers).

gonochoristic species—those species in which any individual spends its entire adult life as either a male or a female, but not as both.

gradient, stream—the vertical drop exhibited by a stream channel over some specific horizontal distance. Gradient is usually measured as the number of meters drop in elevation per kilometer of distance or the number of feet per mile of distance.

groundwater—water beneath the surface of soils or exposed bedrock.

guild—a group of species in a fish community that exploit the same resources in a similar manner.

gyre—circular ocean currents created by surface winds, rotation of the earth, and the Coriolis effect.

headwaters—the portion of a river basin that occurs at the highest elevations for the watershed; the "source" of a river system.

hermaphroditic species—species in which individuals will function both as a male and a female at some time during the adult phase of their lives.

heterotrophic organisms—organisms that cannot produce their own food; their food comes from an outside source.

hybrids—offspring produced by the mating of parents from two different species or subspecies of fishes.

hypolimnion—deeper, denser waters of lakes and the ocean during periods of stratification. The volume of cold water that occurs below the thermocline in standing bodies of water.

inbreeding depression—the loss of genetic variability within a population that occurs when so few individuals breed that the next generation carries a lower diversity of alleles than would be characteristic of a breeding population with higher numbers of participants.

incidental catch—see *bycatch.*

infauna—invertebrates living within the substrates of aquatic habitats.

intertidal—occurring between the high and low tide lines.

introgressive hybridization—cross-breeding between two species of fishes that results in extensive backcrossing (interbreeding) between the hybrids and individuals from either parent species, and between hybrids themselves.

iteroparity—life history in which a specific individual may reproduce in more than one reproductive season during its life.

juvenile—post-metamorphosis stage of a fish's life cycle, starting once all organ systems are essentially fully formed and ending once the individual reaches sexual maturity.

lacustrine—lake ecoystems.

larvae—an immature stage of the life cycle of animals. The larval stage of fishes occurs before an individual metamorphoses into the juvenile/adult body form and life stage.

limited entry—regulation that limits the number of participants in a fishery. It is usually applied in a manner that identifies specific participants.

limnetic zone—in offshore regions of lakes too deep to support rooted aquatic vegetation, the portion of the water column under the surface that has sufficient light to support phytoplankton production.

littoral zone—shoreline region of lakes where light intensity at the submerged substrate is sufficient to allow growth of rooted aquatic vegetation.

mainstem—the main channel of a river basin into which tributaries empty.

marsh—broad, wet lowland at least periodically inundated with water, with substantial stands of herbaceous, terrestrial vegetation such as grasses and rushes.

maximum sustainable yield (MSY)—management of a fish stock that allows the maximum yearly harvest that can be sustained through time.

nitrifying organisms—microscopic organisms that convert ammonia and nitrite nitrogen to nitrate nitrogen, a form of nitrogen highly utilized for the production of protein molecules by green plants. This process occurs during the decomposition of organic nitrogen compounds.

nursery habitat—habitat possessing conditions critical for survival of fish larvae and fry.

nutrients—materials necessary for the production of organic materials. In aquatic systems, this term most often refers to materials required by green plants to carry out photosynthesis.

oligotrophic—aquatic systems with low nutrient concentrations and low levels of biological productivity.

optimum (sustainable) yield (OY or OSY)—managing the harvest of a fishery resource to gain the optimum benefit to society. Differs from MSY in that sociological and economic factors as well as the biological condition and production potential of the fish stock are considered when making management decisions.

otoliths—calcified structures in the inner ears of fishes that are often used in age analyses.

ovoviviparous—species that incubate eggs internally, but yolk reserves provide most or all of the energy available for embryological development.

oxbow—a backwater formed when flowing water erodes across the base of a bend in a river, depositing sediment at the entrance and exit to the bend, cutting it off from direct connection to the river channel.

panne—shallow depression in high salt marsh habitats that holds a permanent pool of water.

passive capture gear—stationary fishing gear that captures fishes when they contact it during their daily movement patterns.

pelagic or pelagial zone—same as *limnetic zone.*

pH—a measure of the logarithm of the reciprocal of the concentration of free hydrogen ions.

photic zone—the zone from the surface to the greatest depth at which sufficient sunlight penetrates the water column to support photosynthetic organisms.

phytoplankton—bouyant photosynthesizing microorganisms that are suspended in the water column, typically in the photic zone.

plankton—microscopic organisms that are suspended within the water column, usually drifting passively with currents.

pool—section of a stream or river where some portion of the water is not noticeably in motion.

population—group of organisms that interbreeds freely among themselves, but is isolated, often due to geographic barriers, from interbreeding with other groups of the same species.

primary production—the production of biomass by plants through photosynthesis.

production—the creation of biomass within a population through reproduction and body growth.

profundal zone—the portion of the water column under the pelagial zone and immediately above the substate; the substrate does not support rooted aquatic vegetation in the profundal.

protogynous—type of hermaphroditism in which all individuals are female upon reaching sexual maturity and may become males at some later time.

quota—level of harvest that cannot be exceeded by fishing fleets participating in the fishery.

recruitment—the addition of new individuals into some life stage or size range of a population. Most often, recruitment is referenced to sexual maturity (that is, recruitment into the spawning stock) or to the size range that is vulnerable to fishing gear used in a specific fishery (recruitment to a fishery).

riffles—reaches of a stream where the surface of the water is obviously turbulent rather than flowing in a laminar manner.

riparian zone—habitats occurring along the banks of river systems or the shorelines of lakes.

salinity—a measure of the concentration of dissolved ions in water. Measured in grams of dissolved salts per kilogram of water.

seines—small-mesh nets that are used to encircle schools of fishes.

semelparity—life history in which all individuals of a population spawn during one spawning season only, then die.

shellfishes—species of invertebrates that support fisheries.

size limits—regulations that restrict harvest within a fishery to some specified size range of fishes.

stock—group of individuals of the same species that share common production characteristics, and support the same basic fisheries. Stocks are often managed as single groups of organisms, even though they may be comprised of individuals from more than one population of a species.

stratification, thermal—the layering of the water column of lakes and the ocean, in most systems due to temperature and density differentials between the surface and deeper waters. Warm, less dense waters near the surface mix due to wave-generated currents, but deeper waters do not mix upward because they are significantly colder, and denser, than water at the surface.

subsistence harvesters—those who fish to put protein on their family dinner table, or for local barter for goods.

subtidal—permanently flooded habitats below the low tide line.

targeted catch—same as *directed catch.*

tailwater—a section of river downstream from a dam in which habitat conditions are modified from their natural state due to the presence of the upstream dam and reservoir.

TED (turtle-excluding device)—mechanism placed in shrimp trawls that reduces the incidental capture of sea turtles.

thermocline—the portion of the water column that exhibits rapid vertical change in water temperature during stratification. The thermocline marks the area of physical separation between warm, less dense surface waters (the *epilimnion*) and cold, denser deep waters (the *hypolimnion*) during summer stratification.

tidal amplitude—a measure of the difference between low and high tide lines in a specific region.

tidal rip—when rising or falling tides flow unidirectionally through some constricted space, such as the area between a submerged barrier beach and the shoreline, or a narrow channel connecting a large estuary to the coastline.

trap net—passive capture gear with an entrapment area consisting of a series of funnels into which fishes wander and do not escape.

trawl—cone-shaped net, closed at the posterior end, that is towed through the water by fishing vessels.

tributaries—small channels within a river basin that funnel water toward the mainstem.

trophic depression—the decline in production seen in reservoirs as they age. This decline is largely due to a reduction in leaching of soluble nutrients from soils of the watershed into the reservoir and from the loss of nutrients within the reservoir due to its high flushing rate.

trophic upsurge—initial pulse of high levels of productivity in a newly filled reservoir, due in large part to an influx of nutrients leaching from newly bared and flooded soils, and from the decomposition of substantial amounts of organic debris left when the reservoir basin was deforested.

turbidity—the level of opaqueness caused by particles suspended in the water column.

turnover—the movement of surface waters to the substrate, and deeper waters toward the surface in standing bodies of water such as lakes and the ocean. In many systems turnover occurs in the spring and fall due to differences in density between the surface and deeper waters, and due to vertical currents created by wave action.

"two-story" lakes or reservoirs—systems that have a substantial volume of deep water below the hypolimnion that remains suitable habitat for coldwater fish species throughout the year, as well as surface water habitats that support warmwater and coolwater species.

upwelling—the upward flow of water that occurs when ocean currents strike a continental coastline and deflect laterally. This deflection causes water to rise upward from the ocean depths to the surface.

viviparous fishes—species in which embryos are incubated within the reproductive tract of the female, and are nourished through a capillary attachment between female and embryo.

weir—permanent, passive capture gear set up as a maize into which fishes wander and do not find their way out.

wetlands—seasonally or permanently flooded areas with emergent herbaceous vegetation, shrubs and/or trees.

yield—harvested portion of a population. Sustainable yield is that portion of a population that can be removed by fishing and replaced by the following year through reproduction and body growth.

zooplankton—microscopic animals that are suspended in the water column. Although many are capable of locomotion, they will drift horizontally in the water column with prevailing currents.

Literature Cited

Acheson, J. M. 1980. Attitudes towards limited entry among finfishermen in northern New England. *Fisheries* 5 (6):20–25.

Acheson, J. M. 1988. *The lobster gangs of Maine.* University Press of New England. Hanover, New Hampshire.

AFS (American Fisheries Society). 1991. *Common and scientific names of fishes from the United States and Canada.* American Fisheries Society. Bethesda, Maryland.

AFS (American Fisheries Society). 1993. Reauthorization of the Magnuson Act. *Fisheries* 18 (10):20–26.

Aggus, L. R., and G. V. Elliott. 1975. Effects of cover and food on year-class stength of largemouth bass. Pages 317–322 in: R. H. Stroud and H. Clepper (eds.), *Black bass biology and management.* Sport Fishing Institute. Washington, D.C.

Allendorf, F. W., and R. F. Leary. 1988. Conservation and distribution of genetic variation in a polytypic species, the cutthroat trout. *Conservation Biology* 2:170–184.

Alm, G. 1959. *The connection between maturity, size and age in fishes.* Institute of Freshwater Research, Drottingholm. Report 40.

Alperin, I. M. 1987. Management of migratory Atlantic coast striped bass: A historical perspective, 1938–1986. *Fisheries* 12 (6):2–3.

Anderson, E. E. 1989. Economic benefits of habitat restoration. Seagrass and the Virginia hard-shell blue crab fishery. *North American Journal of Fisheries Management* 9:140–149.

Anderson, J. J. 1988. Diverting migrating fish past turbines. *Northwest Environmental Journal* 4:109–128.

Anderson, R. M., and R. B. Nehring. 1984. Effects of a catch-and-release regulation on a wild trout population in Colorado and its acceptance by anglers. *North American Journal of Fisheries Management* 4:257–265.

Anthony, V. C., and M. J. Fogarty. 1985. Environmental effects on recruitment, growth and vulnerability of Atlantic herring (*Clupea harengus harengus*) in the Gulf of Maine region. *Canadian Journal of Fisheries and Aquatic Sciences* 42:158–173.

Anthony, V. C., and S. A. Murawski. 1986. *Managing multi-species fisheries with catch quota regulations—the ICNAF experience.* Cooperative Research Report Number 139:42–57. Fifth Dialogue Meeting, ICES, Copenhagen.

Armour, C. L., D. A. Duff, and W. Elmore. 1991. AFS position statement on the effects of livestock grazing on riparian and stream ecosystems. *Fisheries* 16(1):12–17.

Armstrong, N. E. 1987. *The ecology of open-bay bottoms of Texas: A community profile.* U.S. Fish and Wildlife Service Biological Report 85 (7.12).

Arnett, G. R. 1983. Introduction to the session—user group demands on the environment. Pages 11–13 in: J. W. Reintjes (ed.), *Improving multiple use of coastal and marine resources.* American Fisheries Society. Bethesda, Maryland.

Arnold, D. E., W. D. Skinner, and D. E. Spotts. 1988. Evaluation of three experimental low technology approaches to acid mitigation in headwater streams. *Water Air and Soil Pollution* 41:385–406.

ASA (American Sportfishing Association). 1994. ESA and the sportfishing community: Rebuilding bridges? *American Sportfishing Association Bulletin* 454:1–2.

Austen, D. J., and D. J. Orth. 1984. Angler catches from New River, Virginia and West Virginia, in relation to minimum length limit regulations. *Proceedings of the Annual Conference of the Southeastern Association of Fish and Wildlife Agencies* 38:520–531.

Axon, J. R., and D. K. Whitehurst. 1985. Striped bass management in lakes with emphasis on management problems. *Transactions of the American Fisheries Society* 114:8–11.

Babey, G. J., and C. R. Berry, Jr. 1989. Post-stocking performance of three strains of rainbow trout in a reservoir. *North American Journal of Fisheries Management* 9:309–315.

Backiel, T., and E. D. LeCren. 1967. Some density relationships for fish population parameters. Pages 261–293 in: S. Gerking (ed.), *The biological basis of freshwater fish production.* Wiley. New York.

Bahr, L. M., and W. P. Lanier. 1981. *The ecology of intertidal oyster reefs of the South Atlantic coast: A community profile.* U.S. Fish and Wildlife Service, Office of Biological Services, Washington D.C. FWS/OBS-81/15.

Bailey, K. M. 1981. Larval transport and recruitment of Pacific hake *Merluccius productus. Marine Ecology Progress Series* 6:1–9.

Bailey, W. M. 1975. An evaluation of striped bass introductions in the southeastern United States. *Proceedings of the Annual Conference of the Southeastern Association of Game and Fish Commissioners* 28:54–68.

Bain, M. B., and S. E. Boltz. 1992. Effect of aquatic plant control on the microdistribution and population characteristics of largemouth bass. *Transactions of the American Fisheries Society* 121:94–103.

Bain, M. B., J. T. Finn, and H. E. Booke. 1988. Streamflow regulation and fish community structure. *Ecology* 69:382–392.

Baird, S. F. 1873. *Report on the condition of the sea fisheries of the south coast of New England in 1871 and 1872.* Part 1. U.S. Government Printing Office. Washington, D.C.

Balon, E. K. 1975. Reproductive guilds of fishes: A proposal and definition. *Journal of the Fisheries Research Board of Canada* 32:821–864.

Barber, W. E., and J. N. Taylor. 1990. The importance of goals, objectives, and values in the fisheries management process and organization: A review. *North American Journal of Fisheries Management* 10:365–373.

Barnhart, G. A., and V. Engstrom-Heg. 1984. A synopsis of some New York experiences with catch and release management of wild salmonids. Pages 91–101 in: F. Richardson and R. H. Hamre (eds.), *Wild trout III.* Trout Unlimited, Inc.

Barnhart, R. A. 1975. Pacific slope steelhead trout management. Pages 7–11 in: W. King (ed.), *Wild trout management.* Trout Unlimited, Inc.

Baughman, J. 1984. Catch–and-release may be the answer—now, what was the question? Pages 102–104 in: F. Richardson and R. H. Hamre (eds.), *Wild trout III.* Trout Unlimited, Inc.

Baylis, J. R. 1981. The evolution of parental care in fishes, with reference to Darwin's rule of male sexual selection. *Environmental Biology of Fishes* 6:223–251.

Beamish, R. J. 1993. Climate and exceptional fish production off the West Coast of North America. *Canadian Journal of Fisheries and Aquatic Sciences* 50:2270–2291.

Beamish, R. J., and D. R. Bouillon. 1993. Pacific salmon production trends in relation to climate. *Canadian Journal of Fisheries and Aquatic Sciences* 50:1002–1016.

Bean, M. J. 1986. The endangered species program. Pages 347–371 in: R. L. DiSylvestro (ed.), *Audubon wildlife report 1986.* Donnelley and Sons.

Beattie, W. D., and P. T. Clancey. 1991. Effects of *Mysis relicta* on the zooplankton community and kokanee population of Flathead Lake, Montana. Pages 39–48 in: T. P. Nesler and E. P. Bergersen (eds.), *Mysids in fisheries: hard lessons from headlong introductions.* American Fisheries Society Symposium 9.

Becker, C. D., D. A. Neitzel, and C. S. Abernethy. 1983. Effects of dewatering on chinook salmon redds: Tolerance of four development phases to one-time dewatering. *North American Journal of Fisheries Management* 3:373–382.

Becker, G. 1975. Fish toxification: Biological sanity or insanity? Pages 41–53 in: P. H. Eschmeyer (ed.), *Rehabilitation of fish populations with toxicants: A symposium.* American Fisheries Society Special Publication 4. Bethesda, Maryland.

Beeton, A. M. 1969. Changes in the environment and biota of the Great Lakes. Pages 150–187 in: G. A. Rohlich (symposium chairman), *Eutrophication: Causes, consequences, correctives.* Proceedings of a symposium. National Academy of Sciences. Washington, D.C.

Behnke, R. J. 1990. Summary of progress in wild trout management: 1974–1989. Pages 12–17 in: F. Richardson and R. Hamre (eds.), *Wild trout IV.* U.S. Government Printing Office 774-173/25037.

Bell, F. T. 1940. *Report of the U.S. Commissioner of Fisheries for the fiscal year 1938.* U.S. Government Printing Office. Washington, D.C.

Bell, F. W. 1980. Fisheries economics. Pages 197–218 in: R. T. Lackey and L. A. Nielsen (eds.), *Fisheries management.* Halsted Press. New York.

Bennett, G. W. 1970. *Management of lakes and ponds.* Van Nostrand Reinhold. New York.

Berggren, T. J., and J. J. Filardo. 1993. An analysis of variables influencing the migration of juvenile salmonids in the Columbia River basin. *North American Journal of Fisheries Management* 13:48–63.

Berkman, H. E., and C. F. Rabeni. 1987. Effects of siltation on stream fish communities. *Environmental Biology of Fishes* 18:285–294.

Berlin, W. H., R. J. Hesselberg, and J. J. Mac. 1981. *Growth and mortality of fry of Lake Michigan lake trout during chronic exposure to PCBs and DDE.* U. S. Fish and Wildlife Service Technical Paper 105:11–22.

Bertness, M. D. 1991. Interspecific interactions among high marsh perennials in a New England salt marsh. *Ecology* 72:125–137.

Beschta, R. L., R. E. Bilby, G. W. Brown, L. B. Holtby, and T. W. Hofstra. 1987. Stream temperature and aquatic habitat: Fisheries and forestry interactions. Pages 191–232 in: E. O. Salo and T. W. Cundy (eds.), *Streamside management: Forestry and fishery interactions.* Institute of Forest Resources, University of Washington, Seattle.

Bettoli, P. W., M. J. Maceina, R. L. Noble, and R. K. Betsill. 1992. Piscivory in largemouth bass as a function of aquatic vegetation abundance. *North American Journal of Fisheries Management* 12:509–516.

Bettoli, P. W., M. J. Maceina, R. L. Noble, and R. K. Betsill. 1993. Response of a reservoir fish community to aquatic vegetation removal. *North American Journal of Fisheries Management* 13:110–124.

Beverton, R. J. H., and S. J. Holt. 1957. *On the dynamics of exploited fish populations.* Fisheries Investigations of the Ministry of Agriculture, Fisheries and Food (Great Britain) Series II 19.

Bigelow, H. B., and W. C. Schroeder. 1953. *Fishes of the Gulf of Maine.* U.S. Fish and Wildlife Service Bulletin 74, Volume 53. U.S. Government Printing Office.

Bilby, R. E., and J. W. Ward. 1991. Characteristics and function of large woody debris in streams draining old-growth, clear-cut, and second-growth forests in southwestern Washington. *Canadian Journal of Fisheries and Aquatic Sciences* 48:2499–2508.

Bisson, P. A., R. E. Bilby, M. D. Bryant, C. A. Dolloff, G. B. Grette, R. A. House, M. L. Murphy, L. V. Koski, and J. R. Sedell. 1987. Large woody debris in forested streams in the Pacific Northwest: Past, present, and future. Pages 143–190 in: E. O. Salo and T. W. Cundy (eds.), *Streamside management: Forestry and fishery interactions.* Institute of Forest Resources, University of Washington, Seattle.

Bohnsack, J. A., and D. L. Sutherland. 1985. Artificial reef research: A review with recommendations for future priorities. *Bulletin of Marine Science* 37:11–39.

Booth, G. M., C. D. Wren, and J. M. Gunn. 1993. Efficacy of shoal liming for rehabilitation of lake trout populations in acid-stressed lakes. *North American Journal of Fisheries Management* 13:766–774.

Boreman, J. 1991. *Improving habitat quality vs. reducing fishing mortality to restore depleted populations of winter flounder.* Report to the Atlantic States Marine Fisheries Commission, Winter Flounder Scientific and Statistical Committee.

Boreman, J., and H. M. Austin. 1985. Production and harvest of anadromous striped bass stocks along the Atlantic coast. *Transactions of the American Fisheries Society* 114:3–7.

Bowen, J. T. 1970. A history of fish culture as related to the development of fishery programs. Pages 71–93 in: N. G. Benson (ed.), *A century of fisheries in North America.* American Fisheries Society Special Publication Number 7.

Bowers, G. M. 1901. *Report of the U.S. Commissioner of Fisheries for the fiscal year 1910.* U.S. Government Printing Office. Washington, D.C.

Bowles, E. C., B. E. Rieman, G. R. Mauser, and D. H. Bennett. 1991. Effects of introductions of *Mysis relicta* on fisheries in northern Idaho. Pages 65–74 in: T. P. Nesler and E. P. Bergersen (eds.), *Mysids in fisheries: Hard lessons from headlong introductions.* American Fisheries Society Symposium 9.

Boxrucker, J. 1986. Evaluation of supplemental stocking of largemouth bass as a management tool in small impoundments. *North American Journal of Fisheries Management* 6:391–396.

Boyd, C. E. 1981a. Solubility of granular inorganic fertilizers for fish ponds. *Transactions of the American Fisheries Society* 110:451–454.

Boyd, C. E. 1981b. Comparison of five fertilization programs for fish ponds. *Transactions of the American Fisheries Society* 110:541–545.

Boyd, C. E., and Y. Musig. 1981. Orthophosphate uptake by phytoplankton and sediment. *Aquaculture* 22:165–173.

Boyd, C. E., Y. Musig, and L. Tucker. 1981. Effects of three phosphorus fertilizers on phosphorus concentrations and phytoplankton production. *Aquaculture* 22:175–180.

Boyd, C. E., and J. W. Sowles. 1978. Nitrogen fertilization of ponds. *Transactions of the American Fisheries Society* 107:737–741.

Braun, J. L., and H. Kincaid. 1982. Survival, growth and catchability of rainbow trout of four strains. *North American Journal of Fisheries Management* 2:1–10.

Brice, J. J. 1898. *Report of the Commissioner for the year ending June 30, 1897.* U.S. Commission of Fish and Fisheries. Government Printing Office. Washington, D.C.

Bricklemeyer, E. C., Jr., S. Iudicello, and H. J. Hartmann. 1989. Discarded catch in U.S. commercial marine fisheries. Pages 259–296 in: W. J. Chandler (ed.), *Audubon wildlife report 1989–1990.* Academic Press. New York.

Britt, N. W., J. T. Addis, and R. Engel. 1973. Limmnological studies of the island area of western Lake Erie. *Bulletin of the Ohio Biological Survey-New Series* 4(3).

Broches, C. F. 1983. Fish, politics and treaty rights: Who protects salmon resources in Washington state? *British Columbia Studies* 57:86–98.

Brock, R. E. 1985. Preliminary study of the feeding habits of pelagic fish around Hawaiian fish aggregation devices, or can fish aggregation devices enhance local fisheries productivity? *Bulletin of Marine Science* 37:40–49.

Brodeur, R. D., and W. G. Pearcy. 1987. Daily feeding chronology, gastric evacuation and estimated daily ration of juvenile coho salmon, *Oncorhynchus kisutch* (Walbaum), in the coastal marine environment. *Journal of Fish Biology* 31:465–477.

Brown, A. M. 1986. Modifying reservoir fish habitat with artificial structures. Pages 98–102 in: G. E. Hall and M. J. Van Den Ayvle (eds.), *Reservoir fisheries management: Strategies for the 80's.* The American Fisheries Society. Bethesda, Maryland.

Brown, G. W., and J. T. Krygier. 1970. Effects of clear-cutting on stream temperature. *Water Resources Research* 6:1133–1140.

Brown, L. R., P. B. Moyle, and R. M. Yashiyama. 1994. Historical decline and current status of coho salmon in California. *North American Journal of Fisheries Management* 14:237–261.

Bryant, M. D. 1983. The role and management of woody debris in west coast salmonid nursery streams. *North American Journal of Fisheries Management* 3:322–330.

Buckley, J. 1989. *Species profiles: Life histories and environmental requirements of coastal fishes and invertebrates (North Atlantic)—winter flounder.* U.S. Fish and Wildlife Service Biological Report (11.87). U.S. Army Corps of Engineers, TR EL-82-4.

Buckley, R. M. 1989. Habitat alterations as a basis for enhancing marine fisheries. *California Cooperative Oceanic Fisheries Investigations Report* 30: 40–45.

Budgen, G. L., B. T. Hargrave, M. M. Sinclair, C. L. Tang, J. C. Therriault, and P. A. Yeats. 1982. Freshwater runoff effects in the marine environment: The Gulf of St. Lawrence example. *Canadian Technical Report Fisheries and Aquatic Sciences* 1078:1–71.

Bull, J. J., and R. Shine. 1979. Iteroparous animals that skip opportunities for reproduction. *American Naturalist* 114:296–303.

Burress, R. M. 1961. Fishing pressure and success in areas of flooded standing timber in Bull Shoals Reservoir, Missouri. *Proceedings of the 15th Annual Conference of the Southeastern Association of Game and Fish Commissions.* Pages 296–298.

Cada, G. F., J. M. Loar, and M. J. Sale. 1987. Evidence of food limitation of rainbow and brown trout in southern Appalachian soft-water streams. *Transactions of the American Fisheries Society* 116:692–702.

Cada, G. F., and M. J. Sale. 1993. Status of fish passage facilities at non-federal hydropower projects. *Fisheries* 18(7):4–13.

Cairns, M. A., and R. T. Lackey. 1992. Biodiversity and management of natural resources: The issues. *Fisheries* 17 (3):6–10.

Calhoun, A. 1966. The importance of considering the strain of trout stocked. In: A. Calhoun (ed.), *Inland fisheries management.* California Department of Fish and Game. Sacramento.

Campbell, W. J. 1989. Too long at sea for too few fish. Reprint of three-day series, published July 9–11, in the *Hartford Courant.* Hartford, Connecticut.

Canfield, C., and A. Garber. 1994. A way of life at risk. *Portland Press Herald.* 21 September. Pages 1A and 6A–7A.

Carl, L. M. 1982. Natural reproduction of coho salmon and chinook salmon in some Michigan streams. *North American Journal of Fisheries Management* 2:375–380.

Carlander, K. D. 1969. *Handbook of freshwater fishery biology.* Volume I. Iowa State University Press. Ames.

Carlander, K. D. 1977. *Handbook of freshwater fishery biology.* Volume 2. Iowa State University Press. Ames.

Carlander, K. D., and P. M. Payne. 1977. Yearclass abundance, population, and production of walleye (*Stizostedion vitreum vitreum*) in Clear Lake, Iowa, 1948–1974, with varied fry stocking rates. *Journal of the Fisheries Research Board of Canada* 34:1792–1799.

Carlander, K. D., and R. R. Whitney. 1961. Age and growth of walleye in Clear Lake, Iowa, 1935–57. *Transactions of the American Fisheries Society* 90:130–138.

Carline, R. F., and S. P. Klosiewski. 1985. Responses of fish populations to mitigation structures in two small channelized streams in Ohio. *North American Journal of Fisheries Management* 5:1–11.

Carlson, C. A., and R. T. Muth. 1993. Endangered species management. Pages 355–381 in: C. C. Kohler and W. A. Hubert (eds.), *Inland fisheries management in North America.* American Fisheries Society. Bethesda, Maryland.

Carmichael, G. J., J. N. Hanson, M. E. Schmidt, and D. C. Morizot. 1993. Introgression among Apache, cutthroat, and rainbow trout in Arizona. *Transactions of the American Fisheries Society* 122:121–130.

Carr, J. F., and J. K. Hiltunen. 1965. Changes in the bottom fauna of western Lake Erie from 1930 to 1961. *Limnology and Oceanography* 10:551–569.

Cederholm, C. J., and L. M. Reid. 1987. Impacts of forest management on coho salmon (*Oncorhynchus kisutch*) populations of the Clearwater River, Washington: A project summary. Pages 373–398 in: E. O. Salo and T. W. Cundy (eds.), *Streamside management: Forestry and fishery interactions.* Institute of Forest Resources, University of Washington, Seattle.

Chamberlin, T. W., R. D. Harr, and F. H. Everest. 1991. Timber harvesting, silviculture and watershed processes. Pages 181–205 in: W. R. Meehan (ed.), *Influences of forest and rangeland management on salmonid fishes and their habitats.* American Fisheries Society Special Publication 19. Bethesda, Maryland.

Chandler, A. D. 1988. The National Marine Fisheries Service. Pages 3–98 in: W. J. Chandler (ed.), *Audubon wildlife report 1988/1989.* Academic Press. New York.

Chandler, W. J. 1985. Inland fisheries management. Pages 92–129 in: R. L. DiSilvestro (ed.), *Audubon wildlife report 1985.* National Audubon Society. New York.

Chang, S. 1993. Analysis of fishery resources: Potential risk from sewage sludge dumping at the deepwater dumpsite off New Jersey. *Fishery Bulletin* 91:594–610.

Chapman, D. W. 1988. Critical review of variables used to define effects of fines in redds of large salmonids. *Transactions of the American Fisheries Society* 117:1–22.

Chapman, D. W., and R. Demory. 1963. Seasonal changes in the food ingested by aquatic insect larvae and nymphs in two Oregon streams. *Ecology* 44:140–146.

Chapman, D. W., and E. Knudsen. 1980. Channelization and livestock impacts on salmonid habitat and biomass in western Washington. *Transactions of the American Fisheries Society* 109:357–363.

Charnov, E. L., and W. M. Schaffer. 1973. Life history consequences of natural selection: Cole's result revisited. *American Naturalist* 107:791–793.

Cheek. T. E., M. J. Van den Avyle, and C. C. Coutant. 1985. Influences of water quality on distribution of striped bass in a Tennessee river impoundment. *Transactions of the American Fisheries Society* 114:67–76.

Childers, W. F. 1967. Hook and line yield of largemouth bass and redear X green sunfish hybrids in a one-acre pond. *Progressive Fish Culturist* 29:27–35.

Childress, W. M. 1988. Catch-and-release mortality of white and black crappie. Pages 175–186 in: R. A. Barnhart and T. D. Roelofs (eds.), *Catch-and-release fishing—a decade of experience.* California Cooperative Fishery Research Unit, Humboldt State University, Arcata.

Clapp, D. F., and R. D. Clark, Jr. 1989. Hooking mortality of smallmouth bass caught on live minnows and artificial spinners. *North American Journal of Fisheries Management* 9:81–85.

Clark, R. D., Jr., and G. R. Alexander. 1984. Effects of a slotted size limit on a brown trout fishery, Au Sable River, Michigan. Pages 74–84 in: F. Richardson and R. H. Hamre (eds.), *Wild trout III.* Trout Unlimited.

Clarke, T. A. 1970. Territorial behavior and population dynamics of a pomacentrid fish, the garibaldi, *Hypsypops rubicunda. Ecological Monographs* 40:180–212.

Clune, T., and D. Dauble. 1991. Yakima/Klickitat fisheries project: A strategy for supplementation of anadromous salmonids. *Fisheries* 16(5):28–34.

Coble, D. W., R. E. Bruesewitz, T. W. Fratt, and J. W. Scheirer. 1990. Lake trout, sea lampreys, and overfishing in the upper Great Lakes: A review and reanalysis. *Transactions of the American Fisheries Society* 119:985–995.

Cochran, P. A., and I. R. Adelman. 1982. Seasonal aspects of daily ration and diet of largemouth bass, *Micropterus salmoides,* with an evaluation of gastric evacuation rates. *Environmental Biology of Fishes* 7:265–275.

Colle, D. E., J. V. Shireman, W. T. Haller, J. C. Joyce, and D. E. Canfield, Jr. 1987. Influence of hydrilla on harvestable sport-fish populations, angler use, and angler expenditures at Orange Lake, Florida. *North American Journal of Fisheries Management* 7:410–417.

Conover, D. O., and B. E. Kynard. 1984. Field and laboratory observations of spawning periodicity and behavior of a northern population of the Atlantic silverside, *Menidia menidia* (Pisces Atherinidae). *Environmental Biology of Fishes* 11:161–171.

Contreras-B, S., and M. L. Lozano-V. 1994. Water, endangered fishes, and development perspectives in arid lands of Mexico. *Conservation Biology* 8:379–387.

Cooper, E. L. 1952. Rate of exploitation of wild eastern brook trout and brown trout populations in the Pigeon River, Otsego County, Michigan. *Transactions of the American Fisheries Society* 81:224–234.

Courtenay, S. C. 1985. Simultaneous multinesting by the fourspine stickleback, *Apeltes quadracus* (Mitchill). *Canadian Field Naturalist* 99:360–363.

Courtenay, W. R., Jr., D. A. Hensley, J. N. Taylor, and J. A. McCann. 1984. Distribution of exotic fishes in the continental United States. Pages 41–77 in: W. R. Courtenay, Jr., and J. R. Stauffer, Jr. (eds.), *Distribution, biology, and management of exotic fishes.* John Hopkins University Press. Baltimore, Maryland.

Courtenay, W. R., Jr., and P. B. Moyle. 1992. Crimes against biodiversity: The lasting legacy of fish introductions. Transactions of the 57th North American Wildlife and Natural Resources Conference.

Coutant, C. C. 1985. Striped bass, temperature, and dissolved oxygen: A speculative hypothesis for environmental risk. *Transactions of the American Fisheries Society* 114:31–61.

Coutant, C. C., and D. L. Benson. 1990. Summer habitat suitability for striped bass in Chesapeake Bay: Reflections on a population decline. *Transactions of the American Fisheries Society* 119:757–778.

Cowardin, L. M., V. Carter, F. C. Golet, and E. T. LaRoe. 1979. *Classification of wetlands and deepwater habitats of the United States.* U.S. Fish and Wildlife Service FWS/OBS-79/31.

Crecco, V. A., and T. Savoy. 1984. Effects of fluctuations in hydrographic conditions on year-class strength of American shad (*Alosa sapidissima*) in the Connecticut River. *Canadian Journal of Fisheries and Aquatic Sciences* 41:1216–1223.

Crecco, V. A., T. Savoy, and L. Gunn. 1983. Daily mortality rates of larval and juvenile American shad (*Alosa sapidissima*) in the Connecticut River with changes in year-class strength. *Canadian Journal of Fisheries and Aquatic Sciences* 40:1719–1728.

Cronan, C. S., and C. L. Schofield. 1979. Aluminum leaching response to acid precipitation: Effects on high-elevation watersheds in the northeast. *Science* 204:304–305.

Csanady, G. T., and B. A. Magnell. 1987. Mixing processes. Pages 163–169 in: R. H. Backus (ed.), *Georges Bank.* MIT Press. Cambridge, Massachusetts.

Cushing, D. H. 1977. The problems of stock and recruitment. Pages 116–133 in: J. A. Gulland (ed.), *Fish population dynamics.* John Wiley and Sons. New York.

Dadswell, M. J. 1979. Biology and population characteristics of the shortnose sturgeon, *Acipenser brevirostrum* LeSueur 1818 (Osteichthyes:Acipenseridae), in the Saint John River estuary, New Brunswick, Canada. *Canadian Journal of Zoology* 57:2186–2210.

Dahl, T. E., and C. E. Johnson. 1991. *Status and trends of wetlands in the conterminous United States, mid–1970s to mid-1980s.* U.S. Department of the Interior, Fish and Wildlife Service. Washington, D.C.

Daley, W. J. 1993. The use of fish hatcheries: Polarizing the issue. *Fisheries* 18 (3):4–5.

Daniel, J. 1993. Dance of denial. *Sierra* 78(2):64–73.

Davies, W. D. 1983. Sampling with toxicants. Pages 199–214 in: L. A. Nielsen and D. L. Johnson (eds.), *Fisheries techniques.* American Fisheries Society. Bethesda, Maryland.

Davis, J. R., and J. S. Hughes. 1971. Effects of standing timber on fish populations and fisherman success in Bussey Lake, Louisiana. Pages 255–264 in: G. E. Hall (ed.), *Reservoir fisheries and limnology.* Special Publication 8, American Fisheries Society. Bethesda, Maryland.

Davis, R. A., Jr. 1987. *Oceanography: An introduction to the marine environment.* William C. Brown. Dubuque, Iowa.

Dawson, C. P., and B. T. Wilkins. 1981. Motivations of New York and Virginia marine boat anglers and their preferences for potential fishing constraints. *North American Journal of Fisheries Management* 1:151–158.

Daye, P., and E. Garside. 1976. Histopathologic changes in surficial tissues of brook trout, *Salvelinus fontinalis* (Mitchill), exposed to acute and chronic levels of pH. *Canadian Journal of Zoology* 54:2140–2155.

Deacon, J. E. 1988. The endangered woundfin and water management in the Virgin River, Utah, Arizona, Nevada. *Fisheries* 13 (1):18–24.

Deacon, J. E., and W. L. Minckley. 1991. Western fishes and the real world: The enigma of "endangered species" revisited. Pages 404–413 in: W. L. Minckley and J. E. Deacon (eds.), *Battle against extinction: Native fish management in the American West.* University of Arizona Press. Tucson.

Dent, R. J., Jr. 1986. Results of a black bass 15-inch minimum length limit on Pomme de Terre Lake, Missouri. Pages 309–310 in: G. E. Hall and M. J. Van Den Avyle (eds.), *Reservoir fisheries management: Strategies for the 80's.* American Fisheries Society. Bethesda, Maryland.

Detenbach, N. E., P. W. DeVore, G. J. Niemi, and A. Lima. 1992. Recovery of temperate-stream fish communities from disturbance: A review of case studies and synthesis of theory. *Environmental Management* 16:33–53.

Dey, W. P. 1981. Mortality and growth of young-of-the-year striped bass in the Hudson River estuary. *Transactions of the American Fisheries Society* 110:151–157.

Ditton, R. B., D. K. Loomis, and S. Choi. 1992. Recreation specialization: Re-conceptualization from a social world's perspective. *Journal of Leisure Research* 24:33–51.

Dobbins, D. A., and C. E. Boyd. 1976. Phosphorus and potassium fertilization of sunfish ponds. *Transactions of the American Fisheries Society* 105:536–540.

Doeringer, P. B., P. I. Moss, and D. G. Terkla. 1986. *The New England fishing economy.* University of Massachusetts Press. Amherst.

Dolloff, C. A. 1986. Effects of stream cleaning on juvenile coho salmon and dolly varden in southeast Alaska. *Transactions of the American Fisheries Society* 115:743–755.

Dowling, T. E., and B. D. Demarais. 1993. Evolutionary significance of introgressive hybridization in cyprinid fishes. *Nature* 362:444–446.

Dugas, R., V. Guillory, and M. Fischer. 1979. Oil rigs and offshore sport fishing in Louisiana. *Fisheries* 4 (6):2–10.

Durocher, P. P., W. C. Provine, and J. E. Kraai. 1984. Relationship between abundance of largemouth bass and submerged vegetation in Texas reservoirs. *North American Journal of Fisheries Management* 4:84–88.

Dutil, J.-D. 1986. Energetic constraints and spawning interval in the anadromous Arctic charr (*Salvelinus alpinus*). *Copeia* 1986:945–955.

Eaglin, G. S., and W. A. Hubert. 1993. Effects of logging and roads on substrate and trout in streams of the Medicine Bow National Forest, Wyoming. *North American Journal of Fisheries Management* 13:844–846.

Ebert, D. J., K. W. Shirley, and J. J. Farwick. 1987. Evaluation of *Morone* hybrids in a small, shallow, warmwater impoundment. *Proceedings of the Annual Conference of the Southeastern Association of Fish and Wildlife Agencies* 41:55–62.

Echelle, A. A. 1991. Conservation genetics and genic diversity in freshwater fishes of western North America. Pages 141–153 in: W. L. Minckley and J. E. Deacon (eds.), *Battle against extinction: Native fish management in the American west.* University of Arizona Press. Tucson.

Edwards, C. J., B. L. Griswold, R. A. Tubb, E. C. Weber, and L. C. Woods. 1984. Mitigating effects of artificial riffles and pools on the fauna of a channelized warmwater stream. *North American Journal of Fisheries Management* 4:194–203.

Ehrlich, P. R., and A. H. Ehrlich. 1981. *Extinction: The causes and consequences of the disappearance of species.* Random House. New York.

Elgar, M. A. 1990. Evolutionary compromise between a few large and many small eggs: Comparative evidence in teleost fish. *Oikos* 59:283–287.

Elliott, S. T. 1986. Reduction of a dolly varden population and macrobenthos after removal of logging debris. *Transactions of the American Fisheries Society* 115:392–400.

Emlen, J. M., T. A. Strekal, and C. C. Buchanan. 1993. Probabilistic projections for recovery of the endangered cui-ui. *North American Journal of Fisheries Management* 13:467–474.

EPRI (Electric Power Research Institute). 1986. *Assessment of downstream migrant fish protection technologies for hydroelectric application.* Electric Power Research Institute Project 2694–1.

EPRI (Electric Power Research Institute). 1992. *Fish entrainment and turbine mortality review and guidelines.* Electric Power Research Institute Research Project 2694–01.

Ernst, J. P. 1991. Federalism and the Act. Pages 98–113 in: K. A. Kohm (ed.), *Balancing on the brink of extinction: The Endangered Species Act and lessons for the future.* Island Press. Washington, D.C.

Evans, D. O., G. J. Warren, and V. C. Cairns. 1990. Assessment and management of fish community health in the Great Lakes: Synthesis and recommendations. *Journal of Great Lakes Research* 16:639–669.

Everest, F. H., R. L. Beschta, J. C. Scrivener, K. V. Koski, J. R. Sedell, and C. J. Cederholm. 1987. Fine sediment and salmonid production: A paradox. Pages 98–142 in: E. O. Salo and T. W. Cundy (eds.), *Streamside management: Forestry and fishery interactions.* Institute of Forest Resources, University of Washington, Seattle.

Everhart, W. H., A. E. Eipper, and W. D. Youngs. 1975. *Principles of fishery science.* Cornell University Press. Ithaca, New York.

Farquhar, B. W. 1988. Evaluation of fertilizers to increase plankton abundance in reservoir coves. *Proceedings of the Annual Conference of Southeastern Associations of Fish and Wildlife Agencies* 42:193–199.

Fausch, K. D., and R. J. White. 1981. Competition between brook trout (*Salvelinus fontinalis*) and brown trout (*Salmo trutta*) for positions in a Michigan stream. *Canadian Journal of Fisheries and Aquatic Sciences* 38:1220–1227.

Fay, C. W., and G. B. Pardue. 1986. Harvest, survival, growth, and movement of five strains of hatchery-reared rainbow trout in Virginia streams. *North American Journal of Fisheries Management* 6:569–579.

Fedler, A. J., and R. B. Ditton. 1994. Understanding angler motivations in fisheries management. *Fisheries* 19:6–13.

Fetterholf, C. M., Jr. 1984. Lake trout futures in the Great Lakes. Pages 163–170 in: F. Richardson and R. H. Hamre (eds.), *Wild trout III.* Trout Unlimited.

Filipek, S. P., and M. D. Gibson. 1986. Evaluation of supplemental stocking of fingerling largemouth bass in Lake Coronado, Arkansas. Page 313 in: G. E. Hall and M. J. Va Den Avyle (eds.), *Reservoir fisheries management: Strategies for the 80's.* American Fisheries Society. Bethesda, Maryland.

Finkelstein, S. L. 1969. Age at maturity of scup from New York waters. *New York Fish and Game Journal* 16:224–237.

Finkelstein, S. L. 1971. Migration, rate of exploitation and mortality of scup from inshore waters of eastern Long Island. *New York Fish and Game Journal* 18:97–111.

Fisheries 8(2):2–3. *Florida receives artificial reef.* 1983.

Fisheries News. 1991. "America's most wanted fish?" August. American Fisheries Society.

Fiss, F. C., and R. F. Carline. 1993. Survival of brook trout embryos in three episodically acidified streams. *Transactions of the American Fisheries Society* 122:268–278.

Fletcher, R. I., and R. B. Deriso. 1988. Fishing in dangerous waters: Remarks on a controversial appeal to spawner-recruit theory for long-term impact assessment. *American Fisheries Society Monograph* 4:232–244.

Flick, W. A., and D. A. Webster. 1975. Movement, growth and survival in a stream population of wild brook trout (*Salvelinus fontinalis*) during a period of removal of non-trout species. *Journal of the Fisheries Research Board of Canada* 32:1359–1367.

Flickinger, S. A., and F. J. Bulow. 1993. Small impoundments. Pages 469–492 in: C. C. Kohler and W. A. Hubert (eds.), *Inland fisheries management in North America.* American Fisheries Society. Bethesda, Maryland.

Fogarty, M. J., A. A. Rosenberg, and M. P. Sissenwine. 1992. Fisheries risk assessment, sources of uncertainty: a case study of Georges Bank haddock. *Environmental Science and Technology* 26:440–447.

Forney, J. L. 1977. Evidence of inter- and intra-specific competition as factors regulating walleye (*Stizostedion vitreum vitreum*) biomass in Oneida Lake, New York. *Journal of the Fisheries Research Board of Canada* 34:1812–1820.

Forsgren, H., and A. J. Loftus. 1993. Rising to a greater future: Forest Service fisheries program accountability. *Fisheries* 18(5):15–21.

Forshage, A. A., W. D. Harvey, K. E. Kulzer, and L. T. Fries. 1986. Natural reproduction of white bass X striped bass hybrids in a Texas reservoir. *Proceedings of the Annual Conference of Southeastern Fish and Wildlife Agencies* 40:9–14.

Foster, C. H. W. 1991. Yankee salmon: The Atlantic salmon of the Connecticut River. CIS. Cambridge, Massachusetts.

Foster, M. S., and D. R. Schiel. 1985. *The ecology of giant kelp forests in California: A community profile.* U.S. Fish and Wildlife Service Biological Report 85 (7.02).

Frank, K. T., and W. C. Leggett. 1981. Wind regulation of emergence times and early larval survival in capelin (*Mallotus villosus*). *Canadian Journal of Fisheries and Aquatic Sciences* 38:215–223.

Fromm, P. 1980. A review of some physiological and toxicological responses of freshwater fish to acid stress. *Environmental Biology of Fishes* 5:79–93.

Funk, J. L. 1970. Warm-water streams. Pages 141–152 in: N. G. Benson (ed.), *A century of fisheries in North America.* Special Publication Number 7, American Fisheries Society. Bethesda, Maryland.

Gabriel, W. L., M. P. Sissenwine, and W. J. Overholtz. 1989. Analysis of spawning stock biomass per recruit: An example for Georges Bank haddock. *North American Journal of Fisheries Management* 9:383–391.

Gadowski, D. M., and J. A. Hall-Griswold. 1992. Predation by northern squawfish on live and dead juvenile chinook salmon. *Transactions of the American Fisheries Society* 121:680–685.

Gallaway, B. J., and G. S. Lewbel. 1982. *The ecology of petroleum platforms in the northwestern Gulf of Mexico: A community profile.* Bureau of Land Management, Gulf of Mexico OCS Regional Office. FWS/OBS-82/27.

Garber, A., and C. Canfield. 1994a. Growing fleet takes toll. *Portland Press Herald.* September 19. Pages 1A and 9A.

Garber, A., and C. Canfield. 1994b. Regulators ignored depletion warnings. *Portland Press Herald.* September 20. Pages 1A and 14A.

Gatewood, J. B., and B. J. McCay. 1990. Comparison of job satisfaction in six New Jersey fisheries: Implications for management. *Human Organization* 49:14–25.

Gessel, M. H., J. G. Williams, D. A. Brege, R. F. Krcma, and D. R. Chambers. 1991. Juvenile salmonid guidance at the Bonneville Dam second powerhouse, Columbia River, 1983–1989. *North American Journal of Fisheries Management* 11:400–412.

Gilliland, E. R., and J. Whitaker. 1989. Introgression of Florida largemouth bass introduced into northern largemouth bass populations in Oklahoma reservoirs. *Proceedings of the Annual Conference of Southeastern Fish and Wildlife Agencies* 43:182–190.

Gladwin, C. H., and J. Butler. 1982. Gardening: A survival strategy for the small, part-time Florida farm. *Proceedings of the Florida State Horticultural Society* 95:264–268.

Glass, R. D. 1984. Angler compliance with length limits on largemouth bass in an Oklahoma reservoir. *North American Journal of Fisheries Management* 4:457–458.

Glebe, B. D., and W. C. Leggett. 1981. Latitudinal differences in energy allocation and use during the freshwater migrations of American shad (*Alosa sapidissima*) and their life history consequences. *Canadian Journal of Fisheries and Aquatic Sciences* 38:806–820.

Golden, M. F., and C. E. Twilley. 1976. Fisheries investigations of a channelized stream: Big Muddy Creek, watershed, Kentucky. *Transactions of the Kentucky Academy of Science* 37:85–90.

Golet, F. C., A. J. K. Calhoun, W. R. DeRagon, D. J. Lowry, and A. J. Gold. 1993. *Ecology of red maple swamps in the glaciated northeast: A community profile.* U.S. Fish and Wildlife Service, Biological Report 12.

Gosselink, J. G. 1984. *The ecology of delta marshes of coastal Louisiana: A community profile.* U.S. Fish and Wildlife Service FWS/OBS-84/09.

Gowan, C., M. K. Young, K. D. Fausch, and S. C. Riley. 1994. Restricted movement in resident stream salmonids: A paradigm lost? *Canadian Journal of Fisheries and Aquatic Sciences* 51:2626–2637.

Graefe, A. R., and A. J. Fedler. 1986. Situational and subjective determinants of satisfaction in marine recreational angling. *Leisure Sciences* 8:275–295.

Graff, D. R., and L. Sorensen. 1970. The successful feeding of a dry diet to esocids. *Progressive Fish Culturist* 32:31–35.

Graham, H. W. 1970. Management of the groundfish fisheries of the northwest Atlantic. Pages 249–261 in: N. G. Benson (ed.), *A century of fisheries in North America*. American Fisheries Society, Special Publication Number 7.

Griffin, C. R. 1989. Protection of wildlife habitat by state wetland regulations: The Massachusetts initiative. *Transactions of the 54th North American Wildlife and Natural Resources Conference* 54:22–31.

Griffith, J. S. 1993. Coldwater streams. Pages 405–425 in: C. C. Kohler and W. A. Hubert (eds.), *Inland fisheries management in North America*. American Fisheries Society. Bethesda, Maryland.

Groen, C. L., and T. A. Schroeder. 1978. Effects of water level management on walleye and other coolwater fishes in Kansas reservoirs. Pages 278–283 in: R. L. Kendall (ed.), *Selected coolwater fishes of North America*. Special Publication 11. The American Fisheries Society. Bethesda, Maryland.

Gross, M. R., and E. L. Charnov. 1980. Alternative male life histories in bluegill sunfish. *Proceedings of the National Academy of Sciences USA* 77:6937–6940.

Grosslein, M. D., R. W. Langton, and M. P. Sissenwine. 1980. Recent fluctuations in pelagic fish stocks of the Northwest Atlantic, Georges Bank region, in relation to species interactions. *Rapports et Proces-Verbaux des Reunions, Conseil International pour l'Exploration de la Mer* 177:374–404.

Guest, W. C. 1984. Trihybrid sunfishes: Their growth, catchability, and reproductive success compared to parentals and hybrids. *Proceedings of the Annual Conference of Southeastern Fish and Wildlife Agencies* 38:421–435.

Gunderson, D. R. 1968. Floodplain use related to stream morphology and fish populations. *Journal of Wildlife Management* 32:507–514.

Gunn, J. M., J. G. Hamilton, G. M. Booth, C. D. Wren, G. L. Beggs, H. J. Rietveld, and J. R. Munro. 1990. Survival, growth and reproduction of lake trout (*Salvelinus namaycush*) and yellow perch (*Perca flavescens*) after neutralization of an acidic lake near Sudbury, Ontario. *Canadian Journal of Fisheries and Aquatic Sciences* 47:446–453.

Haines, T. A. 1981. Acid precipitation and its consequence for aquatic ecosystems: A review. *Transactions of the American Fisheries Society* 110:669–707.

Hall, J. D., G. W. Brown, and R. L. Lantz. 1987. The Alsea watershed study: A retrospective. Pages 399–416 in: E. O. Salo and T. W. Cundy (eds.), *Streamside management: Forestry and fishery interactions*. Institute of Forest Resources, University of Washington, Seattle.

Hall, L. W., A. E. Pinkney, L. O. Horseman, and S. E. Finger. 1985. Mortality of striped bass larvae in relation to contaminants and water quality in a Chesapeake Bay tributary. *Transactions of the American Fisheries Society* 114:861–868.

Hall-Arber, M. 1993. "They" are the problem: Assessing fisheries management in New England. *Nor'easter* Fall-Winter: 17–21.

Hansen, D. F., G. W. Bennett, R. J. Webb, and J. M. Lewis. 1960. Hook-and-line catch in fertilized and unfertilized ponds. *Illinois Natural History Survey Bulletin* 27:345–390.

Hard, J. 1995. Science, education, and the fisheries scientist. *Fisheries* 20(3):10–16.

Harper, J. L., and H. E. Namminga. 1986. Fish population trends in Texoma Reservoir following establishment of striped bass. Pages 156–165 in: G. E. Hall and J. Van Den Avyle (eds.), *Reservoir fisheries management: Strategies for the 80's*. American Fisheries Society. Bethesda, Maryland.

Harrell, R. M. 1987. Catch and release mortality of striped bass caught with artificial lures and baits. *Proceedings of the Annual Conference of Southeastern Fisheries and Wildlife Agencies* 41:70–75.

Harris, D. D., W. A. Hubert, and T. A. Wesche. 1991. Brown trout population and habitat response to enhanced minimum flow in Douglas Creek, Wyoming. *Rivers* 2:285–294.

Hart, J. L. 1973. *Pacific fishes of Canada*. Fisheries Research Board of Canada Bulletin 180.

Hartel, K. E. 1992. Non-native fishes known from Massachusetts freshwaters. *Occasional Reports of the MCZ Fish Department* 2:1–9.

Hartley, P. H. T. 1948. Food and feeding relationships in a community of fresh-water fishes. *Journal of Animal Ecology* 17:1–14.

Hartman, W. L. 1972. Lake Erie: Effects of exploitation, environmental changes and new species on the fishery resources. *Journal of the Fisheries Research Board of Canada* 29:899–912.

Hartzler, J. R. 1983. The effects of half-log covers on angler harvest and standing crop of brown trout in McMichaels Creek, Pennsylvania. *North American Journal of Fisheries Management* 3:228–238.

Harville, J. P. 1983. Obsolete petroleum platforms as artificial reefs. *Fisheries* 8(2):4–6.

Hassler, T. J. 1970. Environmental influences on early development and year-class strength of northern pike in Lakes Oahe and Sharpe, South Dakota. *Transactions of the American Fisheries Society* 99:369–380.

Hayes, M. L. 1983. Active fish capture methods. Pages 123–146 in: L. A. Nielsen and D. L. Johnson (eds.), *Fisheries techniques*. American Fisheries Society. Bethesda, Maryland.

Hayward, R. S., and F. J. Margraf. 1987. Eutrophication effects upon prey size and food available to yellow perch in Lake Erie. *Transactions of the American Fisheries Society* 116:210–223.

Hebda, A. J., G. M. Jones, and L. J. Hinks. 1990. *Smallmouth bass in Nova Scotia: Biology and options for management*. ERDA Report 24. Nova Scotia Department of Fisheries. Halifax, Canada.

Heidinger, R. C. 1993. Stocking for sport fisheries enhancement. Pages 309–333 in: C. C. Kohler and W. A. Hubert (eds.), *Inland fisheries management in North America*. American Fisheries Society. Bethesda, Maryland.

Heman, M. L., R. S. Campbell, and L. C. Redmond. 1969. Manipulations of fish populations through reservoir drawdown. *Transactions of the American Fisheries Society* 98:293–304.

Hennemuth, R. C., and S. Rockwell. 1987. History of fisheries conservation and management. Pages 430–446 in: R. H. Backus and D. W. Bourne (eds.), *Georges Bank*. MIT Press. Cambridge, Massachusetts.

Herdendorf, C. E. 1987. *The ecology of the coastal marshes of western Lake Erie: A community profile*. U.S. Fish and Wildlife Service, Biological Report 85 (7.9).

Hesse, L. W., and W. Sheets. 1993. The Missouri River hydrosystem: Recovery of lost functions is the best approach toward restoration. *Fisheries* 18(5):5–14.

Hestand, R. S., and C. C. Carter. 1978. Comparative effects of grass carp and selected herbicides on macrophyte and phytoplankton communities. *Journal of Aquatic Plant Management* 16:43–50.

High, B. 1989. Commercial fishing. Pages 225–240 in: F. G. Johnson and R. R. Stickney (eds.), *Fisheries: Harvesting life from water*. Kendall/Hunt. Dubuque, Iowa.

Hilborn, R. 1992. Hatcheries and the future of salmon in the Northwest. *Fisheries* 17 (1):5–8.

Hindar, K., N. Ryman, and F. Utter. 1991. Genetic effects of cultured fish on natural fish populations. *Canadian Journal of Fisheries and Aquatic Sciences* 48:945–957.

Hocutt, C. H., and J. R. Stauffer. 1975. Influence of gradient on the distribution of fishes in Conowingo Creek, Maryland and Pennsylvania. *Chesapeake Science* 16:143–147.

Holden, P. B. 1991. Ghosts of the Green River: Impacts of Green River poisoning on management of native fishes. Pages 43–54 in: W. L. Minckley and J. E. Deacon (eds.), *Battle against extinction: Native fish management in the American west*. University of Arizona Press. Tucson.

Holland, S. M., and R. B. Ditton. 1992. Fishing trip satisfaction: A typology of anglers. *North American Journal of Fisheries Management* 12:28–33.

Howell, P., D. Simpson, and G. Matlezos. 1984. Effects of a length limit on the recreational catch of scup (*Stenotomus chrysops*) in Connecticut. *Transactions of the Northeast Fish and Wildlife Conference* 41:239.

Howell, W. H., and R. Langan. 1992. Discarding of commercial groundfish species in the Gulf of Maine shrimp fishery. *North American Journal of Fisheries Management* 12:568–58.

Hubert, W. A. 1983. Passive capture techniques. Pages 95–122 in: L. A. Nielsen and D. A. Johnson (eds.), *Fisheries Techniques*. American Fisheries Society. Bethesda, Maryland.

Hubert, W. A., R. P. Lanka, T. A. Wesche, and F. Stabler. 1985. Grazing management influences on two brook trout streams in Wyoming. Pages 290–294 in: R. R. Johnson et al. (technical coordinators), *Riparian ecosystems and their management: Reconciling conflicting uses.* First North American Riparian Conference. USDA Forest Service General Technical Report RM-120.

Hudy, M., and C. R. Berry. 1983. Performance of three strains of rainbow trout in a Utah reservoir. *North American Journal Fisheries Management* 3:136–141.

Hulett, P. L., and S. A. Leider. 1990. Genetic interactions of hatchery and wild steelhead trout: Findings and implications of research at Kalama River, Washington. Pages 76–82 in: F. Richardson and R. Hamre (eds.), *Wild trout IV.* U.S. Government Printing Office 774-173/25037.

Hunt, R. 1975. Angling regulations in relation to wild trout management. Pages 66–74 in: W. King (ed.), *Wild trout management.* Trout Unlimited.

Hunt, R. L. 1988. *A compendium of 45 trout stream habitat development evaluations in Wisconsin during 1953–1985.* Wisconsin Department of Natural Resources Technical Bulletin 162.

Hunter, J. R. 1972. Swimming and feeding behaviour of larval anchovy, *Engraulis mordax. Fishery Bulletin* 70:821–838.

Huntsman, G. R., and W. E. Schaaf. 1994. Simulation of the impact of fishing on reproduction of a protogynous grouper, the graysby. *North American Journal of Fisheries Management* 14:41–52.

Hynes, H. B. N. 1970. *The ecology of running waters.* University of Toronto Press. Canada.

Ingmanson, D. E., and W. J. Wallace. 1979. *Oceanography: An introduction.* Wadsworth. Belmont, California.

IPHC (International Pacific Halibut Commission). 1987. *The Pacific halibut: Biology, fishery and management.* Technical Report Number 22. International Pacific Halibut Commission. Seattle, Washington.

IPHC (International Pacific Halibut Commission). 1992. Annual Report. International Pacific Halibut Commission. Seattle, Washington.

Iselin, C. 1955. Coastal currents and the fisheries. Papers Marine Biology Ocean, Deep-Sea Research. Supplement to Volume 3.

Isley, J. J., R. L. Noble, J. B. Koppelman, and D. P. Philipp. 1987. Spawning period and first-year growth of northern, Florida, and intergrade stocks of largemouth bass. *Transactions of the American Fisheries Society* 116:757–762.

Jaap, W. C. 1984. *The ecology of the south Florida coral reefs: A community profile.* U.S. Fish and Wildlife Service, Office of Biological Services. FWS/OBS-82/08.

Jackson, H. W., and R. E. Tiller. 1952. Preliminary observations on spawning potential in the striped bass (*Roccus saxatilis* Walbaum). *Chesapeake Biological Laboratory Publication* 93:1–16.

Janicki, A., and H. S. Greening. 1988. The effects of stream liming on water chemistry and anadromous yellow perch spawning success in two Maryland coastal plain streams. *Water, Air and Soil Pollution* 41:359–383.

Jantzen, R. A. 1986. Keynote address. Pages 1–4 in: G. E. Hall and J. J. Van den Avyle (eds.), *Reservoir fisheries management: Strategies for the 80's.* Reservoir Committee, Southern Division, American Fisheries Society. Bethesda, Maryland.

Jearld, Jr., A. 1983. Age determination. Pages 301–324 in: L. A. Nielsen and D. L. Johnson (eds.), *Fisheries techniques.* American Fisheries Society. Bethesda, Maryland.

Jenkins, R. M. 1970. Reservoir fish management. Pages 173–182 in: N. G. Nelson (ed.), *A century of fisheries in North America.* American Fisheries Society Special Publication 7.

Jennings, C. A., E. A. Gluesing, and R. J. Muncy. 1986. *Proceedings of the Annual Conference of Southeastern Association of Fish and Wildlife Agencies* 40:127–137.

Johnson, D. M., R. J. Behnke, D. A. Harpman, and R. G. Walsh. 1995. Economic benefits and costs of stocking catchable rainbow trout: A synthesis of economic analysis in Colorado. *North American Journal of Fisheries Management* 15:26–32.

Johnson, F. G. 1989. Fisheries: Harvesting life from water. Pages 1–9 in: F. G. Johnson and R. R. Stickney (eds.), *Fisheries: Harvesting life from water.* Kendall/Hunt. Dubuque, Iowa.

Johnson, J. E. 1987. Protected fishes of the United States and Canada. American Fisheries Society. Bethesda, Maryland.

Johnson, J. E., and B. L. Jensen. 1991. Hatcheries for endangered freshwater fishes. Pages 199–218 in: W. L. Minckley and J. E. Deacon (eds.), *Battle against extinction: Native fish management in the American west.* University of Arizona Press. Tucson.

Johnson, J. E., and J. N. Rinne. 1982. The Endangered Species Act and southwest fishes. *Fisheries* 7(3):2–7.

Johnson, R. R. 1978. The lower Colorado River: A western system. Pages 41–55 in: R. R. Johnson and J. F. McCormick (technical coordinators), *Strategies for protection and management of floodplain wetlands and other riparian ecosystems.* Proceedings of the symposium. Forest Service (USDA). Washington D.C.

Johnson, R. R., and S. W. Carothers. 1987. External threats: The dilemma of resource management on the Colorado River in Grand Canyon National Park, USA. *Environmental Management* 11:99–107.

Jones, A. R. 1982. The "two-story" rainbow trout fishery at Laurel River Lake, Kentucky. *North American Journal of Fisheries Management* 2:132–137.

Jones, R. D. 1984. Ten years of catch-and-release in Yellowstone Park. Pages 105–108 in: F. Richardson and R. H. Hamre (eds.), *Wild trout III.* Trout Unlimited.

Jones, S. R. 1994. Endangered Species Act battles. *Fisheries* 19 (1):22–25.

Jonsson, B., K. Hindar, and T. G. Northcote. 1984. Optimal age at sexual maturity of sympatric and experimentally allopatric cutthroat trout and Dolly Varden charr. *Oecologia* 61:319–325.

Jordan, D. S., and B. W. Evermann. 1900. *The fishes of North and Middle America: A descriptive catalogue.* U.S. Government Printing Office. Washington, D.C.

Joseph, J., W. Klawe, and P. Murphy. 1988. *Tuna and billfish—fish without a country.* Inter-American Tropical Tuna Commission. LaJolla, California.

Kallemeyn, L. W. 1987. Correlations of regulated lake levels and climatic factors with abundance of young-of-the-year walleye and yellow perch in four lakes in Voyageurs National Park. *North American Journal of Fisheries Management* 7:513–521.

Kapuscinski, A. R., and L. W. Jacobson. 1987. *Genetic guidelines for fisheries management.* Minnesota Sea Grant Research Report Number 17. University of Minnesota. St. Paul.

Karr, J. R., and I. J. Schlosser. 1978. Water resources and the land-water interface. *Science* 201:229–234.

Kauffman, J. 1980. Effect of a mercury-induced consumption ban on angling pressure. *Fisheries* 5(1):10–12.

Kauffman, J. 1983. Effects of a smallmouth bass minimum size limit on the Shenandoah River sport fishery. *Proceedings of the Annual Conference of the Southeastern Association of Fish and Wildlife Agencies* 37:459–467.

Keast, A. 1965. Resource subdivisions amongst cohabiting fish species in a bay, Lake Opinicon, Ontario. Pages 106–132 in: *Proceedings of the 8th Conference of Great Lakes Research.* University of Michigan.

Keast, A. 1966. Trophic interrelationships in the fish fauna of a small stream. Pages 51–79 in: *Proceedings of the 9th Conference on Great Lakes Research.* University of Michigan.

Keast, A., and D. Webb. 1966. Mouth and body form relative to feeding ecology in the fish fauna of a small lake, Lake Opinicon, Ontario. *Journal of the Fisheries Research Board of Canada* 23:1845–1874.

Keith, W. E. 1986. A review of introduction and maintenance stocking in reservoir fisheries management. Pages 144–148 in: G. E. Hall and M. J. Van Den Avyle (eds.), *Reservoir fisheries management: Strategies for the 80's.* American Fisheries Society. Bethesda, Maryland.

Keller, C. R., and K. P. Burnham. 1982. Riparian fencing, grazing, and trout habitat preference on Summit Creek, Idaho. *North American Journal of Fisheries Management* 2:53–59.

Kennedy, J. J., and P. J. Brown. 1976. Attitudes and behavior of fishermen in Utah's Uinta Primitive Area. *Fisheries* 1 (6):15–17, 30–31.

Ketcheson, G. L., and W. F. Megahan. 1993. Sediment deposition on slopes below roads in the Idaho batholith. EOS, *Transactions of the American Geophysical Union* 24(43):315.

Keup, L. E. 1979. Fisheries in lake restoration. *Fisheries* 4(1):7–9, 20.

Kimmel, B. L., and A. W. Groeger. 1986. Limnological and ecological changes associated with reservoir aging. Pages 103–109 in: G. E. Hall and J. J. Van Den Avyle (eds.), *Reservoir fisheries management: Strategies for the 80's*. The American Fisheries Society. Bethesda, Maryland.

Kincaid, H. L. 1983. Inbreeding in fish populations used for aquaculture. *Aquaculture* 33:215–227.

Kircheis, F. W., and J. G. Stanley. 1981. Theory and practice of forage-fish management in New England. *Transactions of the American Fisheries Society* 110:729–737.

Klauda, R. J., and R. E. Palmer. 1987. Responses of blueback herring eggs and larvae to pulses of acid and aluminum. *Transactions of the American Fisheries Society* 116:561–569.

Klein, W. D. 1965. Mortality of rainbow trout caught on single and treble hooks and released. *Progressive Fish-Culturist* 27:231–235.

Kohler, C. C., R. J. Sheehan, and J. J. Sweatman. 1993. Largemouth bass hatching success and first-winter survival in two Illinois reservoirs. *North American Journal of Fisheries Management* 13:125–133.

Kohm, K. A. 1991. The act's history and framework. Pages 10–24 in: K. A. Kohm (ed.), *Balancing on the brink of extinction: The Endangered Species Act and lessons for the future*. Island Press. Washington, D.C.

Koppelman, J. B., K. P. Sullivan, and P. J. Jeffries, Jr. 1992. Survival of three sizes of genetically marked walleye stocked into two Missouri impoundments. *North American Journal of Fisheries Management* 12: 291–298.

Kraai, J. E., W. C. Provine, and J. A. Prentice. 1983. Case histories of three walleye stocking techniques with cost-to-benefit considerations. *Proceedings of the Annual Conference of Southeastern Association of Fish and Wildlife Agencies* 37:395–400.

Kretser, W. A., and J. R. Colquhoun. 1984. Treatment of New York's Adirondack lakes by liming. *Fisheries* 9(1):36–41.

Krieger, D. A., and S. Puttman. 1986. Evaluation of supplemental stocking of yearling largemouth bass in Chatfield Reservoir, Colorado. Page 311 in: G. E. Hall and M. J. Van Den Avyle (eds.), *Reservoir fisheries management: Strategies for the 80's*. American Fisheries Society. Bethesda, Maryland.

Krueger, C. C., A. J. Gharrett, T. R. Dehring, and F. W. Allendorf. 1981. Genetic aspects of fisheries rehabilitation programs. *Canadian Journal of Fisheries and Aquatic Sciences* 38:1877–1881.

Krueger, C. C., and B. May. 1991. Ecological and genetic effects of salmonid introductions in North America. *Canadian Journal of Fisheries and Aquatic Sciences* 48 (Supplement 1):66–77.

Kutkuhn, J. H. 1980. Great lakes lake trout: Have we really lost what we are trying to restore? Pages 15–20 in: W. King (ed.), *Wild trout II*. Trout Unlimited.

Kwain, W. and G. A. Rose. 1986. Growth of brook trout *Salvelinus fontinalis* subject to sudden reductions of pH during their early life history. *Transactions of the American Fisheries Society* 114:564–570.

Laarman, P. N. 1978. Case histories of stocking walleye in inland lakes, impoundments and the Great Lakes—100 years with walleyes. *American Fisheries Society Special Publication* 11:254–260.

Lagler, K. F. 1978. Capture, sampling and examination of fishes. Pages 7–47 in: T. Bagenal (ed.), *Methods for assessment of fish production in fresh waters*. Third Edition. Blackwell Scientific. Oxford, Great Britain.

Lagler, K. F., John E. Bardach, Robert R. Miller, and Dora R. May Passino. 1977. *Ichthyology*. John Wiley and Sons. New York.

LaJeone, L. J., T. W. Bowzer, and D. L. Bergerhouse. 1992. Supplemental stocking of fingerling walleyes in the upper Mississippi River. *North American Journal of Fisheries Management* 12: 307–312.

Lake Erie Task Group. 1994. *Report of the Lake Erie Task Group to the Great Lakes Fishery Commission*.

Lamb, B. L., and B. A. K. Coughlan. 1993. Legal considerations in inland fisheries management. Pages 77–104 in: C. C. Kohler and W. A. Hubert (eds.), *Inland fisheries management in North American*. American Fisheries Society. Bethesda, Maryland.

Lantz, B. 1994. Trials of a fisherman. *Cape Cod Times*. 30 October. Pages B1 and B3.

Larkin, P. A. 1956. Interspecific competition and population control in fresh-water fish. *Journal of the Fisheries Research Board of Canada* 13:327–342.

Larkin, P. A., and T. G. Northcote. 1969. Fish as indices of eutrophication. Pages 256–273 in: Rohlich, G. A. (symposium chairman), *Eutrophication: Causes, consequences, correctives*. Proceedings of a symposium. National Academy of Sciences. Washington, D.C.

Lasker, R., H. M. Feder, G. H. Theilacker, and R. C. May. 1970. Feeding, growth and survival of *Engraulis mordax* larvae reared in the laboratory. *Marine Biology* 5:345–353.

Laurence, G. C. 1974. Growth and survival of haddock, *Melanogrammus aeglefinus*, larvae in relation to plankton prey concentration. *Journal of the Fisheries Research Board of Canada* 31:1415–1419.

Lawson, P. W. 1993. Cycles in ocean productivity, trends in habitat quality, and the restoration of salmon runs in Oregon. *Fisheries* 18(8):6–10.

Leggett, W. C., and J. E. Carscadden. 1978. Latitudinal variation in reproductive characteristics of American shad (*Alosa sapidissima*): Evidence for population specific life history strategies in fish. *Journal of the Fisheries Research Board of Canada* 35:1469–1478.

Leggett, W. C., K. T. Frank, and J. E. Carscadden. 1984. Meteorological and hydrographic regulation of year-class strength in capelin (*Mallotus villosus*). *Canadian Journal of Fisheries and Aquatic Sciences* 41:1193–1201.

Li, H. W., and P. B. Moyle. 1993. Management of introduced fishes. Pages 287–307 in: C. C. Kohler and W. A. Hubert (eds.), *Inland fisheries management in North America*. American Fisheries Society. Bethesda, Maryland.

Lichtkoppler, F., and C. E. Boyd. 1977. Phosphorus fertilization of sunfish ponds. *Transactions of the American Fisheries Society* 106:634–636.

Loftus, A. J., W. W. Taylor, and M. Keller. 1988. An evaluation of lake trout (*Salvelinus namaycush*) hooking mortality in the upper Great Lakes. *Canadian Journal of Fisheries and Aquatic Sciences* 46:2153–2156.

Loh-Lee Low (ed.). 1991. *Status of living marine resources off Alaska as assessed in 1991*. U.S. Department of Commerce, NOAA Technical Memorandum NMFS F/NWC-211.

Loomis, D. K., and R. B. Ditton. 1987. Analysis of motive and participation differences between saltwater sport and tournament fishermen. *North American Journal of Fisheries Management* 7:482–487.

Lynch, W. E., Jr., D. L. Johnson, and S. A. Schell. 1982. Survival, growth and food habits of walleye X sauger hybrids (saugeye) in ponds. *North American Journal of Fisheries Management* 2:381–387.

Mabbott, L. B. 1991. Artificial habitat for warmwater fish in two reservoirs in southern Idaho. Pages 62–66 in: *Warmwater Fisheries Symposium I*. USDA Forest Service. General Technical Report RM-207.

Maceina, M. J., P. W. Bettoli, W. G. Klussmann, R. K. Betsill, and R. L. Noble. 1991. Effect of aquatic macrophyte removal on recruitment and growth of black crappies and white crappies in Lake Conroe, Texas. *North American Journal of Fisheries Management* 11:556–563.

Maceina, M. J., B. R. Murphy, and J. J. Isely. 1988. Factors regulating Florida largemouth bass stocking success and hybridization with northern largemouth bass in Aquilla Lake, Texas. *Transactions of the American Fisheries Society* 117:221–231.

Magnusson, K. B., and O. K. Palsson. 1989. On the trophic ecological relationships of Iceland cod. *Rapports et Proces-Verbaux des Reunions, Conseil International pour l'Exploration de la Mer* 188:206–224.

Marcus, M. D. 1988. Differences in pre- and post–treatment water qualities for twenty limed lakes. *Water, Air and Soil Pollution* 41:279–291.

Maril, R. L. 1983. *Texas shrimpers: Community, capitalism, and the sea*. Texas A & M University Press. College Station.

Marking, L. L. 1992. Evaluation of toxicants for the control of carp and other nuisance fishes. *Fisheries* 17 (6):6–13.

Marsden, J. E. 1993. Responding to aquatic pest species: Control or management? *Fisheries* 18 (1):4–5.

Marsden, J. E., C. C. Krueger, and B. May. 1989. Identification of parental origins of naturally produced lake trout in Lake Ontario: Application of mixed-stock analysis to a second generation. *North American Journal of Fisheries Management* 9:257–268.

Marsh, J. H., and J. H. Johnson. 1985. The role of Stevens treaty tribes in the management of anadromous fish runs in the Columbia basin. *Fisheries* 10 (4):2–5.

Martin, C., and T. B. Hess. 1986. The use of a "slot" length limit regulation for largemouth bass on a new Georgia reservoir. Page 306 in: G. E. Hall and M. J. Van Den Avyle (eds.), *Reservoir fisheries management: Strategy for the 80's.* American Fisheries Society. Bethesda, Maryland.

Martin, J., J. Webster, and G. Edwards. 1992. Hatcheries and wild stocks: Are they compatible? *Fisheries* 17(1):4.

Martinez, P. J., and E. P. Bergersen. 1991. Interactions of zooplankton, *Mysis relicta,* and kokanees in Lake Granby, Colorado. Pages 49–64 in: Nesler, T. P. and E. P. Bergesen, *Mysids in fisheries: Hard lessons from headlong introductions.* American Fisheries Society Symposium 9.

Matlock, G. C., L. W. McEachron, J. A. Dailey, P. A. Unger, and P. Chai. 1993. Short-term hooking mortalities of red drums and spotted seatrout caught on single-barb and treble hooks. *North American Journal of Fisheries Management* 13:186–189.

Matthews, W. J., L. G. Hill, and S. M. Schellhaass. 1985. Depth distribution of striped bass and other fish in Lake Texoma (Oklahoma-Texas) during summer stratification. *Transactions of the American Fisheries Society* 114:84-91.

Mayer, K. S., F. L. Mayer, and A. Witt, Jr. 1985. Waste transformer oil and PCB toxicity to rainbow trout. *Transactions of the American Fisheries Society* 114:869–886.

McCarthy, D. T. 1985. The adverse effects of channelization and their amelioration. In: J. S. Alabaster (ed.), *Habitat modification and freshwater fisheries.* Butterworths. London.

McGinnis, S. M. 1984. *Freshwater fishes of California.* California Natural History Guide Number 49. University of California Press. Berkeley.

McGurrin, J. 1989. An assessment of Atlantic artificial reef development. *Fisheries* 14(4):19–25.

McHugh, J. L. 1970. Trends in fisheries research. Pages 25–56 in: N. G. Benson (ed.), *A century of fisheries in North America.* American Fisheries Society Special Publication Number 7.

McKiernan, D. J. 1995. Mass. anglers resist changes to striped bass rules. *DMF News.* Newsletter of the Massachusetts Division of Marine Fisheries. April-June 1995:3.

McWilliams, R. H., and J. G. Larscheid. 1992. Assessment of walleye fry and fingerlings and factors affecting their success in Lower Lake Oahe, South Dakota. *North American Journal of Fisheries Management* 12:329–335.

Meador, K. L., and A. W. Green. 1986. Effects of a minimum size limit on spotted seatrout recreational harvest. *North American Journal of Fisheries Management* 6:509–518.

Meador, M. R. 1992. Inter-basin water transfer: Ecological concerns. *Fisheries* 17(2):17–22.

Meehan, W. R., and W. S. Platts. 1978. Livestock grazing and the aquatic environment. *Journal of Soil and Water Conservation* 33(b):274–278.

Megahan, W. F. 1993. Sediment production from granitic cutslopes on forest roads in Idaho. EOS, *Transactions of the American Geophysical Union* 24(43):140.

Merna, J. W. 1986. Contamination of stream fishes with chlorinated hydrocarbons from eggs of Great Lakes salmon. *Transactions of the American Fisheries Society* 115:69–74.

Metzger, R. J., and C. E. Boyd. 1980. Liquid ammonium polyphosphate as a fish pond fertilizer. *Transactions of the American Fisheries Society* 109:563–570.

Miller, M. L., and F. G. Johnson. 1989. Fish and people. Pages 10–23 in: F. G. Johnson and R. R. Stickney (eds.), *Harvesting life from water.* Kendall/Hunt. Dubuque, Iowa.

Miller, R. R., J. D. Williams, and J. E. Williams. 1989. Extinctions of North American fishes during the past century. *Fisheries* 14 (6):22–38.

Minckley, W. L., and J. E. Deacon (eds.). 1991. *Battle against extinction: Native fish management in the American West.* University of Arizona Press. Tucson.

Minckley, W. L., and M. E. Douglas. 1991. Discovery and extinction of western fishes: A blink of the eye in geologic time. Pages 7–18 in: W. L. Minckley and J. E. Deacon (eds.), *Battle against extinction: Native fish management in the American West.* University of Arizona Press. Tucson.

Minckley, W. L., G. K. Meffe, and D. L. Soltz. 1991. Conservation and management of short-lived fishes: The cyprinodontoids. Pages 247–282 in: W. L. Minckley and J. E. Deacon (eds.), *Battle against extinction: Native fish management in the American West.* University of Arizona Press. Tucson.

Miranda, L. E., and W. D. Hubbard. 1994. Length-dependent winter survival and lipid composition of age-0 largemouth bass in Bay Springs Reservoir, Mississippi. *Transactions of the American Fisheries Society* 123:80–87.

Mitchell, J. M., and K. K. Sellers. 1989. Effects of two alternative minimum-length and creel limits on a largemouth bass population. *Proceedings of the Annual Conference of the Southeastern Fisheries and Wildlife Agencies* 43:164–171.

Mitton, J. B., and W. M. Lewis, Jr. 1989. Relationships between genetic variability and life-history features of bony fishes. *Evolution* 43:1712–1723.

Mitzner, L. 1992. Evaluation of walleye fingerling and fry stocking in Rathbun Lake, Iowa. *North American Journal of Fisheries Management* 12:321–328.

Mix, M. C. 1986. Cancerous diseases in aquatic animals and their association with environmental pollutants: A critical literature review. *Marine Environmental Research* 20:1–141.

Modde, T. 1980. State stocking policies for small warmwater impoundments. *Fisheries* 5(5):13–17.

Moeller, G. H., and J. Engelken. 1972. What fishermen look for in a fishing experience. *Journal of Wildlife Management* 36:1253–1257.

Moller, H. 1984. Reduction of a larval herring population by jellyfish predator. *Science* 224:621–622.

Moore, K. M. S., and S. V. Gregory. 1988. Response of young-of-the-year cutthroat trout to manipulation of habitat structure in a small stream. *Transactions of the American Fisheries Society* 117:162–170.

Morin, R., J. D. Dodson, and G. Power. 1982. Life history variations of anadromous cisco (*Coregonus artedii*), lake whitefish (*Coregonus clupeaformis*), and round whitefish (*Prosopium cylindraceum*) populations of eastern James-Hudson Bay. *Canadian Journal of Fisheries and Aquatic Sciences* 39:958–967.

Moring, J. R. 1982. An efficient strain of rainbow trout for stocking Oregon streams. *North American Journal of Fisheries Management* 2:209–215.

Moring, J. R. 1993. Anadromous stocks. Pages 553–580 in: C. C. Kohler and W. A. Hubert (eds.), *Inland fisheries management in North America.* American Fisheries Society. Bethesda, Maryland.

Moss, B. 1988. *Ecology of fresh waters.* Blackwell Scientific Publications. London.

Moss, J. L. 1985. Summer selection of thermal refuges by striped bass in Alabama reservoirs and tailwaters. *Transactions of the American Fisheries Society* 114:77–83.

Moyle, P. B. 1977a. In defense of sculpins. *Fisheries* 2(1):20–23.

Moyle, P. B. 1977b. Are coarse fish a curse? *Fly Fisherman* (early season):35–39.

Moyle, P. B. 1993. *Fish: An enthusiast's guide.* University of California Press. Berkeley.

Moyle, P. B., and J. J. Cech, Jr. 1996. *Fishes: An introduction to ichthyology.* Prentice Hall. Upper Saddle River, New Jersey.

Moyle, P. B., B. Vondracek, and G. D. Grossman. 1983. Responses of fish populations in the North Fork of the Feather River, California, to treatments with fish toxicants. *North American Journal of Fisheries Management* 3:48–60.

Moyle, P. B., and J. E. Williams. 1990. Biodiversity loss in the temperate zone: Decline of the native fish fauna of California. *Conservation Biology* 4:275–284.

Muoneke, M. I. 1992. Seasonal hooking mortality of bluegills caught on natural baits. *North American Journal of Fisheries Management* 12:645–649.

Murawski, S. A., and J. S. Idoine. 1992. Multispecies size composition: A conservative property of exploited fishery systems? *Journal of Northwest Atlantic Fishery Science* 14:79–85.

Murphy, G. I. 1968. Pattern in life history and the environment. *American Naturalist* 102:390–404.

Murphy, G. I. 1977. Clupeoids. Pages 283–308 in: J. A. Gulland (ed.), *Fish population dynamics.* Wiley. London.

Murphy, M. L., and K. V. Koski. 1989. Input and depletion of weedy debris in Alaska streams and implications for streamside management. *North American Journal of Fisheries Management* 9: 427–436.

Murray, J. D., J. J. Bahen, and R. A. Rulifson. 1992. Management considerations for by-catch in the North Carolina and Southeast shrimp fishery. *Fisheries* 17(1):21–26.

Muth, R. M. In press. *Use of fisheries resources for subsistence: Contemporary patterns in Southeast Alaska.* USA Proceedings, World Fisheries Congress. Volume III. Oxford and IBH Publishing Company. New Dehli, India.

Muth, R. M., D. E. Ruppert, and R. J. Glass. 1987. Subsistence use of fisheries resources in Alaska: Implications for Great Lakes fisheries management. *Transactions of the American Fisheries Society* 116:510–518.

Myers, N. 1985. A look at the present extinction spasm and what it means for the future evolution of species. Pages 47–57 in: R. J. Hoage (ed.), *Animal extinctions: What everyone should know.* Smithsonian Institution Press. Washington, D.C.

NEFSC. 1992. *Status of fishery resources off the northeastern United States for 1991.* NOAA Technical Memorandum NMFS-F/NEC-86.

Nehlsen, W., J. E. Williams, and J. A. Lichatowich. 1991. Pacific salmon at the crossroads: Stocks at risk from California, Oregon, Idaho and Washington. *Fisheries* 16(2):4–21.

Nehring, R. B., and R. Anderson. 1984. Catch and release management in Colorado—what works? How, when, where, why? Pages 109–112 in: F. Richardson and R. H. Hamre (eds.), *Wild trout III.* Trout Unlimited.

Nelson, W. R., and J. Bodle. 1990. *Ninety years of salmon culture at Little White Salmon National Fish Hatchery.* U.S. Fish and Wildlife Service Biological Report 90(17).

Nelson, W. R., M. C. Ingham, and W. E. Schaaf. 1977. Larval transport and year class strength of Atlantic menhaden, *Brevoortia tyrannus. Fishery Bulletin* 75:23–42.

Nelson, W. R., and C. H. Walburg. 1977. Population dynamics of yellow perch (*Perca flavescens*), sauger (*Stizostedion canadense*), and walleye (*S. vitreum vitreum*) in four mainstem Missouri River reservoirs. *Journal of the Fisheries Research Board of Canada* 34:1748–1763.

Nesler, T. P., and E. P. Bergersen. 1991. Mysids and their impacts on fisheries: An introduction to the 1988 mysid-fisheries symposium. Pages 1–4 in: T. P. Nesler and E. P. Bergersen (eds.), *Mysids in fisheries: Hard lessons from headlong introductions.* American Fisheries Society Symposium 9.

Neves, R. J. 1993. A state-of-the-unionids address. Pages 1–10 in: K. S. Cummings, A. C. Buchanan, and L. M. Koch (eds.), *Conservation and management of freshwater mussels.* Proceedings of a UMRCC symposium.

Ney, J. J. 1981. Evolution of forage-fish management in lakes and reservoirs. *Transactions of the American Fisheries Society* 110:725–728.

Nickelson, T. E. 1986. Influences of upwelling, ocean temperature, and smolt abundance on marine survival of coho salmon (*Oncorhynchus kisutch*) in the Oregon production area. *Canadian Journal of Fisheries and Aquatic Sciences* 43:527–535.

Nickelson, T. E., M. F. Solazzi, and S. L. Johnson. 1986. Use of hatchery coho salmon (*Oncorhynchus kisutch*) presmolts to rebuild wild populations in Oregon coastal streams. *Canadian Journal of Fisheries and Aquatic Sciences* 43:2443–2449.

Nickelson, T. E., M. F. Solazzi, S. L. Johnson, and J. D. Rodgers. 1992. Effectiveness of selected stream improvement techniques to create suitable summer and winter rearing habitat for juvenile coho salmon (*Oncorhynchus kisutch*) in Oregon coastal streams. *Canadian Journal of Fisheries and Aquatic Sciences* 49:790–794.

Nielsen, L. A. 1976. The evolution of fisheries management philosophy. *Marine Fisheries Review* 38(12):15–23.

Nielsen, L. A. 1993. History of inland fisheries management in North America. Pages 3–32 in: C. C. Kohler and W. A. Hubert (eds.), *Inland fisheries management in North America.* American Fisheries Society. Bethesda, Maryland.

Nikolsky, G. V. 1963. *The ecology of fishes.* Academic Press. New York.

Nixon, S. W. 1982. *The ecology of New England high salt marshes: A community profile.* U.S. Fish and Wildlife Service, Office Biological Services, Washington D.C. FWS/OBS-81/55.

NMFS (National Marine Fisheries Service). 1994. *Fisheries of the United States, 1993.* National Marine Fisheries Service Current Fishery Statistics Number 9300.

NOAA (National Oceanic and Atmospheric Administration). 1991. *Our living oceans.* NOAA Technical Memorandum NMFS-F/SPO-1.

NOAA (National Oceanic and Atmospheric Administration). 1993. *Our living oceans.* NOAA Technical Memorandum NMFS-F/SPO-15.

Noble, R. L. 1980. Management of lakes, reservoirs, and ponds. Pages 265–295 in: R. T. Lackey and L. A. Nielsen (eds.), *Fisheries management.* Halsted Press. New York.

Noble, R. L. 1981. Management of forage fishes in impoundments of the southern United States. *Transactions of the American Fisheries Society* 110:738–750.

Norcross, J. J., S. L. Richardson, E. H. Massman, and E. V. Joseph. 1974. Development of young bluefish (*Pomatomus saltatrix*) and distribution of eggs and young in Virginian coastal waters. *Transactions of the American Fisheries Society* 103:477–497.

Northcote, T. G. 1991. Success, problems, and control of introduced mysid populations in lakes and reservoirs. Pages 5–16 in: T. P. Nesler and E. P. Bergersen (eds.), *Mysids in fisheries: Hard lessons from headlong introductions.* American Fisheries Society Symposium 9.

Northwest Energy News. 1994. "Washingtonians willing to pay for salmon restoration." *Fisheries* 19(11):47.

Novinger, G. D. 1984. Observations on the use of size limits for black basses in large impoundments. *Fisheries* 9(4):2–5.

Novinger, G. D. 1986. Effects of a 15-inch minimum length limit on largemouth bass and spotted bass fisheries in Tale Rock Lake, Missouri. Pages 308–309 in: G. E. Hall and M. J. Van Den Avyle (eds.), *Reservoir fisheries management: Strategies for the 80's.* American Fisheries Society. Bethesda, Maryland.

Novinger, G. D. 1987. Evaluation of a 15.0-inch minimum length limit on largemouth bass and spotted bass catches at Table Rock Lake, Missouri. *North American Journal of Fisheries Management* 7:260–272.

Nuhfer, A. J., and G. R. Alexander. 1992. Hooking mortality of trophy-sized wild brook trout caught on artificial lures. *North American Journal of Fisheries Management* 12:634–644.

ODNR (Ohio Department of Natural Resources). 1994. *Ohio Strategic Plan for Lake Erie walleye.* Ohio Department of Natural Resources. Columbus.

Odum, E. P. 1980. The status of three ecosystem-level hypotheses regarding salt marsh estuaries: tidal subsidy, outwelling, and detritus-based food chains. Pages 485–495 in: V. S. Kennedy (ed.), *Estuarine perspectives.* Academic Press. New York.

Officer, C. B., R. B. Biggs, J. L. Taft, L. E. Cronin, M. A. Tyler, and W. R. Boynton. 1984. Chesapeake Bay anoxia: Origin, development, and significance. *Science* 223:22–27.

Oliver, M. L. 1984. The rainbow trout fishery in the Bull Shoals-Norfolk Tailwaters, Arkansas, 1971–81. *Proceedings of the Annual Conference of the Southeastern Association of Fish and Wildlife Agencies* 38:549–561.

Orbach, M. K. 1980. The human dimension. Pages 149–166 in: R. L. Lackey and L. A. Nielsen (eds.), *Fisheries management.* Halsted Press. New York.

Owen, C. R., and H. M. Jacobs. 1992. Wetland protection as land-use planning: The impact of Section 404 in Wisconsin, USA. *Environmental Management* 16:345–353.

Paragamian, V. L. 1982. Catch rates and harvest records under a 14.0-inch minimum length limit for largemouth bass in a new Iowa impoundment. *North American Journal of Fisheries Management* 2:224–231.

Paragamian, V. L., and R. Kingery. 1992. A comparison of walleye fry and fingerling stockings in three rivers in Iowa. *North American Journal of Fisheries Management* 12:313–320.

Paragamian, V. L., and M. J. Wiley. 1987. Effects of variable streamflows on growth of smallmouth bass in the Maquoketa River, Iowa. *North American Journal of Fisheries Management* 7:357–362.

Pardue, G. B. 1973. Production response of the bluegill sunfish, *Lepomis macrochirus* Rafinesque, to added attachment surface for fish food organisms. *Transactions of the American Fisheries Society* 102:622–626.

Patriarche, M. H., and R. S. Campbell. 1957. The development of the fish population in a new flood-control reservoir in Missouri, 1948 to 1954. *Transactions of the American Fisheries Society* 86:240–258.

Payer, R. D., R. B. Pierce, and D. L. Periera. 1989. Hooking mortality of walleyes caught on live and artificial baits. *North American Journal of Fisheries Management* 9:188–192.

Pearcy, W. 1962. Ecology of an estuarine population of winter flounder, *Pseudopleuronectes americanus* (Walbaum). *Bulletin of the Bingham Oceanographic Collection.* Yale University. 18:1–78.

Peden, A. E., and C. A. Corbett. 1973. Commensalism between a liparid fish, *Careproctus* sp., and the lithodid box crab, *Lopholithodes foraminatus. Canadian Journal of Zoology* 51:555-556.

Pella, J. J., and P. K. Tomlinson. 1969. A generalized stock production model. *Inter-American Tropical Tuna Commission Bulletin* 13(3):421–458.

Pelzman, R. 1980. Impact of Florida largemouth bass, *Micropterus salmoides floridanus,* introductions at selected northern California waters with a discussion of the use of meristics for detecting introgression and for classifying individual fish of intergraded populations. *California Fish and Game* 66:133–162.

Perra, P. 1992. By-catch reduction devices as a conservation measure. *Fisheries* 17(1):28–29.

Peters, J. C. 1982. Effects of river and streamflow alteration on fishery resources. *Fisheries* 7:20–22.

Peters, J. C., and W. Alvord. 1964. Man-made channel alterations in thirteen Montana streams and rivers. *Transactions of the North American Wildlife and Natural Resources Conference* 29:93–102.

Peterson, C. H., and N. M. Peterson. 1979. *The ecology of intertidal flats of North Carolina: A community profile.* U.S. Fish and Wildlife Service, Office of Biological Services. Washington, D.C. FWS/OBS-79/39.

Peterson, R. H., and D. J. Martin-Robichaud. 1986. Growth and major inorganic cation budgets of Atlantic salmon alevins at three ambient acidities. *Transactions of the American Fisheries Society* 115:220–226.

Petrosky, C. E., and T. C. Bjornn. 1988. Response of wild rainbow (*Salmo gairdneri*) and cutthroat trout (*S. clarki*) to stocked rainbow trout in fertile and infertile streams. *Canadian Journal of Fisheries and Aquatic Sciences* 45:2087–2105.

Peven, C. M., and S. G. Hays. 1989. Proportions of hatchery- and naturally produced steelhead smolts migrating past Rock Island Dam, Columbia River, Washington. *North American Journal of Fisheries Management* 9:53–59.

Pfitzer, D. 1975. Tailwater trout fisheries with special reference to the southeastern states. Pages 23–27 in: W. King (ed.), *Wild trout management.* Trout Unlimited.

Philipp, D. P., and G. S. Whitt. 1991. Survival and growth of northern, Florida, and reciprocal F_1 hybrid largemouth bass in central Illinois. *Transactions of the American Fisheries Society* 120-58-64.

Phillips, G. L., D. Eminson, and B. Moss. 1978. A mechanism to account for macrophyte decline in progressively eutrophicated freshwaters. *Aquatic Botany* 4:103–126.

Phillips, G. L., W. D. Schmid, and J. C. Underhill. 1982. *Fishes of the Minnesota region.* University of Minnesota Press. Minneapolis.

Phillips, R. C. 1984. *The ecology of eelgrass meadows in the Pacific northwest: A community profile.* FWS/OBS-84/24.

Phinney, L. A. 1986. Chinook salmon of the Columbia River basin. Pages 715–742 of: R. L. DiSilvestro (ed.), *Audubon wildlife report 1986.* National Audubon Society.

Pimentel, R., and R. V. Bulkley. 1983. Concentrations of total dissolved solids preferred or avoided by endangered Colorado River fishes. *Transactions of the American Fisheries Society* 112:595–600.

Pister, E. P. 1991. The desert fishes council: Catalyst for change. Pages 55–68 in: W. L. Minckley and J. E. Deacon (eds.), *Battle against extinction: Native fish management in the American West.* University of Arizona Press. Tucson.

Pitcher, T. J., and P. J. B. Hart. 1982. *Fisheries ecology.* Avi Press. Westport, Connecticut.

Plan Development Team. 1990. *The potential of marine fishery reserves for reef fish management in the U.S. Southern Atlantic.* NOAA Technical Memorandum NMFS-SEFC-261.

Platts, W. S. 1991. Livestock grazing. Pages 389–423 in: W. R. Meehan (ed.), *Influences of forest and range-land management on salmonid fishes and their habitats*. American Fisheries Society Special Publication 19. Bethesda, Maryland.

Platts, W. S., and S. B. Martin. 1980. Livestock grazing and logging effects on trout. Pages 34–46 in: W. King (ed.), *Wild trout II*. Trout Unlimited.

Platts, W. S., and M. L. McHenry. 1988. *Density and biomass of trout and char in western streams*. U.S. Forest Service General Technical Report INT-241.

Ploskey, G. R. 1986. Effects of water-level changes on reservoir ecosystems, with implications for fisheries management. Pages 86–97 in: G. E. Hall and J. J. Van Den Avyle (eds.), *Reservoir fisheries management: Strategies for the 80's*. American Fisheries Society. Bethesda, Maryland.

Price, K. S., D. A. Flemer, J. L. Taft, G. B. Mackiernan, W. Nehlsen, R. B. Biggs, N. H. Burger, and D. A. Blaylock. 1985. Nutrient enrichment of Chesapeake Bay and its impact on the habitat of striped bass: A speculative hypothesis. *Transactions of the American Fisheries Society* 114:97–106.

Primack, R. B. 1993. *Essentials of conservation biology*. Sinauer Associates. Sunderland, Massachusetts.

Prince, E. D., and O. E. Maughan. 1978. Freshwater artificial reefs: Biology and economics. *Fisheries* 3 (1):5–9.

Quinn, S. 1993. Bass seasons: Conservation measure or needless regulation. *In-Fisherman* 18(6):31–38.

Quinn, S. P., and M. R. Ross. 1985. Non-annual spawning in the white sucker, *Catostomus commersoni*. *Copeia* 1985:613–618.

Rabeni, C. F. 1993. Warmwater streams. Pages 427–443 in: C. C. Kohler and W. A. Hubert (eds.), *Inland fisheries management in North America*. American Fisheries Society. Bethesda, Maryland.

Radonski, G. C., and R. G. Martin. 1986. Fish culture is a tool, not a panacea. Pages 7–13 in: R. H. Stroud (ed.), *Fish culture in fisheries management*. American Fisheries Society, Fish Culture and Fisheries Administrators Section. Bethesda, Maryland.

Radonski, G. C., N. S. Prosser, R. G. Martin, and R. H. Stroud. 1984. Exotic fishes and sport fishing. Pages 313–321 in: W. R. Courtenay, Jr., and J. R. Stauffer, Jr. (eds.), *Distribution, biology and management of exotic fishes*. John Hopkins University Press. Baltimore, Maryland.

Rago, P. J., and C. P. Goodyear. 1987. Recruitment mechanisms of striped bass and Atlantic salmon: Comparative liabilities of alternative life histories. Pages 402–416 in: Dadswell et al. (eds.), *Common strategies of anadromous and catadromous fishes*. AFS Symposium 1. Bethesda, Maryland.

Rainwater, W. C., and A. Houser. 1975. Relations of physical and biological variables to black bass crops. Pages 306–309 in: R. H. Stroud and H. Clepper (eds.), *Black bass biology and management*. Sport Fishing Institute. Washington, D.C.

Raustron, R. R. 1977. Effects of a reduced bag limit and later planting date on the mortality, survival, and cost and yield of three domestic strains of rainbow trout at Lake Berryessa and Merle Collins Reservoir 1971–1974. *California Fish and Game* 63:219–227.

Raymond, H. L. 1979. Effects of dams and impoundments on migrations of juvenile chinook salmon and steelhead from the Snake River, 1966 to 1975. *Transactions of the American Fisheries Society* 108:505–529.

Redmond, L. C. 1986. Management of reservoir fish populations by harvest regulation. Pages 186–195 in: G. E. Hall and M. J. Van Den Avyle (eds.), *Reservoir fisheries management: Strategies for the 80's*. American Fisheries Society. Bethesda, Maryland.

Reed, D. J. 1992. Effects of weirs on sediment deposition in Louisiana coastal marshes. *Environmental Management* 16:55–65.

Reeves, G. H., J. D. Hall, T. D. Roelofs, T. L. Hickman, and C. O. Baker. 1991. Rehabilitating and modifying stream habitats. Pages 519–557 in: W. R. Meehan (ed.), *Influences of forest and rangeland management on salmonid fishes and their habitats*. American Fisheries Society Special Publication 19. Bethesda, Maryland.

Reeves, W. C., and F. R. Harders. 1983. Liquid fertilization of public fishing lakes in Alabama. *Proceedings of the Annual Conference of the Southeastern Association of Fish and Wildlife Agencies* 37:371–375.

Regier, H. A. 1962. On the evolution of bass-bluegill stocking policies and management recommendations. *Progressive Fish Culturist* 24:99–111.

Reid, G. K., and R. D. Wood. 1976. *Ecology of inland waters and estuaries.* D. Van Nostrand. New York.

Reinert, R. E., B. A. Knuth, M. A. Kamrin, and Q. J. Stober. 1991. Risk assessment, risk management, and fish consumption advisories in the United States. *Fisheries* 16(6):5–12.

Reisenbichler, R. R., and S. R. Phelps. 1989. Genetic variation in steelhead (*Salmo gairdneri*) from the north coast of Washington. *Canadian Journal of Fisheries and Aquatic Sciences* 46:66–73.

Reiser, D. W., J. P. Ramey, S. Beck, T. R. Lambert, and R. E. Geary. 1989. Flushing flow recommendations for maintenance of salmonid spawning gravels in a steep, regulated stream. *Regulated Rivers: Research and Management* 3:267–275.

Reiser, D. W., and R. G. White. 1990. Effects of streamflow reduction on chinook salmon egg incubation and fry quality. *Rivers* 1:110–118.

Richards, K. 1986. Evaluation of a 15-inch minimum length limit for black bass at Lake of the Ozarks, Missouri. Page 309 in: G. E. Hall and M. J. Van Den Avyle (eds.), *Reservoir fisheries management: Strategies for the 80's.* American Fisheries Society. Bethesda, Maryland.

Ricker, W. E. 1975. *Computation and interpretation of biological statistics of fish populations.* Fisheries Research Board of Canada Bulletin 191.

Ricker, W. E. 1981. Changes in the average size and average age of Pacific salmon. *Canadian Journal of Fisheries and Aquatic Sciences* 38:1636–1656.

Riehle, M. D., B. L. Parker, and J. S. Griffith. 1990. Rainbow trout populations in Silver Creek, Idaho, following a decade of catch-and-release regulations. Pages 153–162 in: F. Richardson and R. Hamre (eds.), *Wild trout IV.* U.S. Government Printing Office 774-173/25037.

Rieman, B. E., and R. C. Beamesderfer. 1990. Dynamics of a northern squawfish population and the potential to reduce predation on juvenile salmonids in a Columbia River reservoir. *North American Journal of Fisheries Management* 10:228–241.

Rieman, B. E., R. C. Beamesderfer, S. Vigg, and T. P. Poe. 1991. Estimated loss of juvenile salmonids to predation by northern squawfish, walleyes, and smallmouth bass in John Day Reservoir, Columbia River. *Transactions of the American Fisheries Society* 120:448–458.

Riley, S. C., and K. D. Fausch. 1995. Trout population response to habitat enhancement in six northern Colorado streams. *Canadian Journal of Fisheries and Aquatic Sciences* 52:34–53.

Rinne, J. N., and P. R. Turner. 1991. Reclamation and alteration as management techniques, and a review of methodology in stream renovation. Pages 219–246 in: W. L. Minckley (eds.), *Battle against extinction: Native fish management in the American West.* University of Arizona Press. Tucson.

Roff, D. A. 1982. Reproductive strategies in flatfish: A first synthesis. *Canadian Journal of Fisheries and Aquatic Sciences* 39:1686–1698.

Rohde, F. C., R. G. Arndt, D. G. Lindquist, and J. F. Parnell. 1994. *Freshwater fishes of the Carolinas, Virginia, Maryland and Delaware.* The University of North Carolina Press. Chapel Hill.

Rolston, H., III. 1991. Fishes in the desert: Paradox and responsibility. Pages 93–108 in: W. L. Minckley and J. E. Deacon (eds.), *Battle against extinction: Native fish management in the American West.* University of Arizona Press. Tucson.

Rood, S. B., and J. M. Mahoney. 1990. Collapse of riparian poplar forests downstream from dams in western prairies: Probable causes, and prospects for mitigation. *Environmental Management* 14:451–464.

Roseboom, D. 1993. What can the feds do to help control nonpoint pollution? Fisheries 18 (9):32–33.

Roseboom, D. and K. Russell. 1985. Riparian vegetation reduces stream bank and row crop flood damages. Pages 241–244 in: R. Johnson, C. Ziebell, D. Patton, P. Folliott, and R. Hamre (technical coordinators), *Riparian ecosystems and their management.* USDA Forest Service. General Technical Report RM-120.

Ross, D. R. 1990. Ten year recovery plan for the western banded killifish *Fundulus diaphanus menona.* Unpublished recovery plan, Ohio Department of Natural Resources.

Ross, M. R. 1977. Aggression as a social mechanism in the creek chub (*Semotilus atromaculatus*). *Copeia* 1977:393–397.

Ross, M. R. 1983. The frequency of nest construction and satellite male behavior in the fallfish minnow. *Environmental Biology of Fishes* 9:65–70.

Ross, M. R. 1991. *Recreational fisheries of coastal New England.* University of Massachusetts Press. Amherst.

Ross, M. R., and F. P. Almeida. 1986. Density-dependent growth of silver hakes. *Transactions of the American Fisheries Society* 115:548–554.

Ross, M. R., and T. M. Cavender. 1981. Morphological analyses of four experimental intergeneric cyprinid hybrid crosses. *Copeia* 1981:377–387.

Ross, M. R., and G. A. Nelson. 1992. Influences of stock abundance and bottom-water temperature on growth dynamics of haddock and yellowtail flounder on Georges Bank. *Transactions of the American Fisheries Society* 121:578–587.

Ross, M. R., and R. J. Reed. 1978. The reproductive behavior of the fallfish *Semotilus corporalis. Copeia* 1978:215–221.

Rothschild, B. J. 1986. *Dynamics of marine fish populations.* Harvard University Press. Cambridge, Massachusetts.

Rottman, R. 1977. Management of weedy lakes and ponds with grass carp. *Fisheries* 2(5):8–14.

Roussou, G. 1957. Some considerations concerning sturgeon spawning periodicity. *Journal of the Fisheries Research Board of Canada* 14:553–572.

Royce, W. F. 1984. Introduction to the practice of fishery science. Academic Press. New York.

Royce, W. F. 1987. *Fishery development.* Academic Press. New York.

Royce, W. F. 1989. A history of marine fishery management. *Review of Aquatic Sciences* 1:27–44.

Rulifson, R. A., J. D. Murray, and J. J. Bahen. 1992. Finfish catch reduction in south Atlantic shrimp trawls using three designs of by-catch reduction devices. *Fisheries* 17(1):9–20.

Russell, E. S. 1942. *The overfishing problem.* Cambridge University Press. Cambridge, England.

Ruttner, F. 1974. *Fundamentals of limnology.* University of Toronto Press. Canada.

Saiki, M. K. 1984. Environmental conditions and fish faunas in low elevation rivers of the irrigated San Joaquin valley floor, California. *California Fish and Game* 70:145–157.

Saiki, M. K., and C. J. Schmitt. 1985. Population biology of bluegills, *Lepomis macrochirus,* in lotic environments on the irrigated San Joaquin valley floor. *California Fish and Game* 71:225–244.

Saunders, J. W., and M. W. Smith. 1962. Physical alteration of stream habitat to improve brook trout production. *Transactions of the American Fisheries Society* 91:185–188.

Savino, J. F., M. G. Henry, and H. L. Kincaid. 1993. Factors affecting feeding behavior and survival of juvenile lake trout in the Great Lakes. *Transactions of the American Fisheries Society* 122:366–377.

Savino, J. F., and R. A. Stein. 1982. Predator-prey interactions between largemouth bass and bluegills as influenced by simulated submersed vegetation. *Transactions of the American Fisheries Society* 111:255–266.

Schaefer, W. F. 1989. Hooking mortality of walleyes in a northwestern Ontario lake. *North American Journal of Fisheries Management* 9:193–194.

Schaffer, W. M. 1974. Optimal reproductive effort in fluctuating environments. *American Naturalist* 108:783–790.

Schaffer, W. M., and P. E. Elson. 1975. The adaptive significance of variations in life history among local populations of Atlantic salmon in North America. *Ecology* 56:577–590.

Schill, D. J., J. S. Griffith, and R. E. Gresswell. 1986. Hooking mortality of cutthroat trout in a catch-and-release segment of the Yellowstone River, Yellowstone National Park. *North American Journal of Fisheries Management* 6:226–232.

Schloesser, D. W., and T. F. Nalepa. 1994. Dramatic decline of unionid bivalves in offshore waters of western Lake Erie after infestation by the zebra mussel, *Dreissena polymorpha. Canadian Journal of Fisheries and Aquatic Sciences* 51: 2234–2242.

Schofield, C. L., S. P. Gloss, B. Plonski, and R. Spateholts. 1989. Production and growth efficiency of brook trout (*Salvelinus fontinalis*) in two Adirondack mountain (New York) lakes following liming. *Canadian Journal of Fisheries and Aquatic Sciences* 46:333–341.

Schreck, C. B. 1980. Research perspectives for management of wild steelhead trout. Pages 26–32 in: W. King (ed.), *Wild trout II.* Trout Unlimited.

Schwiewe, M. H., D. D. Weber, M. S. Meyers, F. J. Jacques, W. L. Reichert, C. A. Krone, D. C. Malins, B. B. McCain, S. Chan, and U. Varanasi. 1991. Induction of foci of cellular alteration and other hepatic lesions in English sole (*Parophrys vetulus*) exposed to an extract of an urban marine sediment. *Canadian Journal of Fisheries and Aquatic Sciences* 48:1750–1760.

Scoppetone, G. C., and G. Vinyard. 1991. Life history and management of four endangered lacustrine suckers. Pages 359–378 in: W. L. Minckley and J. E. Deacon (eds.), *Battle against extinction: Native fish management in the American West.* University of Arizona Press. Tucson.

Scoppettone, G. C., G. A. Wedemeyer, M. Coleman, and H. Burge. 1983. Reproduction by the endangered cui-ui in the lower Truckee River. *Transactions of the American Fisheries Society* 112:788–793.

Scott, W. B., and E. J. Crossman. 1973. *Freshwater fishes of Canada.* Fisheries Research Board of Canada Bulletin 184.

Scott, W. B., and M. G. Scott. 1988. *Atlantic fishes of Canada.* Canadian Bulletin of Fisheries and Aquatic Sciences 219.

Seelbach, P. W., and G. E. Whelan. 1988. Identification and contribution of wild and hatchery steelhead stocks in Lake Michigan tributaries. *Transactions of the American Fisheries Society* 117:444–451.

Seliskar, D. M., and J. L. Gallagher. 1983. *The ecology of tidal marshes of the Pacific Northwest coast: A community profile.* U.S. Fish and Wildlife Service, Division of Biological Services. Washington, D.C. FWS/OBS-82/32.

SFI (Sport Fishing Institute). 1990. The changing face of America's anglers. *Sport Fishing Institute Bulletin* 418:1–2.

SFI (Sport Fishing Institute). 1991. Sport fishing license sales. *Sport Fishing Institute Bulletin* 424:1–2.

SFI (Sport Fishing Institute). 1992. Red drum stocking program shows promising results. *Sport Fishing Institute Bulletin* 439:3.

SFI (Sport Fishing Institute). 1993. Sport fishing: Past, present and future. *Sport Fishing Institute Bulletin* 450:1–4.

SFI (Sport Fishing Institute). 1994. The political correctness of sport fishing. *Sport Fishing Institute Bulletin* 451:1–4.

Shafland, P. L. 1986. A review of Florida's efforts to regulate, assess, and manage exotic fishes. *Fisheries* 11(2): 20–25.

Shannon, L., R. G. Biggins, and R. E. Hylton. 1993. Freshwater mussels in peril: Perspective of the U.S. Fish and Wildlife Service. Pages 66–68 in: K. S. Cummings, A. C. Buchanan, and L. M. Koch (eds.), *Conservation and management of freshwater mussels.* Proceedings of a UMRCC symposium.

Sheehan, R. J., and J. L. Rasmussen. 1993. Large rivers. Pages 445–468 in: C. C. Kohler and W. A. Hubert (eds.), *Inland fisheries management in North America.* American Fisheries Society. Bethesda, Maryland.

Shephard, B. B., S. A. Leathe, T. M. Weaver, and M. D. Enk. 1984. Monitoring levels of fine sediment within tributaries to Flathead Lake, and impacts of fine sediment on bull trout recruitment. Pages 146–156 in: F. Richardson and R. H. Hamre (eds.), *Wild trout III.* Trout Unlimited.

Shetter, G. A., and G. R. Alexander. 1962. Effects of a flies-only restriction on angling and on fall trout populations in Hunt Creek, Montmorency County, Michigan. *Transactions of the American Fisheries Society* 91:295–302.

Shields, J. T. 1958. Experimental control of carp reproduction through water drawdowns in Fort Randall Reservoir, South Dakota. *Transactions of the American Fisheries Society* 87:23–33.

Shireman, J. V. 1984. Control of aquatic weeds with exotic fishes. Pages 302–312 in: W. R. Courtenay, Jr., and J. R. Stauffer, Jr. (eds.), *Distribution, biology, and management of exotic fishes.* John Hopkins University Press. Baltimore, Maryland.

Shoemaker, T. G. 1988. Wildlife and water projects on the Platte River. Pages 285–334 in: W. J. Chandler (ed.), *Audubon wildlife report* 1988/1989. Academic Press. New York.

Siddens, L. K., W. K. Siem, L. R. Curtis, and G. A. Chapman. 1986. Comparisons of continuous and episodic exposure to acidic, aluminum-contaminated waters of brook trout (*Salvelinus fontinalis*). *Canadian Journal of Fisheries and Aquatic Sciences* 43:2036–2040.

Siewert, H. F., and J. B. Cave. 1990. Survival of released bluegill, *Lepomis macrochirus,* caught on artificial flies, worms, and spinner lures. *Journal of Freshwater Ecology* 5:407–411.

Simenstad, C. A. 1983. *The ecology of estuarine channels of the Pacific Northwest coast: A community profile.* U.S. Fish and Wildlife Service, Office of Biological Services. FWS/OBS-83/05.

Smith, C. L. 1985. *The inland fishes of New York State.* New York State Department of Environmental Conservation.

Smith, E. V., and H. S. Swingle. 1940. Effect of organic and inorganic fertilizers on plankton production and bluegill bream carrying capacity of ponds. *Transactions of the American Fisheries Society* 69:257–262.

Smith, E. V., and H. S. Swingle. 1942. The use of fertilizer for controlling several submerged aquatic plants in ponds. *Transactions of the American Fisheries Society* 71:94–101.

Smith, R. P., and J. L. Wilson. 1982. Growth comparison of two subspecies of largemouth bass in Tennessee ponds. *Proceedings of the Southeastern Association of Fish and Wildlife Agencies* 34:25–30.

Smith, R. W., and J. S. Griffith. 1994. Survival of rainbow trout during their first winter in the Henrys Fork of the Snake River, Idaho. *Transactions of the American Fisheries Society* 123:747–756.

Smith, S. H. 1972. Factors of ecologic succession in oligotrophic fish communities of the Laurentian Great Lakes. *Journal of the Fisheries Research Board of Canada* 29:717–730.

Snow, H. E. 1974. *Effects of stocking northern pike in Murphy's Flowage.* Wisconsin Department of Natural Resources Technical Bulletin Number 79.

Sparks, R. E. 1995. Need for ecosystem management of large rivers and their floodplains. *BioScience* 45:168–182.

Stearns, S. C. 1992. *The evolution of life histories.* Oxford University Press. New York.

Stevens, D. E. 1977. Striped bass (*Morone saxatilis*) year class strength in relation to river flow in the Sacramento-San Joaquin estuary, California. *Transactions of the American Fisheries Society* 106:34–42.

Stevens, D. E., and H. K. Chadwick. 1979. Sacramento-San Joaquin estuary—biology and hydrology. *Fisheries* 4(4):2–6.

Stevens, D. E., D. W. Kohlhurst, and L. W. Miller. 1985. The decline of striped bass in the Sacramento-San Joaquin Estuary, California. *Transactions of the American Fisheries Society* 114:12–30.

Stevens, D. E., and L. W. Miller. 1983. Effects of river flow on abundance of young chinook salmon, American shad, longfin smelt, and delta smelt in the Sacramento-San Joaquin River system. *North American Journal of Fisheries Management* 3:425–437.

Stevens, W. K. 1994. Dwindling salmon spur West to save rivers. Science Times of the *New York Times.* 15 November 1994. Pages C1 and C11.

Stewart, D. J., J. F. Kitchell, and L. B. Crowder. 1981. Forage fishes and their salmonid predators in Lake Michigan. *Transactions of the American Fisheries Society* 110:751–763.

Stoffle, R. W., F. V. Jensen, and D. L. Rasch. 1987. Cultural basis of sport anglers' response to reduced lake trout catch limits. *Transactions of the American Fisheries Society* 116:503–509.

Stokesbury, K. D. E., and J. J. Dadswell. 1991. Mortality of juvenile clupeids during passage through a tidal, low-head hydroelectric turbine at Annapolis Royal, Nova Scotia. *North American Journal of Fisheries Management* 11:149–154.

Stolte, L. 1981. *The forgotten salmon of the Merrimack.* U.S. Government Printing Office. Washington, D.C.

Stone, R. B. 1978. Artificial reefs and fishery management. *Fisheries* 3(1):2–4.

Stout, J. P. 1984. *The ecology of irregularly flooded salt marshes of the northeastern Gulf of Mexico: A community profile.* U.S. Fish and Wildlife Service Biological Report 85 (7.1).

Stowe, K. S. 1979. *Ocean science.* John Wiley and Sons. New York.

Stromberg, J. C., and D. T. Patten. 1990. Riparian vegetation instream flow requirements: A case study from a diverted stream in the eastern Sierra Nevada, California USA. *Environmental Management* 14:185–194.

Stroud, R. H., and R. G. Martin. 1968. *Fish conservation highlights 1963–67.* Sport Fishing Institute. Washington, D.C.

Stuber, R. J. 1985. Trout habitat, abundance, and fishing opportunities in fenced versus unfenced riparian habitat along Sheep Creek, Colorado. Pages 310–314 in: R. R. Johnson et al. (technical coordinators), *Riparian ecosystems and their management: Reconciling conflicting uses.* First North American Riparian Conference. USDA Forest Service. General Technical Report RM-120.

Summerfelt, R. C. 1986. Summarization of the symposium. Pages 314–327 in: G. E. Hall and M. J. Van den Avyle (eds.), *Reservoir fisheries management: Strategies for the 80's.* Reservoir Committee, Southern Division, American Fisheries Society. Bethesda, Maryland.

Summers, G. L. 1988. Largemouth bass population changes following implementation of a slot length limit. *Proceedings of the Annual Conference of the Southeastern Association of Fish and Wildlife Agencies* 42:209–217.

Sutter, F. C. 1980. Reproductive biology of anadromous rainbow smelt, *Osmerus mordax,* in the Ipswich Bay area, Massachusetts. Unpublished master's thesis, University of Massachusetts, Amherst.

Swanson, F. J., L. E. Benda, S. H. Duncan, G. E. Grant, W. F. Megahan, L. M. Reid, and R. R. Ziemer. 1987. Mass failures and other processes of sediment production in Pacific Northwest forest landscapes. Pages 9–38 in: E. O. Salo and T. W. Cundy (eds.), *Streamside management: Forestry and fishery interactions.* Contribution Number 57. Institute of Forest Resources, University of Washington, Seattle.

Swingle, H. S. 1970. History of warmwater pond culture in the United States. Pages 95–106 in: N. G. Benson (ed.), *A century of fisheries in North America.* Special Publication 7, American Fisheries Society. Bethesda, Maryland.

Swingle, H. S., B. C. Gooch, and H. R. Rabanal. 1963. Phosphate fertilization of ponds. *Proceedings of the Annual Conference of the Southeastern Association of Fish and Game and Fish Commissions* 17:213–218.

Swingle, H. S., and E. V. Smith. 1939. Fertilizer for increasing the natural food for fish in ponds. *Transactions of the American Fisheries Society* 68:126–133.

Swingle, H. S., and E. V. Smith. 1942. *Management of farm fish ponds.* Alabama Agricultural Experiment Station, Auburn University Bulletin 254.

Swingle, H. S., and E. V. Smith. 1947. *Management of farm ponds.* Alabama Polytechnical Institute Aricultural Experiment Station Bulletin 254.

Swink, W. D. 1983. Nonmigratory salmonids and tailwaters—a survey of stocking practices in the United States. *Fisheries* 8(3):5–9.

Tam, W. H., and P. D. Payson. 1986. Effects of chronic exposure to sublethal pH on growth, egg production and ovulation on brook trout *Salvelinus fontinalis. Canadian Journal of Fisheries and Aquatic Sciences* 43:275–280.

Taubert, B. D. 1980. Reproduction of shortnose sturgeon (*Acipenser brevirostrum*) in Holyoke Pool, Connecticut River, Massachusetts. *Copeia* 1980:114–117.

Taylor, J. N., W. R. Courtenay, Jr., and J. A. McCann. 1984. Known impacts of exotic fishes in the continental United States. Pages 322–373 in: W. R. Courtenay, Jr., and J. R. Stauffer, Jr. (eds.), *Distribution, biology, and management of exotic fishes.* John Hopkins University Press. Baltimore, Maryland.

Teal, J. M. 1986. *The ecology of regularly flooded salt marshes of New England: A community profile.* U.S. Fish and Wildlife Service, Biological Services Program, Washington D.C. FWS/OBS-81/01.

Thayer, G. W., W. J. Kenworthy, and M. S. Fonseca. 1984. *The ecology of eelgrass meadows of the Atlantic coast: A community profile.* U.S. Fish and Wildlife Service, Biological Services Program, Washington D.C. FWS/OBS-84/02.

Thurow, R. 1984. Wild steelhead trout populations in Idaho. Pages 38–40 in: F. Richardson and R. H. Hamre (eds.), *Wild trout III.* Proceedings of the symposium. Trout Unlimited.

Tiner, R. W., Jr. 1984. *Wetlands of the United States: Current status and recent trends.* National Wetlands Inventory. U.S. Fish and Wildlife Service, Department of the Interior.

Toneys, M. L., and D. W. Coble. 1979. Size-related, first winter mortality of freshwater fishes. *Transactions of the American Fisheries Society* 108:415–419.

Trandahl, A. 1978. Preface. Pages ix–x in: R. L. Kendall (ed.), *Selected coolwater fishes of North America*. Special Publication Number 11. American Fisheries Society. Bethesda, Maryland.

Trautman, M. B. 1942. Fish distribution and abundance correlated with stream gradients as a consideration in stocking programs. *Transactions of the North American Wildlife Conference* 7:211–223.

Trautman, M. B. 1981. *The fishes of Ohio*. Ohio State University Press. Columbus.

Trippel, E. A. 1993. Relations of fecundity, maturation, and body size of lake trout, and implications for management in northwestern Ontario lakes. *North American Journal of Fisheries Management* 13:64–72.

Trippel, E. A., and H. H. Harvey. 1989. Missing opportunities to reproduce: An energy dependent or fecundity gaining strategy in white sucker (*Catostomus commersoni*)? *Canadian Journal of Zoology* 67: 2180–2188.

Trotter, P. C. 1990. Preserve, protect and perpetuate: The status of wild cutthroat trout stocks in the West. Pages 83–95 in: *Wild trout IV*. U.S. Government Printing Office 774-173/25037.

Tyler, A. V., and V. F. Gallucci 1980. Dynamics of fish stocks. Pages 111–147 in: R. T. Lackey and L. A. Nielsen (eds.), Fisheries management. John Wiley and Sons. New York.

Tyus, H. M. 1991. Ecology and management of Colorado squawfish. Pages 379–402 in: W. L. Minckley and J. E. Deacon (eds.), *Battle against extinction: Native fish management in the American West*. University of Arizona Press. Tucson.

U.S. Department of Interior and U.S. Department of Commerce (USDI and USDC). 1993. *1991 National Survey of fishing, hunting, and wildlife-associated recreation*. U.S. Department of Interior and U.S. Department of Commerce. U.S. Government Printing Office. Washington, D.C.

U.S. Fish and Wildlife Service (USFWS). 1989. *Fish and egg distribution report of the national fish hatchery system*. Report Number 24. U.S. Fish and Wildlife Service, Department of Interior.

U.S. Government Accounting Office (GAO). 1993. *Potential economic costs of further protection for Columbia River Salmon*. Report to Congressional Requesters. General Accounting Office. GAO/RCED-93-41.

Vallentyne, J. R. 1957. Principles of modern limnology. *American Scientist* 45:218–244.

Van Den Avyle, M. J. 1993. Dynamics of exploited fish populations. Pages 105–135 in: C. C. Kohler and W. A. Hubert (eds.), *Inland fisheries management in North America*. American Fisheries Society. Bethesda, Maryland.

Varanasi, U., B. B. McCain, J. E. Stein, and S. Chan. 1993. Effects of coastal pollution on living marine resources. *Transactions, Proceedings of 58th North American Wildlife and Natural Resources Conference*. Washington, D.C.

Vickerman, S. E. 1989. State wildlife protection efforts: The non-game programs. Pages 67–96 in: G. Mackintosh (ed.), *Preserving communities and corridors*. Defenders of Wildlife. Washington, D.C.

Vincent, E. R. 1984. Effect of stocking hatchery rainbow trout on wild stream-dwelling trout. Pages 48–52 in: F. Richardson and R. H. Hamre (eds.), *Wild trout III*. Trout Unlimited.

Vincent, E. R. 1987. *Effects of stocking catchable-size hatchery rainbow trout on two wild trout species in the Madison River and O'Dell Creek, Montana*.

Voorhees, D. A., J. F. Witzig, M. F. Osborn, M. C. Holliday, and R. J. Essig. undated. *Marine recreational fishery statistics survey, Atlantic and Gulf coasts, 1990–1991*. Current Fisheries Statistics Number 9204. NOAA.

Walters, C. J., M. Stocker, A. V. Tyler, and S. J. Westrheim. 1986. Interactions between Pacific cod (*Gadus macrocephalus*) and herring (*Clupea harengus pallasi*) in the Hecate Strait, British Columbia. *Canadian Journal of Fisheries and Aquatic Sciences* 43:830–837.

Ware, D. M. 1985. Life history characteristics, reproductive value, and resilience of Pacific herring (*Clupea harengus pallasi*). *Canadian Journal of Fisheries and Aquatic Sciences* 42 (Supplement 1):127–137.

Warner, R. R. 1975. The adaptive significance of sequential hermaphroditism in animals. *American Naturalist* 109:61–82.

Warner, R. R. 1984. Mating behavior and hermaphroditism in coral reef fishes. *American Scientist* 72:128–136.

Warner, W. W. 1983. *Distant water: The fate of the North Atlantic fisherman*. Little Brown. Boston.

Warren, M. L., and B. M. Burr. 1994. Status of freshwater fishes of the United States: Overview of an imperiled fauna. *Fisheries* 19(1):6–18.

Waters, T. F. 1983. Replacement of brook trout by brown trout over 15 years in a Minnesota stream: Production and abundance. *Transactions of the American Fisheries Society* 112:137–146.

Weatherley, A. H. 1972. Growth and ecology of fish populations. Academic Press. New York.

Weber, M. 1985. Marine mammal protection. Pages 180–211 in: R. L. Di Silvestro (ed.), *Audubon wildlife report 1985*. National Audubon Society. New York.

Weber, M. 1986. Federal marine fisheries management. Pages 267–346 in: A. Eno (ed.), *Audubon wildlife report 1986*. National Audubon Society. New York.

Weber, M. 1987. Marine mammal protection. Pages 163–178 in: R. L. Di Silvestro (ed.), *Audubon wildlife report 1987*. National Audubon Society. New York.

Weber, P. 1994. *Net loss: Fish, jobs, and the marine environment*. Worldwatch Paper 120. Worldwatch Institute.

Wege, G. J., and R. O. Anderson. 1979. Influence of artificial structures on largemouth bass and bluegills in small ponds. Pages 59–69 in: D. L. Johnson and R. A. Stein (eds.), *Response of fish to habitat structures in standing water*. North Central Division, American Fisheries Society, Special Publication 6.

Weis, J. S., and P. Weis. 1989. Effects of environmental pollutants on early fish development. *Reviews in Aquatic Sciences* 1:45–73.

Weithman, A. S. 1993. Socioeconomic benefits of fisheries. Pages 159–180 in: C. C. Kohler and W. A. Hubert (eds.), *Inland fisheries management in North America*. American Fisheries Society. Bethesda, Maryland.

Weisberg, S. B., and W. H. Burton. 1993. Enhancement of fish feeding and growth after an increase in minimum flow below the Conowingo dam. *North American Journal of Fisheries Management* 13:103–109.

Welsh, B. L. 1980. Comparative nutrient dynamics of a marsh-mudflat ecosytem. *Estuarine and Coastal Marine Science* 10:143–164.

Werner, E. E., and D. J. Hall. 1976. Niche shifts in sunfishes: Experimental evidence and significance. *Science* 191:404–406.

Werner, E. E., and D. J. Hall. 1977. Competition and habitat shift in two sunfishes (Centrarchidae). *Ecology* 58:869–876.

Werner, R. G., and J. H. S. Blaxter. 1980. Growth and survival of larval herring, *Clupea harengus*, in relation to prey density. *Canadian Journal of Fisheries and Aquatic Sciences* 37:1063–1069.

Wesche, T. A. 1993. Watershed management and land-use practices. Pages 181–203 in: C. C. Kohler and W. A. Hubert (eds.), *Inland Fisheries Management*. American Fisheries Society. Bethesda, Maryland.

Wesche, T. A., C. M. Goertler, and C. B. Frye. 1987. Contribution of riparian vegetation to trout cover in small streams. *North American Journal of Fisheries Management* 7:151–153.

Westin, D. T., C. E. Olney, and B. A. Rogers. 1985. Effects of parental and dietary organochlorines on survival and body burdens of striped bass larvae. *Transactions of the American Fisheries Society* 114:125–136.

Wetzel, R. G. 1983. *Limnology*. Saunders College Publishing. Philadelphia.

Wharton, C. H., W. M. Kitchens, and E. C. Pendleton. 1982. *The ecology of bottomland hardwood swamps of the southeast: A community profile*. U.S. Fish and Wildlife Service, Biological Services Program, Washington, D.C. FWS/OBS-81/37.

White, R. 1975. In-stream management. Pages 48–58 in: W. King (ed.), *Wild trout management*. Trout Unlimited.

White, R. J. 1992. Why wild fish matter: Balancing ecological and aquacultural fishery management. *Trout* 1992(Autumn):17–48.

Whitlach, R. B. 1982. *The ecology of New England tidal flats: A community profile*. U. S. Fish and Wildlife Service, Biological Services Program, Washington, D.C. FWS/OBS-81/01.

Wildlife Laws News Quarterly. Fall 1994. Chippewas retain right to hunt and fish. Center for Wildlife Law, Institute for Public Law. University of New Mexico School of Law. Albuquerque. Page 3.

Wiley, M. J., R. W. Gorden, S. W. Waite, and T. Powless. 1984. The relationship between aquatic macrophytes and sport fish production in Illinois ponds: A simple model. *North American Journal of Fisheries Management* 4:111–119.

Wiley, R. W., and J. W. Mullan. 1975. Philosophy and management of the Fontenelle Green River tailwater trout fisheries. Pages 28–31 in: W. King (ed.), *Wild trout management*. Trout Unlimited.

Wiley, R. W., R. A. Whaley, J. B. Satake, and M. Fowden. 1993. Assessment of stocking hatchery trout: A Wyoming perspective. *North American Journal of Fisheries Management* 13:160–170.

Wiley, R. W., and R. S. Wydoski. 1993. Management of undesirable fish species. Pages 335–354 in: C. C. Kohler and W. A. Hubert (eds.), *Inland fisheries management in North America*. American Fisheries Society. Bethesda, Maryland.

Wilkin, D. C., and S. J. Hebel. 1982. Erosion, redeposition and delivery of sediments to midwestern streams. *Water Resources Research* 18:1278–1282.

Williams, C. D. 1994. Aquatic resources and the Endangered Species Act. *Fisheries* 19(1):19–21.

Williams, C. D., and J. E. Deacon. 1991. Ethics, federal legislation, and litigation in the battle against extinction. Pages 109–121 in: W. L. Minckley and J. E. Deacon (eds.), *Battle against extinction: Native fish management in the American West.* University of Arizona Press. Tucson.

Williams, J. D., J. L. Warren, Jr., K. S. Cummings, J. L. Harris, and R. J. Neves. 1993. Conservation status of freshwater mussels of the United States and Canada. *Fisheries* 18(9):6–22.

Williams, J. E. 1991. Preserves and refuges for native western fishes: History and management. Pages 171–189 in: W. L. Minckley and J. E. Deacon (eds.), *Battle against extinction: Native fish management in the American West.* University of Arizona Press. Tucson.

Williams, J. E., J. E. Johnson, D. A. Hendrickson, S. Contreras-Balderas, J. D. Williams, M. Navarro-Mendoza, D. E. McAllister, and J. E. Deacon. 1989. Fishes of North America endangered, threatened, or of special concern: 1989. *Fisheries* 14(6):2–20.

Williams, L. L., and J. P. Geisy. 1992. Relationships among concentrations of individual polychlorinated biphenyl (PCB) congeners, 2,3,7,8-tetrachlorodibenzo-p-dioxin equivalents (TCDD-EQ), and rearing mortality of chinook salmon (*Oncorhynchus tshawytscha*) eggs from Lake Michigan. *Journal of Great Lakes Research* 18:108–124.

Williams, T. 1994. Tongass salmon lose again. *Fly Rod and Reel* November-December 1994:13–17.

Willis, D. W. 1986. Review of water level management on Kansas reservoirs. Pages 110–114 in: G.E. Hall and M. J. Van Den Avyle (eds.), *Reservoir fisheries management: Strategies for the 80's.*

Winger, P. V. 1981. Physical and chemical characteristics of warmwater streams. Pages 32–44 in: L. A. Krumholtz (ed.), *The warmwater stream symposium.* Southern Division, American Fisheries Society. Bethesda, Maryland.

Winston, M. R., C. M. Taylor, and J. Pigg. 1991. Upstream extirpation of four minnow species due to damming of a prairie stream. *Transactions of the American Fisheries Society* 120:98–105.

Witzig, J. F., M. C. Holliday, R. J. Essig, and D. L. Sutherland. 1992. *Marine recreational fishery statistics survey, Pacific coast, 1987–1989.* Current Fisheries Statistics Number 9205. NOAA.

Wood, C. A. 1993. Implementation and evaluation of the water budget. *Fisheries* 18(11):6–17.

Woodhead, A. 1979. Senescence in fishes. Pages 179–205 in: P. J. Miller (ed.), *Fish phenology: Anabolic adaptiveness in teleosts.* Symposium of the Zoological Society of London 44.

Woodward, D. F., W. G. Brumbaugh, A. J. DeLonay, E. E. Little, and C. E. Smith. 1994. Effects on rainbow trout fry of a metals-contaminated diet of benthic invertebrates from the Clark Fork River, Montana. *Transactions of the American Fisheries Society* 123:51–62.

Wootton, R. J. 1990. *Ecology of teleost fishes.* Chapman and Hall. New York.

Wright, G. L. 1991. Results of a water level management plan on largemouth bass recruitment in Lake Eufaula, Oklahoma. Pages 126–130 in: *Warmwater Fisheries Symposium I.* USDA Forest Service. General Technical Report RM-207.

Wright, G. L., and G. W. Wigtil. 1982. Comparison of growth, survival, and catchability of Florida, northern, and hybrid bass in a new Oklahoma reservoir. *Proceedings of the Annual Conference of the Southeastern Association of Fish and Wildlife Agencies* 34:31–38.

Wydoski, R. S., and D. H. Bennett. 1981. Forage species in lakes and reservoirs of the western United States. *Transactions of the American Fisheries Society* 110:764–771.

Wydoski, R. S., and J. Hamill. 1991. Evolution of a cooperative recovery program for endangered fishes in the upper Colorado River basin. Pages 123–135 in: W. L. Minckley and J. E. Deacon (eds.), *Battle against extinction: Native fish management in the American West.* University of Arizona Press. Tucson.

Zaret, T. M., and A. S. Rand. 1971. Competition in tropical stream fishes: Support for the competitive exclusion principle. *Ecology* 52:336–342.

Zedler, J. B. 1982. *The ecology of southern California coastal salt marshes: A community profile.* U.S. Fish and Wildlife Service, Office of Biological Services. FWS/OBS-81/54.

Zieman, J. C. 1982. *The ecology of the seagrasses of South Florida: A community profile.* U.S. Fish and Wildlife Services, Office of Biological Services. Washington, D.C. FWS/OBS-82/25.

Zurbuch, P. E. 1984. Neutralization of acidified streams in West Virginia. *Fisheries* 9(1):42–47.

Index